T0188895

Genetically Engineered Crops
Interim Policies, Uncertain Legislation

Iain E. P. Taylor, PhD
Editor

Routledge
Taylor & Francis Group
New York London

First published 2007 by The Haworth Press, Inc.

Published 2021 by Routledge
605 Third Avenue, New York, NY 10017
2 Park Square, Milton Park, Abingdon, Oxon OX14 4RN

Routledge is an imprint of the Taylor & Francis Group, an informa business

Copyright © 2007 Taylor & Francis.

PUBLISHER'S NOTE
The development, preparation, and publication of this work has been undertaken with great care. However, the Publisher, employees, editors, and agents of The Haworth Press are not responsible for any errors contained herein or for consequences that may ensue from use of materials or information contained in this work. The Haworth Press is committed to the dissemination of ideas and information according to the highest standards of intellectual freedom and the free exchange of ideas. Statements made and opinions expressed in this publication do not necessarily reflect the views of the Publisher, Directors, management, or staff of The Haworth Press, Inc., or an endorsement by them.

Cover design by Lora Wiggins.

Library of Congress Cataloging-in-Publication Data

Genetically engineered crops: Interim policies, uncertain legislation/Iain E. P. Taylor, editor.
   p. cm.
Includes bibliographical references and index.
ISBN: 978-1-56022-988-9 (case : alk. paper)
ISBN: 978-1-56022-989-6 (soft : alk. paper)
   1. Plant biotechnology—Risk assessment. 2. Transgenic plants—Risk assessment. 3. Plant genetic engineering. 4. Agricultural biotechnology—Law and legislation. 5. Plant biotechnology—Research—Law and legislation. 6. Genetically modified foods. I. Taylor, Iain E. P.

SB106.B56G46 2007
631.5'233-dc22   2006022968

ISBN 13: 978-1-56022-989-6 (pbk)

CONTENTS

PART III: CHALLENGES TO CIVIL SOCIETY

ABOUT THE EDITOR

Iain E. P. Taylor, PhD, holds BSc and doctorate degrees from the University of Liverpool. He joined the faculty of the University of British Columbia Botany Department in 1968. His research interests in plant cell walls have recently been paralleled by research on ethics in science, scholarly publishing, and professionalism. He retired and was appointed Professor Emeritus in December 2003. He was editor-in-chief of the *Canadian Journal of Botany* from 1989 to 1999, and has served as assistant editor in chief of the NRC of Canada Research Journals since 1991. He is associated with the W. Maurice Young Centre for Applied Ethics at UBC and is the research director at the university's Botanical Garden.

Genetically Engineered Crops
© 2007 by The Haworth Press, Inc. All rights reserved.
doi:10.1300/5880_a

xi

CONTRIBUTORS

Elisabeth Abergel, PhD, is an assistant professor in the International Studies Department and is on the faculty of environmental studies at York University. She obtained her doctorate in environmental studies in which she examined the environmental risk assessment of herbicide-tolerant canola in Canada. She now teaches science and technology and international relations, global environmental politics, the politics of food, and globalization and international civil society. Her research, which is interdisciplinary, focuses on the international governance of techno-science on the interface between science and global governance regimes for agricultural biotechnology.

Miguel Altieri, PhD, is Professor of Agroecology, at the Division of Insect Biology, UC Berkeley. In addition to teaching agroecology in the Department of Environmental Science, Policy, and Management at UC Berkeley, Altieri has served as scientific advisor to the Latin American Consortium on Agroecology and Development in Santiago, Chile (an NGO network promoting agroecology as a strategy for sustainable rural development in the region). He also served for four years as the general coordinator for the United Nations Development Programme's Sustainable Agriculture Networking and Extension Programme, which aimed at capacity building in agroecology among NGOs and the scaling up of successful local sustainable agriculture initiatives in Africa and Latin America. He was the chairman of the NGO committee of the Consultative Group on International Agriculture Research whose mission was to make sure that the research agenda of international agricultural research centers benefits the poor farmers of the developing world. Currently he is advisor to the FAO-GIAHS program (Globally Indigenous Agricultural Heritage Systems), a program devoted at identifying and dynamically conserving traditional farming systems in the developing world. He is also direc-

Genetically Engineered Crops
© 2007 by The Haworth Press, Inc. All rights reserved.
doi:10.1300/5880_b

tor of the U.S.-Brasil Consortium on Agroecology and Sustainable Rural Development (CASRD), an academic-research exchange program involving students and faculty of UC Berkeley, University of Nebraska, UNICAMP, and Universidad Federal de Santa Catarina. He has written numerous books, including *Agroecology: The Science of Sustainable Agriculture; Biodiversity and Pest Management;* and *Genetic Engineering in Agriculture: The Myths, Environmental Risks and Alternatives,* and more than 250 scientific articles. He is very involved in promoting sustainable agriculture that emphasizes conservation of biodiversity, ensures food security, and empowers local communities. Lately he has become an outspoken scientist on the ecological risks of agricultural biotechnology.

Katherine Barrett holds an MSc in microbiology, and a PhD in botany and applied ethics from the University of British Columbia. At the time of contributing to this book, she was a research associate with the Polis Project in Faculty of Law at the University of Victoria. Katherine has written extensively on the precautionary principle, and the regulation of biotechnology. She currently works with the Government of Canada, in Ottawa.

Paul Billings, MD, PhD, is vice president for Biotechnology and Healthcare Strategy at the Laboratory Corporation of America. Trained as a medical geneticist, Dr. Billings has been on the faculties of Harvard, UCSF, Stanford, and UC-Berkeley, where he is currently a visiting professor of anthropology. He has formerly served as the chief medical officer of the Heart of Texas Veteran's Integrated Health Care System. Dr. Billings is on the boards of several not-for-profits, and also consults on genetic issues for a variety of organizations. He holds a PhD in immunology from Harvard University and an MD from the University of Washington. Dr. Billings is the author of *DNA on Trial: Genetic Information and Criminal Justice* (Cold Spring Harbor Press, 1992).

Conrad Brunk is professor of philosophy and director of the Centre for Studies in Religion and Society at the University of Victoria. His areas of research and teaching include ethical aspects of environmental and health risk management, risk perception and communication, and value aspects of science in public policy. Dr. Brunk is a regular consultant to the Canadian government and international organizations on environmental and health risk management and biotechnology. He served as co-chair of the Royal Society of Canada Expert Panel on the Future of Food Biotechnology and from 2002 to 2004 as a member of the Canadian Biotechnology Advisory Committee. He is a member of the International Forum for TSE and Food Safety. He is co-author of *Value Assumptions in Risk Assessment,* a book exploring how moral and political values influence scientific judgments about technological risks, and author of numerous articles in journals and books on ethical issues in technology, the environment, law, and professional practice. Professor Brunk holds a PhD in philosophy from Northwestern University.

Kristin Dawkins is a graduate in city planning, specializing in international environmental negotiation, from the Massachusetts Institute of Technology. She worked at the Harvard Law School Program on Negotiation, where she was senior writer for their international publication *Consensus.* She joined the Institute for Agriculture and Trade Policy, Minneapolis, Minnesota, in 1991 and she is now senior fellow on global governance. With expertise in trade policy, food security, environmental regulation, intellectual property rights, and institutional governance, Kristin is currently studying democratization of the system of international decision making. She travels widely, representing the Institute at a broad range of international negotiations and conferences.

Kristin is widely published internationally including journals in the United States, Europe, Brazil, Uruguay, Malaysia, India, South Africa, and New Zealand. She is the author of two short books, *Global Governance: The Battle for Planetary Power,* and *Gene Wars: The Politics of Biotechnology,* and a photo essay called *The Ownership of Life: When Patents and Values Clash.*

Josh DuBois works in domestic and international social justice law. He graduated from the University of Minnesota Law School in December 2004, after spending his final semester studying Ghanaian customary law and the University of Ghana Law faculty. Domestically he has done work in the area of affordable housing. He is currently clerking for a Minnesota district court judge.

Brian Ellis received his undergraduate education at the University of New Brunswick in biology and chemistry (1965), and his PhD in plant biochemistry from the University of British Columbia (1969). After postdoctoral training in Germany and Ontario, he accepted a faculty position in the Department of Chemistry and Biochemistry at the University of Guelph and was the acting director of the Centre for Plant Biotechnology (1986/1987).

In 1989, Dr. Ellis became head of the Department of Plant Science at the University of British Columbia, a position that he held until 1998. He is now a professor and associate director of the UBC Michael Smith Laboratories.

Dr. Ellis's current research interests are in the development of genomics resources for poplar and spruce, and the signaling mechanisms through which plants sense and respond to environmental changes. He was co-chair of the Royal Society of Canada Expert Panel on "The Future of Food Biotechnology" (2000/2001) and has served on the "Science and Industry Advisory Committee" of Genome Canada (2001-2005).

Simon Joss is the director of the Centre for the Study of Democracy, University of Westminster, London, and a Fellow of the Royal Society of Arts. His research interests include science and technology policy, technology assessment, and the public accountability and governance of science and technology. He is involved in several multinational research programs on the social dimensions of nanotechnology, brain science, biotechnology, and transport policy. He is a member of several advisory committees.

Michiel Korthals is Professor of applied philosophy at Wageningen University. He studied philosophy, sociology, German, and anthropology at the University of Amsterdam and the Karl Ruprecht Universität in Heidelberg. His academic interests include deliberative theories, American pragmatism, and bioethics and ethical problems concerning food production and environmental issues. Main publications: *Filosofie en intersubjectiviteit,* Alphen a/d Rijn, 1983; *Duurzaamheid en democratie (Sustainability and democracy,* Boom, 1995); *Philosophy of development* (Kluwer, 1996 with Wouter van Haaften and Thomas Wren), *Tussen voeding en medicijn (Between food and medicine),* Utrecht 2001, *Pragmatist ethics for a technological culture* (with Keulartz et al.; Kluwer, 2002), *Before dinner. An introduc-*

tion into food ethics (Kluwer, 2004) and *Ethics for Life Scientists* (with Rob Bogers; Kluwer, 2005).

Sheldon Krimsky is Professor of urban and environmental policy and planning in the School of Arts and Sciences and adjunct professor of public health and family medicine at the School of Medicine at Tufts University. He received his bachelors and masters degrees in physics from Brooklyn College, CUNY and Purdue University respectively, and a masters and doctorate in philosophy at Boston University.

Professor Krimsky's research has focused on the linkages between science/technology, ethics/values, and public policy. He is the author of eight books, most recently *Science in the Private Interest: Has the Lure of Profits Corrupted Biomedical Research?* (Rowman & Littlefield Pub.) 2003; over 140 essays and reviews have appeared in many books and journals. His current book is a co-edited volume titled *Rights and Liberties in the Biotech Age: Why We Need a Genetic Bill of Rights* (Rowman & Littlefield Pub.) 2005.

Professor Krimsky served on the National Institutes of Health's Recombinant DNA Advisory Committee from 1978 to 1981. He was a consultant to the Presidential Commission for the Study of Ethical Problems in Medicine and Biomedical and Behavioral Research and to the Congressional Office of Technology Assessment. Currently he serves on the board of directors for the Council for Responsible Genetics and as a fellow of the Hastings Center on Bioethics. Professor Krimsky was elected fellow of the American Association for the Advancement of Science for "seminal scholarship exploring the normative dimensions and moral implications of science in its social context."

Michelle Marvier received a PhD in Biology from University of California, Santa Cruz in 1996. She is currently associate professor of biology and executive director of the Environmental Studies Institute at Santa Clara University. Her research interests include risk assessment of GE crops and conservation biology.

Kathleen Merrigan is Director of the Agriculture, Food and Environment graduate program at the Friedman School of Nutrition Science and Policy at Tufts University. Previously, she served as administrator of the Agricultural Marketing Service of the U.S. Department of Agriculture (1999-2001), senior analyst at the Henry A. Wallace Institute for Alternative Agriculture (1994-1999), and senior science and technology staff person (1987-1992) to Senator Patrick Leahy, then chairman of the U.S. Senate Agriculture Committee. While in the Senate, Kathleen played a major role in drafting the Biotechnology Risk Assessment Research Program as well as lead congressional oversight of the FDA approval process of bovine somatatropin. While at USDA, she was involved in numerous biotechnology issues from GM-contaminated corn, to debates over GMO-free thresholds, to organic food standards. Most recently, Kathleen participated in a two-year forum of select stakeholders organized by the Pew Initiative on Food and Biotechnology to forge consensus on regulatory reform in agbiotech. Kathleen holds a BA from Williams College, a master's in public affairs from the LBJ School at the University of Texas, and a PhD in environmental planning and policy from MIT.

Marília Regini Nutti obtained an MSc in food science at the State University of Campinas, Brazil (1986), and completed studies in nutritional planning at the University of Gent, Belgium (1980), and in consumer studies at the University of Guelph, Canada (1985). She held a full-time research and lecturer position at the Department of Nutrition and Food Planning, State University of Campinas (1983-1987), and was technical director of Nutricia S.A.—Produtos Dieteticose Nutricionais, leading a group of forty researchers conducting R&D on foods for special dietary uses. She was the general manager of Embrapa Food Technology, a research unit with 140 employees, from 1996 to 2003 and is now a researcher in human nutrition and biosafety at Embrapa Food Technology and coordinator of the Brazilian activities in the HarvestPlus-Breeding Crops for better Nutrition Program.

Her research is in food safety, biosafety, and labeling of genetically modified foods. She has been the Brazilian delegate at the Codex Alimentarius Committee in Food Labelling (Canada) since 1997 and at the Codex Alimentarius ad hoc Task Force on Foods Derived from Biotechnology (Japan) since 2000. She is a member of the FAO/WHO Expert Consultation *"Safety Aspects of Genetically Modified Foods of Plant Origin."*

Carolyn Raffensperger is the executive director of the Science and Environmental Health Network, an environmental think tank. In 1982 she left a career as an archaeologist in the U.S. southwest desert to join the environmental movement. She first worked for the Sierra Club where she addressed an array of environmental issues including forest management, river protection, pesticide pollutants, and disposal of radioactive waste. She began working for SEHN in December 1994. As an environmental lawyer she specializes in the fundamental changes in law and policy necessary for the protection and restoration of public health and the environment.

Carolyn is co-editor of *Protecting Public Health and the Environment: Implementing the Precautionary Principle,* published by Island Press (1999) and co-editor of *Precautionary Tools for Reshaping Environmental Policy,* published by M.I.T Press (2006). In addition, she writes the Science for Lawyers column for the Environmental Law Institute's journal *Environmental Forum.*

Carolyn is at the forefront of developing new models for government that depend on these larger ideas of precaution and ecological integrity. The new models include a vision for the courts of the twenty-first century and the public trust doctrine. Carolyn coined the term "ecological medicine" to encompass the broad notions that both health and healing are entwined with the natural world.

Maria José Amstalden Sampaio earned a degree in agronomy from the University of São Paulo (ESALQ/1975) and a PhD in molecular biochemistry from Dundee University, Scotland, in 1979. She has been involved with the Brazilian Agriculture Research Corporation (Embrapa), linked to the Ministry of Agriculture, since 1976. She is a member of the Brazilian Delegation to the Convention of Biological Diversity—COP meetings and the COP/MOP meetings for the Cartagena Protocol on Biosafety and has worked in all meetings leading to the approval, in 2003, of the FAO International Treaty on Plant Genetic Resources for Food and Agriculture. She is currently involved with the design of the Material Transfer Agreement for its implementation.

She worked at Cornell University/Cornell Research Foundation on intellectual property management in close collaboration with the Agricultural Research Service (ARS)/USDA—(Embrapa/Labex: 1998-2000). Since 2002 she has been joint coordinator of Embrapa's

Biosafety Network, a five-year program to develop GMO risk analysis protocols adapted to Brazilian conditions.

Peter Shorett is a graduate student in the John F. Kennedy School of Government at Harvard University. Formerly the director of programs at the Council for Responsible Genetics, he is the co-editor of *Rights and Liberties in the Biotech Age: Why We Need a Genetic Bill of Rights* (Lanham, MD: Rowman & Littlefield, 2005). Mr. Shorett holds a BA in political science and anthropology from UC-Berkeley.

Armin Spök received a PhD in molecular genetics from the University of Graz (1991) and an MSc in science and technology policy from the University of Sussex (2003). After several years of basic and applied research in molecular genetics he switched to more interdisciplinary approaches in technology assessment and risk analysis and became a researcher and project manager at IFZ—Inter-University Research Centre for Technology, Work and Culture in Graz. Since 1998 he has been head of the research unit in Modern Biotechnology. He also lectures at the University of Klagenfurt, Graz University of Technology, and at the Advanced Technical College Technikum Joanneum in Kapfenberg.

His recent research and consultant work has been on scientific, regulatory, and policy issues in environmental and health risk assessment and management of biotech products, including genetically engineered plants and microorganisms, food, pharmaceuticals, chemicals, cosmetics, and other products derived from them.

Jennifer Thomson received her PhD from Rhodes University in South Africa, and was a postdoctoral fellow at Harvard Medical School. She joined the Genetics Department at the University of the Witwatersrand in Johannesburg and was the director of the CSIR Laboratory for Molecular and Cell Biology before becoming professor and head of the Department of Microbiology at the University of Cape Town. She is currently professor in the Department of Molecular and Cell Biology at UCT. Her research involves the development of transgenic maize resistant to *Maize Streak Virus* and tolerant to drought.

Edson Watanabe is a food engineer who obtained BSc and MSc degrees from the State University of Campinas, Brazil, and a PhD in food science from the University of Reading, United Kingdom. After postdoctoral work with the Biotechnology and Biochemical Engineering Group at the University of Reading, he became a researcher with the Brazilian Agricultural Research Corporation (Embrapa). His current work has been in the area of food safety assessment of genetically modified crops. He is a member of the Brazilian Council for Biotechnology Information (www.cib.org.br).

Sabrina West received a BS in biology from Santa Clara University in 2002. She has since worked as a conservation intern and reserve steward and is currently applying to graduate school in conservation biology.

Brian Wynne is Professor of science studies at Lancaster University. He has a PhD in materials science from Cambridge University, and an MPhil in Sociology of Science from Edinburgh University. He has been research director of the Centre for the Study of Environmental Change at Lancaster, as well as director of the Centre for Science Studies and Science Policy. He is now a senior partner in the ESRC-funded Centre for the Economic and Social Aspects of Genomics, which is a partnership between Lancaster and Cardiff Universities.

Brian has worked since the 1980s on risk science and public attitudes to decisions framed as risk issues, and has published extensively on these topics in energy, biotechnology and other fields. A key issue is scientific uncertainty in its various forms, and how this is understood, as well as the hidden social assumptions in risk assessments. He coordinated a major EU research project on Public Attitudes to Agricultural Biotechnologies in Europe (PABE) completed in 2001, and has written, spoken, and advised on the GM issue in the United Kingdom and Europe. He co-authored an Amicus Curiae brief on risk assessment submitted in April 2004 to the WTO Disputes Panel on the US-EU GMOs case. From 1994 to 2000 Brian was an inaugural member of the Management Committee and Scientific Committee of the European Environment Agency, and currently chairs an EU Expert Group on Science and Governance. He is a joint editor of *Science and Citizens: Globalisation and the Challenge of Engagement,* published in 2005 by Zed Books, and joint author of the Demos London think-tank pamphlet, *The Public Value of Science* (2005).

Acknowledgments

The original conception, planning, and editing of this book was done in 2001 in collaboration with Katherine Barrett, then a postdoctoral fellow at the Polis Project, University of Victoria.

Thanks to the peer reviewers who provided constructive and timely comments on each chapter.

My sincere thanks is extended to the authors, who have provided great support both in their willingness to keep their contributions up-to-date and in responding to the perpetual nagging that comes with contributing to an edited work.

Introduction

Genetic Engineering of Crops: Science Meets Civil Society's Response

Iain E. P. Taylor

Science is inherently an uncertain process. The driving forces are the interests and intellectual backgrounds of individual researchers. The articulation of a research question is motivated by a desire to find the answer, the interest in expanding our understanding and, not least, the hope of recognition. The move from a new discovery to its possible application for broader societal benefit requires not only new, developmental research, but promotion of the benefits that will possibly lead to the funding and marketing of the new product; all of which requires decisions without certain knowledge of the outcomes. The basic research in molecular biology and the developmental work that leads to new inventions in genetic engineering (GE) are no different in their uncertainty, their risk, and their need for precaution from discoveries and inventions of nuclear fission, identification of antibiotics, or the many mechanical developments that led to the internal combustion engine.

The invitation to prepare this book came at a time (2001) when plant GE had clearly arrived and was here to stay, and there were strong currents of approval and concern being expressed in many sectors of civil society. The issues were moving from science and technology to governance, regulation, and impact.

Like many new inventions, the first developments were enormously expensive and the risk takers were venture capitalists. In less than twenty years, the technology has become more effective and ef-

Genetically Engineered Crops
© 2007 by The Haworth Press, Inc. All rights reserved.
doi:10.1300/5880_d

ficient. The emergence of genomics and its spin-off technologies to engineer proteins and cellular biochemistry (proteomics and meta-bolomics), which depend ultimately on genetic information, have ensured that GE will become a long-term contributor to all forms of biotechnology. The successes in the laboratory led to many well-founded as well as more speculative proposed uses. The potential benefits for agricultural production were widely publicized and led to major public relations effort by industry and governments in the in-dustrialized world. The opportunity for massive corporate profit was, as always, hard to resist. Very soon, the benefits became exaggerated and the manipulation of the public in the name of "science-based developments" led to heightened distrust of science and increased cynicism of the regulators and their political masters.

Herbicide and pest resistance are currently the most widely used traits. Resistance to herbicides was carefully and wisely selected as a marker or co-marker character in the early studies on molecular plant development, because carriers of resistance were the survivors after herbicide application and hence carried the desirable trait. There was also obvious value for the industrial seed producer whose farming cus-tomers sought a homogeneous, high-productivity, high-quality crop. The farmer could reasonably expect to cover operating costs (seed, fer-tilizer, herbicide, labor, and equipment) from sale of the crop grown without competition from weeds and, in the case of grain crops, uncon-taminated by weed seeds.

Resistance to disease and pests perhaps is perceived by society in general to be one of the more obvious benefits of GE. High school bi-ology, geography, and social history classes inevitably have a story of plant disease causing major social upheaval and hardship somewhere in the world. The Irish potato famine, the periodic devastations of wheat rust in North America, corn borer, and cotton boll weevil pro-vide examples, but it seems that many of the major pest problems affect largely introduced crop plants for which commercial interests have created massive demands that can only been met by high-inten-sity, monocultural farming practices.

Farming takes place in a relatively uncontrolled environment and inevitably plant ecologists and other researchers began to look for ev-idence that engineered genes could escape into related wild species growing in uncultivated areas around the farms. Several features may distinguish the inventions of GE from other earlier agricultural advances. While the modern plant breeder is the descendent of the earliest agriculturists, the genetic engineer has added what many see as an "unnatural" element to the process of improving crops. This perceived unnaturalness puts a sense of hazard into the issue of safety. Governmental response in North America seems to have assumed that the new GE crops should be regulated in generally the same way as any traditionally bred new crop is regulated. The presumption is that because they look so like the products of traditional breeding, they are substantially equivalent and can be considered as equally safe. The European response, on the other hand, has been driven by a very strong reaction that does not accept regulatory constraints with any real trust. Concerns over safety to humans as well as unknown and unpredicted environmental hazards led to major public demonstrations and proactive campaigns to ban GE crops.

Recently, some countries in the developing world have also banned GE crops, but for rather different reasons. Their concerns are not only to protect local farming and food cultures, but also to their local farms from becoming cheap producers of GE crops for the industrialized world, particularly those whose own agricultural systems are beginning to lose their international competitiveness.

The decisions to develop and grow GE crops, like any other farm commodity, have an impact on the neighbors as well on the market. It is relatively easy to predict potential benefits and to sort the credible short-term disadvantages, but as the technology advances and the societal responses emerge there will certainly be both advantages and disadvantages that were not recognized initially. Most of these goods and bads may emerge in the longer term, but the approval and release of GE crops does differ significantly from nonbiological inventions. Even the first GE that was done with microbes was under strict containment. It is almost certain that once a new gene is released from containment into a farming environment, it can never be recalled should it be found to be faulty.

It is this uncertainty that led to the decision to invite contributors to address the question of "decision-making under uncertainty." The chapters are grouped to address questions about the applied biology, a sample of legal and governance arrangements, and the interactions between the expert scientists and the public. It was particularly pleasing that all those who did accept our invitation were in our first choice group of authors. We have voices from molecular biology, agroecology, environmental testing, governance, international law, public participation in science policy, and the whole issue of whether use of risk assessment can work in this arena.

I used editorial prerogative to use the term "genetically engineered" as far as possible through this book. It directs our thinking to activities that are planned and undertaken to manage and control change. "Genetically modified" (GM) and the uniquely Canadian term "plants with novel traits" (PNT) seem more suitable to accommodate the "gray area" that exists between the emergence and selection of natural mutations, the products of traditional plant breeding, and the products arising from genomic information used to locate suitable genes for selection and possible engineering. The emphasis is on the crop plants in which the gene introduction required skills of engineering, with all its implicit meanings of calculated and precise human intervention. The term GM often seems less artificial, especially in the context of plant domestication and selection that farmers and plant breeders have engaged in for thousands of years.

We selected experts in governance to represent different approaches. The United States and Canada have based their governance on the concept of "substantial equivalence," and, with Argentina, are the major GE crop-producing countries worldwide. Other countries such as China and India have generally followed similar approaches to governance. While "substantial equivalence" is accepted in many European Union (EU) countries, the EU governance models are diverse as well as being strongly influenced by public opinion. Africa provides further examples of diversity in governance, and Brazil has recently moved from a very limited and short-term approval system to passage of legislation that will allow GE crops to become a major part of that country's agriculture.

Reviews such as those contained in the following chapters are individual perspectives based upon rigorous analysis. Inevitably, there is controversy. Just like any other aspect of science, the future of GE in general is uncertain and there are changing and different societal expectations and demands. Historically, the applications that seem to have potential harms for society attracted most attention. In this respect GE plant crops seem less threatening—who can argue with the desire to address worldwide hunger?—but each of the authors has raised challenging issues that both science and civil society must address.

It seems likely that many more GE crops will be developed. Many uses will come as surprises to the original inventors. Many will bring great benefit, some will cause unintended harm, and some will have no effect either way. Whatever the directions, we can only make educated guesses at the outcomes. As Brian Ellis states in his chapter, "It is hoped that this book makes a useful contribution to that global dialogue."

ABBREVIATIONS

AESGP	Association of the European Self-Medication Industry
APEC	Asia Pacific Economic Council
APHIS	The Animal Plant Health Inspection Service
BSE	Bovine Spongiform Encephalopathy
Bt crops	*Bacillus thuringiensis* protein-engineered insect-resistant crops
CA	Competent Authority
CAP	Common Agricultural Policy
CBA	Cost-Benefit-Analysis
CBD	Convention on Biological Diversity
CBP	Cartagena Biosafety Protocol
CCFL	Codex Committee on Food Labelling
CFIA	Canadian Food Inspection Agency
CGIAR	Consultative Group on International Agricultural Research
CoE	Council of Europe
CP	Coat protein
CSR	Corporate social responsibility
DFAIT	Department of Foreign Affairs and International Trade
EC	European Commission
ECPA	European Crop Protection Association
EFSA	European Food Safety Authority
EPA	Environmental Protection Agency
EPO	European Patent Office
ESF	European Science Foundations
EU	European Union
FAO	Food and Agriculture Organization
FDA	Food and Drug Administration
FDCA	Federal Food Drug and Cosmetic Act

Genetically Engineered Crops
© 2007 by The Haworth Press, Inc. All rights reserved.
doi:10.1300/5880_e

FSEs	Farm-scale evaluations
GATT	General Agreement on Trade and Tariffs
GE	Genetically engineered or genetic engineering
GEF	Global Environment Facility
GEO	Genetically engineered organism
GEP	Genetically engineered products or genetically engineered plants
GRASP	Greedy Randomized Adaptive Search Procedures
HRC	Herbicide-resistant crops
HT	Herbicide-tolerant
IPRs	Private Intellectual Property Rights
IRRI	International Rice Research Institute
LMOs	Living modified organisms
NAFTA	North American Free Trade Agenda
NBF	National Biosafety Frameworks
NEA	National Executing Agency
NIH	National Institutes of Health
OECD	Organization for Economic Cooperation and Development
PIPs	Plant incorporated protectants
PNTs	Plants with novel traits
PPP	People, Power, and Profit
r-DNA	recombinant DNA
TBT	Technical Barrier to Trade
TRIPS	Trade-Related Intellectual Property Rights
UNDP	United Nations Development Program
USDA	U.S. Department of Agriculture
WHO	World Health Organization
WTO	World Trade Organization

PART I:
SCIENCE AND THE FUTURE

Chapter 1

The Birth of Synthetic Biology
and the Genetic Mode of Production

Sheldon Krimsky

Biology is as important as the sciences of lifeless matter, and biotechnology will in the long run be more important than mechanical and chemical engineering.[1]

Julian Huxley, 1936

Nearly 400 years ago, the English scientist-philosopher Francis Bacon envisioned a time when the plants and animals on the earth were the starting materials for refashioning biological life forms according to human design.[2] To a degree, agricultural scientists have been fulfilling Bacon's prophecy through crossbreeding of crops and animals. We have seen the results of these genetic experiments in the highly developed domesticated varieties of corn and tomato plants, which have evolved from wild relatives of these plants that would today seem unsuitable to our palette. For example, the North American subsidiary of the Swiss agri-biotechnology company Syngenta recently announced the result of years of consumer research and crossbreeding in its five pound seedless, miniature spherical watermelon.[3]

With crossbreeding, scientists were limited in how much they could modify plants and animals. They could only combine the traits of somewhat similar species with compatible DNA through grafting and cross-fertilization. However, thirty years ago, Bacon's vision that science would eventually exploit the biological resources of the planet

Genetically Engineered Crops
© 2007 by The Haworth Press, Inc. All rights reserved.
doi:10.1300/5880_01

the way it had learned to transform the earth's natural ores, such as copper and iron, to create the industrial revolution, seemed within grasp. In 1973 Stanley Cohen, Herbert Boyer, and others developed recombinant DNA (rDNA) molecules[4] and the following year demonstrated the expression of foreign genes implanted in a bacterium by rDNA methods.[5] These discoveries prepared the way for transporting genes from biological organisms of distant phyla. Thus, the concept of crossing species barriers was introduced. DNA molecules, the fundamental units of inherited traits, could now be redistributed or repositioned according to the desires of the gene engineers.

This chapter discusses the early and later developments in a new field of applied molecular genetics, including the birth of an academic–industrial complex based on the potential of gene splicing, the role of gene technology as a new mode of production in agriculture, drugs, and therapeutics, and the marketing of biotechnology as a green revolution in agricultural genetics.

BIOTECHNOLOGY: EVOLUTION OR REVOLUTION?

Robert Bud has traced the roots of modern biotechnology to zymotechnics, or the fermentation industries in Europe.[6] According to Bud's in-depth review of the history, Karl Ereky, a Hungarian agricultural engineer who worked on the production of food animals, coined the term "biotechnology" in 1917. "His notion of biotechnology was a conception that food animals like the pig were machines converting inputs into human protein."[7] Ereky described a pig as a "Biotechnologische Arbeitsmachine."

Historian Bud viewed the change from classical breeding to modern GE of crops and microbes as a change of degree and not of kind. However, there are reasons for characterizing the tools of molecular genetics as fostering a revolution (abrupt and discontinuous change) in biology. I once characterized it in the following way:

> The discovery of the fungibility of genes forces us to make the next frame shift in our concepts of life. We can no longer accept uncritically the aphorism that "like begets like." It cannot be said that a pig's snout is uniquely of a pig and that there is some-

thing we call "pigness" that is trapped in the evolutionary construct of the pig family of animals. These so-called species
demarcations have been transcended by the discovery that genes
can be shifted from organism to organism and with these shifts
in genes the phenotypic properties of living forms on the planet
can be rearranged as Bacon had foreseen.[8]

In 1984, the U.S. Office of Technology Assessment (OTA) defined
biotechnology as "any technique that uses living organisms (or parts
of organisms) to make or modify products, to improve plants or animals or to develop micro-organisms for specific uses."[9] The OTA distinguished between new and old biotechnology. Modern (post-1973)
or new biotechnology is based on a set of techniques for undertaking
"precision" genetic and cell engineering that include uses of rDNA,
cell fusion, monoclonal antibodies, tissue culture technology, and
novel bioprocessing methods. The OTA emphasized the historical continuum between old and new uses of biological organisms for practical purposes.
    Three discoveries are central to the development of new biotech-
nology as distinguished from traditional fermentation engineering,
and cross breeding of animals and crops. These include the discovery
of new classes of enzymes, DNA sequencing, and methods of trans-
posing genes within and across species.
    A group of enzymes called restriction enzymes were found to cut
DNA at predictable sites. These enzymes initially gave scientists the
tools to isolate DNA sequences that could be reintegrated into other
organisms. Another group of enzymes called ligases were found to
seal the ends of DNA molecules. The ligases are the "chemical glue"
that give scientists the ability to splice together segments of DNA
from different biological organisms. Finally, an enzyme called reverse
transcriptase transcribes single-stranded messenger RNA into dou-
ble-stranded DNA. Mammalian genes contain noncoding regions of
DNA that are split off from the gene when messenger RNA is formed.
Prior to the discovery of reverse transcriptase independently by
Howard Temin and David Baltimore in 1970, it was believed that the
transformation from DNA to RNA was not reversible.[10]
    A second core discovery central to biotechnology's development
was gene sequencing. This is the process by which the precise nucleotide components of a gene are determined. Gene sequencing is

essential to understand which segments of DNA correspond with specific proteins, or how the coding and noncoding regions of DNA differ. Scientists can derive the amino acid sequence from the DNA sequence, but as of yet cannot predict the three-dimensional structure and function of the protein from the gene sequence. According to Gilbert, new theoretical breakthroughs will be required to make that happen. "It is here that a theoretical biology will emerge. It will be a science of pattern recognition—extracting from the genetic sequence the identity of human genes, their interrelationships, and their control elements. This information will be used to predict how the genes and their proteins function."[11]

The third set of discoveries central to the new field of biotechnology are methods for transporting segments of DNA across biological systems. Even prior to the Cohen-Boyer experiments, Paul Berg created recombinant DNA molecules constructed from viruses, which, because they naturally infect cells, can unload their DNA into the cell's chromosome.[12] Simpler, more efficient methods for transporting DNA used circular segments of DNA called plasmids. The plasmids can be cut into segments and can be attached to a foreign piece of DNA, and the spliced segments are annealed at the ends to reestablish the circular plasmid suitable for activation of its genetic components in the cell. Other methods of transporting DNA molecules, where naked DNA is delivered by physical mechanisms to the cell's chromosome, include microinjection or micropropulsion (gene gun).

DNA alone, without the apparatus of the cell, cannot synthesize anything or even replicate itself. Cells provide the environment and biological mechanisms within which the chemical structure of DNA can be "read" and translated into instructions, serving as signals to the cell to undertake the process of protein synthesis. In his *Harper's* essay, titled "Unraveling the DNA Myth," Barry Commoner noted: "Genetic information arises not from DNA alone but through its essential collaboration with protein enzymes."[13] The new gene engineers have learned how to intervene in and co-collaborate with the cellular machinery. In 1980, an editorial/commentary in *Nature* stated: "Genetic manipulation is used by the biotechnologist to enhance the natural genetic repertoire of microorganisms."[14] This was indeed Bacon's vision of expanding the biodiversity of life through human invention.

The cell is the protein factory. The DNA in the cell provides the chemical template for protein production. In the new field of molecular biotechnology, scientists can bring foreign DNA segments into organisms and activate the cellular "production apparatus" to synthesize a protein that the cell(s) had not synthesized before.

FIRST-GENERATION FEARS ABOUT GENE SPLICING: LABORATORY HAZARDS

Concurrent to the discoveries that made the front pages of the national daily newspapers and popular science magazines were expressions of concern about the potential risks of GE. A group of scientists asked: Would these new set of techniques produce organisms that can unexpectedly create harm? The first voices of caution came from young scientists who were poised to apply the new techniques in their own work. They cosigned letters in leading science journals and organized symposia to consider the potential or speculative hazards associated with transplanting genes from mammalian cells (eukaryotes) to bacteria (prokaryotes).

The cautions first raised by scientists soon turned into public debates in dozens of communities where laboratories were being built to accommodate the new U.S. National Institutes of Health containment guidelines for rDNA molecule research. Scientists who viewed the rDNA techniques as a new frontier in biology were fearful of overreaction by the government, which might proscribe or delay fruitful lines of inquiry. The confluence of new science and new fears provided the grist for extensive print media coverage of genetics. There were events that stoked the flames of publicity such as the International Conference at Asilomar, California; the Cambridge, Massachusetts rDNA debates; meetings of the National Institutes of Health Recombinant DNA Molecule Advisory Committee; and U.S. congressional hearings between 1977 and 1980 on the dozen active bills that would regulate gene-splicing experiments. Public fears about the inadvertent release of rDNA organisms were met with unrestrained claims of medical and commercial benefits that the new research methods would bring. The commercial possibilities of rDNA were manifestly obvious to the young scientists who embraced the new research program. The expression "cloning scarce proteins" was among the first clues for potential commercial application. This was a no-brainer for anyone following the science.

By 1974 it was demonstrated that toad genes could be incorporated and expressed in a bacterium.[15] Since bacteria reproduce rapidly, scientists could exploit their cellular mechanism to copy (clone) and express the foreign gene. Consequently, large volumes of the bacteria can yield large quantities of a foreign protein.

The commercial opportunities of cloning foreign genes into bacteria and scaling up for production of human proteins were of great interest to the pharmaceutical industry. In 1977, at a U.S. National Academy of Sciences symposium on rDNA research, a representative of Eli Lilly and Company cited four classes of human proteins as candidates for large-scale production by rDNA techniques: hormones, coagulation factors, hereditary disease replacement enzymes, and immunological factors.[16]

Applications of rDNA to agriculture were also enthusiastically discussed within a brief period after the science became understood. Extrapolating from ideas of traditional plant and animal breeding, scientists began thinking of biological systems as having interchangeable parts, where desirable characteristics could be transferred from one species to another, somewhat like moving Lego™ blocks. Thus, molecular plant scientists began planning research to move nitrogen-fixing genes from bacteria to plants so the latter could become auto-nitrogenous. Four other agricultural applications highlighted in a 1977 National Academy Forum were enhancing photosynthesis and increasing the efficiency of CO_2 fixation, biological pest control, fuel production through bioconversion, and plant breeding.[17] A quarter century after those predictions were made, rDNA applications in biological pest controls and plant breeding have spawned dozens of commercial products such as insect-resistant and herbicide-tolerant crops, including *Bt* corn and glyphosate-tolerant soybeans.[18]

Those outside of the scientific community were distrustful of scientists serving as their own gatekeepers, while the scientists themselves could see it no other way. Citizen groups harkened back to the nuclear industry where well-paid atomic scientists underestimated the risks of radiation hazards. Both the novelty and the perceived powers of rDNA technology created a buzz in business and among public-interest communities but for different reasons. Ironically, the controversy over gene splicing and its accompanying media publicity did not drive away the venture capitalists. Rather, it seemed to pique the interest of investors who demonstrated enthusiasm for becoming players in the very early stages of the technological breakthroughs.[19] As Martin Kenney noted: "Curiously, the debate and publicity about health and safety issues actually attracted the attention of venture capitalists, the potential financial backers; it may also have discouraged established pharmaceutical firms from capturing the technology."[20]

EMERGING INDUSTRIAL SECTORS

The methods described in the aforementioned section defined a new mode of protein production. Within a very short time after the Cohen-Boyer discovery, scientists understood the commercial applications of applied molecular genetics and communicated these to the business sector. Academic biologists, responding to new liberal patent regulations and federal technology transfer inducements for universities, developed for-profit partnerships with companies or set up their own venture capital firms. There were four companies dedicated to biotechnology in the 1970s: Cetus, Biogen, Genex, and Genentech. The period between 1980 and 1984 saw a rapid growth and expansion of the biotechnology business. Over thirty new firms were established in 1980. Nearly seventy more were added in 1981 and twenty-two in 1982.[21] A report by the OTA cited more than 400 dedicated biotechnology companies in operation and 70 major corporations which had invested in biotechnology by 1988.[22]

By 1981, the biotechnology market was worth about $25 million, mostly for reagents and contract research. The biotech market was projected to reach $20 billion by 1990 and jump to $30-34 billion by 2000. The actual growth of biotechnology fell far short of the early estimates with revenues in 1989 at $1.5 billion almost exclusively from pharmaceuticals and diagnostics. In 1992 the revenues from agricultural biotechnology products grew to about $184 million.

By the mid-1980s about 10 percent of the 500 largest U.S. companies reported that they were investing in biotechnology.[23] These companies were from five major commercial sectors: pharmaceuticals, agriculture, environmental, therapeutics, and industrial. The latter included the production of new materials and energy from biomass.

The first major drug to come out of rDNA biotechnology was human insulin. It was marketed in the United States in 1982 and was expected to replace bovine and porcine insulin, which were manufactured by extracting and purifying the protein from the pancreases of cows and pigs.

Toward the end of the 1980s, in addition to insulin, four other proteins produced from GE cells had been approved by the U.S. Food and Drug Administration (FDA). They included human growth hormone, hepatitis B vaccine, alpha interferon (an antiviral agent), and tissue plasminogen activator (tPA), an agent that reduces blood clots. Throughout the 1990s the therapeutic biotechnology companies brought scores of products to clinical trials including anticancer therapies, vaccines, early diagnostic screening tools, and viral vectors designed for human gene therapy.

The agricultural sector embraced biotechnology somewhat more cautiously as debates about releasing GE plants and microorganisms into the environment persisted. The Monsanto Corporation was the first major established company to develop in-house research programs in biotechnology starting as early as 1978. Four years later it was spending about 28 percent of its total R&D budget on biotechnology.[24]

In 1983 the first engineered plant (petunia) was grown using biotechnology. Two years later the first field tests were begun for plants resistant to insects, viruses, and bacteria. By 1986 the Environmental Protection Agency (EPA) approved the release of the first transgenic crop, a gene-altered tobacco plant. This was also the year that the federal government approved the Coordinated Framework for regulation of biotechnology, which allocated agency responsibility among the EPA, the Department of Agriculture (USDA), and the FDA for different aspects of genetically modified organisms and products developed from them. The new framework was introduced without enacting new laws.

From the mid-1980s through the mid-1990s, extensive field trials for transgenic plants were carried out. The USDA and EPA approved more than 2,500 field trials between 1987 and 1995. A little more than 800 of these trials were for herbicide-resistant crops, about 700 for insect resistance, about 600 for plant quality (such as value-added properties), and nearly 400 for disease resistance. By the mid-1990s the first commercial biotech plants entered the marketplace. In 2001 significant percentages of the U.S. production of cotton (69 percent), soybeans (68 percent), canola (55 percent), and corn (26 percent) consisted of genetically modified varieties that were either insect- or herbicide resistant.

The science media echoed the boundless enthusiasm of the investment community that GE crops would revolutionize agriculture by creating more food, a cleaner environment, and more wealth. GE crops were dubbed as the next "Green Revolution." A 1991 report from the World Bank captured the sense of optimism regarding GE plants.

> The great appeal of these techniques is that they can be used to improve the tolerance of both crops and animals to particular stresses, pests, and pathogens, and to increase the efficiency with which plants and livestock use limiting nutrients. They also hold out the promise of relieving the present biological constraints to higher yields. In countries where the new technologies are applied the results should be increased agricultural production, improved comparative advantage in the production of some commodities, new opportunities for the use of marginal lands and a reduced need for agrochemicals.[25]

Between 1996 and 2002 the global acreage planted with GE crops increased from 4.25 million acres to 146.8 million acres, a 35-fold increase. The acreage devoted to transgenic crops represented about 51 percent of the total agricultural acreage. Nearly 100 million acres of transgenic crops were planted in the United States during 2002, an increase of about 15 million acres over the previous year. The estimated global area devoted to GM crops for 2004 was 200 million acres, up from 167 acres in 2003. The number of countries growing transgenic crops also steadily increased from six in 1996, nine in 1998, twelve in 1999, and sixteen and seventeen in 2002 and 2004 respectively. Monsanto's genetically modified seeds accounted for about 118 million acres of transgenic crops or about 81 percent of the world acreage planted with GE products in 2002.

Many European farmers were opposed to GE crops. Nevertheless, the European Community approved the sale of GE soybeans in the 1990s. In total, there were eighteen biotech food products approved by the European Union (EU) prior to June 1999. Since that year there has been a de facto moratorium on additional approvals pending the passage of new regulations. One of Europe's traditionally large U.S. imports affected by the moratorium is corn. Exports of corn from the United States to Europe nosedived from 1.5 million metric tons in 1998 to 23,000 metric tons in 2003.

The EU is developing new labeling and GE food traceability requirements for biotechnology food products and animal feed. Under its preliminary provisions, even highly refined products such as corn and soybean oil, and animal feed produced from biotechnology crops, would have to be labeled. The passage of the new labeling and traceability rules has been linked to the lifting of the EU moratorium.

The acreage dedicated to transgenic crops also increased in the developing countries, which grew from 14 to 27 percent of the global acreage between 1997 and 2002. The developing nations with the largest acreage of transgenic crops in 2004 were Argentina (16.2 million hectares), Brazil (5.0 million hectares), and China (3.7 million hectares). Four crops that dominated the transgenic varieties planted in 2002 are soybeans, corn, cotton and canola, with soybeans (herbicide tolerant) occupying 62 percent of the global acreage.

SECOND-GENERATION CONTROVERSIES: ENVIRONMENTAL RELEASES OF GEOS

During the 1970s, the public angst about rDNA focused mainly on laboratory hazards of genetically modified organisms. The debates were about containment levels, siting of laboratories, worker health and safety, and proscribed experiments. Commercialization had begun but on a small scale. By the 1980s agrochemical companies and small biotech start-ups were beginning to file applications to field-test GE crops, rDNA-produced veterinary hormones, and genetically engineered microorganisms (GEOs). The environmental release of GE fish was also being considered. The second-generation rDNA controversies were about products, rather than techniques, specifically about large-scale releases of GEOs into the environment and their impacts on ecosystems, and on the human health effects of GE crops and rDNA-derived animal hormones.

Three agricultural products introduced during the 1980s and early 1990s, ice-minus bacteria, slow-ripening tomatoes, and recombinant bovine growth hormone (rBGH), were among the first commercialized organisms involving rDNA technology. Ice-minus bacteria, a genetically modified strain of *Pseudomonas syringae,* were the first gene-engineered product released into the environment. After five years of regulatory review, lawsuits, and community protests in Monterey, California, this soil bacterium with its ice-nucleation gene excised was field-tested in northern California in 1987. Ice-minus was designed to be sprayed on crops in the frostbelt when temperatures fell to a few degrees below freezing to prevent damage from ice crystallization. The company that developed ice-minus, Advanced Genetic Resources, merged with DNA Plant Technology in 1989. Research on the genetically modified form of *P. syringae* was halted by the company in 1990.[26]

The Flav'r Savr tomato was a GE product developed in response to consumer interests for a tomato picked ripe from the vine, rather than green, that could still be transported without losing its firmness and freshness. The genetic technology that made this possible is known as "antisense." It involves reversing a DNA sequence in the plant. The chemical ethylene, produced naturally by plants, is an essential part of the ripening process. Plants that exhibit lower levels of ethylene are otherwise the same except that they ripen more slowly. By applying the antisense technology to the gene that synthesizes ethylene, scientists were able to reduce the rate of ethylene produced in the plant. In laboratory trials, delaying ethylene synthesis allowed the tomato to remain firm for as much as six weeks longer than non-transgenic tomatoes. The Flavr Savr tomato was designed to increase consumer use of tomatoes during the off-season and became the first transgenic whole-food product introduced on the market.

In May 1994 the FDA issued a finding that the Flavr Savr tomato was as safe as traditionally bred varieties. It also approved, as a food additive in the tomato, the marker enzyme for the resistance to the antibiotic kanamycin. The gene for that enzyme is part of the genetic alteration of the tomato. Under its policy of "substantial equivalence" the FDA did not require any special labeling of the tomato because it argued that the Flavr Savr maintained the essential characteristics of traditionally bred tomatoes.

There was scarcely a public debate over the marketing of the GE tomato per se, in large part because no new proteins were added to it, although, there was some concern about the spread of antibiotic resistance markers in fresh produce. In addition, the introduction of the first fresh food GE product raised the specter that "genes from different food sources, exchanged and rearranged, might alter the quality, toxicity or nutritional value of food sources."[27]

The Flavr Savr(™) tomato proved to be largely unsuccessful as a consumer product. It was introduced at a time when new foreign non-transgenic hothouse tomato varieties successfully entered the market at competitive prices. Some also attribute the limited commercial success of the Flavr Savr(™) to the fact that the antisense technology was initially used on a poor variety of tomato. By 1995 transgenic delayed-ripening tomatoes had been granted nonregulated status by the USDA. Many of the new transgenic varieties were used primarily for processing. Other tomato varieties were developed with thicker skin, altered pectin, and increased lycopene content.

The third of the first three new rDNA products generated the largest public reaction. Recombinant bovine growth hormone (rBGH) is a veterinary product whose development closely resembles human protein products. The gene for the animal hormone is transferred to a bacterium, which is then grown in large fermentation tanks and induced to express the hormone.

Cows injected with rBGH will increase their lactation and produce 15 to 20 percent more milk. Critics of rBGH cited increased cases of mastitis in cows, the inhumane treatment of animals who are chemically lactated, and uncertainty over the relationship between rBGH and the production of Insulin Growth Factor (a potentially dangerous side product). Moreover, groups such as the Consumers Union and small farmer organizations argued that consumers would receive no benefit from this product. Also, the benefits would accrue disproportionately to large highly mechanized dairy farms.

The ecological side of the second-generation controversies over rDNA technology included the extent to which transgenic crops released into the environment would (1) invade natural habitats through accelerated germination, root growth, and dispersal by acquiring resistance to biotic and abiotic stressers; (2) transfer herbicide tolerance traits from domesticated crops to weeds;[28] (3) reinforce the increased use of chemical herbicides adding to the human and wildlife toxic load;[29] (4) support the use of monocultural herbicide applications increasing the probability of weed resistance; (5) accelerate the growth of resistant traits in insects; (6) harm nontarget insects from the pesticidal properties of the plant; (7) result in the loss of genetic diversity.[30]

The release of transgenic animals into the natural environment also became a contested issue when proposals were made to the U.S. federal government for restocking rivers with genetically modified salmon, enlarged by growth hormone genes.

Bacillus thuriengensis (*Bt*) is a natural bacterium known to be effective against certain insects (lepidoptera) because of its toxic proteins. Some ecologists and environmentalists argued that the overuse of *Bt* would accelerate the onset of resistant strains. They cited evidence that more than 500 species of pests have developed resistance to conventional pesticides.[31]

By the mid-1990s the print media in the United States began reporting research results that confirmed some of the environmental concerns raised by natural resource ecologists. For example, *The New York Times* reported in March 1996: "A field study has shown that a gene inserted into a crop plant can easily be transferred to a close relative, highlighting potentially unseen consequences of the genetic engineering of plants. . . ."[32] Four years later the same paper ran the headline "New study links biotech corn to butterfly deaths" referring to the corn pollen with *Bt* toxins that can be carried to milkweeds, plants that are the food sources for Monarch butterfly *(Danaus plexippus)* caterpillars.[33] The *Times* referred to a field study in which scientists observed the toxic effects of pollen from transgenic plants on milkweed-feeding Monarch larvae.[34] This study came after a series of investigations involving pollen and Monarch butterflies began at Cornell University in 1999. After depositing *Bt* corn pollen on milkweeds and exposing them to Monarch larvae, Losey and his colleagues observed the toxic effects.[35] Although the results of these studies did demonstrate that *Bt* corn pollen could be toxic to Monarch larvae, it did not stop the planting of *Bt* corn. Instead, regulators were more attentive to the concentration of *Bt* pollen and effects on nontarget insects. U.S. regulators promoted the use of buffer zones to separate planted areas from sensitive species.

THIRD-GENERATION BIOTECH CONTROVERSIES: GLOBALIZATION

After almost a decade of negotiations on international trade liberalization, on January 1, 1995 the World Trade Organization (WTO) became a formalized part of the new economic order, receiving support from over 100 nations. It was the culmination of treaties such as the General Agreement on Tariffs and Trade (GATT) and the North American Free Trade Act (NAFTA). The single term that describes this multilateral effort to eliminate trade barriers and create permeable borders for commerce, investment, and, some would say, labor, is "globalization."

While free marketers throughout the world rallied around this concept, many grassroots organizations, unions, agricultural collectives, and small farmers began to question the equity of the radical rearrangement of market flows. Large multinational corporations, operating under enormous economies of scale and low resource labor costs from plants located in developing countries, were poised to drive out local entrepreneurs.

The first products of agricultural biotechnology were reaching the marketplace during the period that globalization was taking hold. The lens of globalization was turned on biotechnology. Many of the critics of globalization chose GE crops as their first example of the downsides of a global market system.

Europe had experienced some severe food contamination crises in the late 1990s including outbreaks of mad cow and hoof and mouth diseases. When U.S. GE crops were ready for European markets, many states wanted more extensive testing and demonstrable proof that these products were safe. In the United Kingdom there was a protracted public controversy involving Arpad Puzstai, a respected scientist working at the Rowett Research Institute in Aberdeen, Scotland, with 270 scientific publications on his resume, who reported that he found intestinal changes in rats fed on GE potatoes.[36] Puzstai's work was actively discredited by other scientists in the United Kingdom and he was relieved of his position at the institute, but the controversy over his findings continued after he published his study in *The Lancet* on October 16, 1999.[37] Meanwhile, U.S. biotech companies who were seeking foreign markets for their transgenic seeds were finding it increasingly difficult to turn Europe into a biotech importer.

Anti-biotech activists formed alliances with anti-free trade activists. Their common concern was that several major chemical–agricultural corporations were gaining a world monopoly over GE seeds. These transnational agribusinesses sought open trade barriers so they could sell seeds cheaply to European farmers. Because the seeds were linked to certain chemical inputs such as herbicides, they could also expand their global herbicide markets. Anti-free trade activists observed U.S. chemical companies investing heavily in biotechnology while buying up seed companies to develop a global distribution network. According to agricultural expert Charles Benbrook, the developing nations in Africa were less than enthusiastic about the first-generation GE crops "created to make pest management simpler on America's large, mechanized farms."[38]

When the biotechnology industry was criticized for ignoring the needs of developing countries and expanding intellectual property ownership over biological entities, the industry released a GE product designed to turn the tide of public opinion, "Golden Rice," a strain of rice that contains beta carotene, which the body turns into vitamin A. People who are deficient in vitamin A from lack of leafy green vegetables and carrots are at risk of becoming blind. The biotech industry supported a research program to develop beta-carotene rice that would provide a person sufficient amount of vitamin A to prevent blindness. Thus, in lieu of addressing the problem of vitamin deficiency by enriching the diversity of the diet in developing countries, the approach chosen through the "genetic mode of production" was to create a single crop with all the essential amino acids and micronutrients. Early prototypes of "Golden Rice" had levels of beta carotene that were too low to reduce blindness when individuals were consuming normal diets of rice. While the media made it sound like we had turned the corner in preventing blindness from vitamin-A deficiency, there was still considerable R&D development left to increase the expression of beta carotene and to assess the acceptability of the orange-colored rice in the developing world where people prize white rice.

Golden Rice sparked a lively debate over biotech's role in improving the quality of life in developing countries. Critics of Golden Rice pointed to the complex set of reasons why the rural poor in the developing world go hungry including poor soil fertility, lack of inexpensive seeds, inferior infrastructure for transporting food and supplies, and lack of agricultural technology. They dismissed the single product solution to rural poverty.

In May 2003 U.S. trade representatives petitioned the WTO to declare illegal the de facto moratorium adopted by the EU on approving new GE crops under the new international trade agreements. The WTO convened a dispute resolution panel in early 2006 to hear the U.S. petition, which claimed $300 million in lost exports resulting from the GE moratorium The WTO resolution panel declared that the EU moratorium was not justified, but since the de jure moratorium was over it took no action. In another action, the U.S. president sent a message to developing nations that America would link foreign aid to a nation's policy on GE foods. In essence, developing nations that refused GE crops in aid would not get privileged foreign aid status. This came in the aftermath of Zambia's 2002 rejection of shipments of U.S. food aid containing GE corn. The Bush administration rebuked those opposed to GE products as "undermining efforts to fight hunger in Africa."[39]

Efforts by the U.S biotech companies and trade representatives to make Europe GE friendly according to American standards had not succeeded by late 2003. In July 2003 the European Parliament approved legislation requiring strict labels for goods made with genetically altered ingredients. This action is consistent with the EU's desire for a standard for labeling and traceability. "This legislation also ensures that genetically modified . . . foodstuffs like grains will be traced from the moment of their inception to their arrival in the European Union through the processing stage and into the supermarket."[40] American farmers and grain processors, who wish to export to European markets, would have to separate GE from non-GE seed to comply with the labeling provision. In another setback for the biotechnology sector, the Cartagena Protocol on Biosafety, first agreed upon by 130 nations in January 2000, came into effect in September 2003 after being formally ratified by the fiftieth state.[41] By July 2006, the Congo became the 134th signatory nation to ratify the biosafety convention. According to the Cartagena Protocol, countries can bar the imports of GE entities (seeds, crops, microbes, or animals) if they believe it would threaten their environment. Like the European Parliament decision, the Cartagena Protocol also calls for labeling of GE products.

CONCLUSION

The biotechnology revolution has passed its thirtieth anniversary, if we mark its beginning with the publication of the first plasmid-mediated gene transplantation experiment. Like other industrial revolutions, it is premised on new forms of production. Scientists have learned the secrets behind the system under which living cells produce proteins and have commandeered that system to either replicate scarce products of nature or to synthesize new ones. GE is as much a revolution in molecular genetics as it is in biology as a whole. Thirty years ago, biology became transformed in ways that chemistry and physics had years before. In Barry Commoner's words: "Biology once was regarded as a languid, largely descriptive discipline, a passive science that was content, for much of its history, merely to observe the natural world rather than change it. No longer."[42]

Since genes are transferable across living things, they can be reassigned to new cellular factories. Thus, a protein typically synthesized in a human cell can be produced more efficiently and in greater quantities in a plant cell. Agriculture becomes a new production system for human proteins called biopharmaceuticals supplanting human tissue culture production.[43]

The "genetic mode of production" has given rise to new products, new methods of producing old products, and new delivery systems (such as vaccines delivered through crops). Moreover, it has created a bridge between universities, small start-up companies, and multinational corporations.

Nearly two decades ago, in his book *Biotechnology: The University-Industry Complex,* Martin Kenney questioned whether biotechnology will survive as a freestanding industry or whether it will provide the tools for and be absorbed by traditional industries.[44] In fact both these developments have taken place. Traditional industries have incorporated the tools of biotechnology into their production and/or service systems. At the same time GE techniques have spawned a new information-based industrial sector. In this sense biotechnology is like the computer revolution. It is both a freestanding industrial sector and a set of tools that have been integrated into other sectors. The most visible signs of the new industry are to be seen in the field of pharmaceuticals and agriculture. Other applications, especially in the field of biomaterials, are also likely to evolve although outside of the intense media limelight. According to the leading trade organization, the Biotechnology Industry Organization (BIO), in 2003 there were 1,457 biotechnology companies in the United States; 342 are publicly held. According to BIO, the revenues in this sector reached $34.8 billion in 2001.[45] According to the twentieth-anniversary edition of *Beyond Borders: The Global Biotechnology Report 2006,* Ernst and Young report that revenues of publicly traded biotechnology companies reached $63.1 billion in 2005, the highest in its thirty-year history.

Previous technological revolutions in the twentieth century, such as the invention of plastics, microelectronics, and computers, have sold themselves. Biotechnology has met numerous forms of public opposition at the outset, and as it matures, it will carry new moral dilemmas and force adjustments within civil society. With the benefits of hindsight we will also be in a better position to distinguish between exaggerated claims and ideologically based criticisms of this new industry, the benefits and liabilities of which have thus far been largely assessed by a prospective rather than a retrospective analysis.

NOTES

1. Huxley, J. 1936. Chairman's introductory address. In Lancelot Hogben. *The retreat from reason.* Conway Memorial Lecture. May 20. London: Watts & Co. p. vii.

2. Bacon, Lord Francis. Orig. pub. 1622. *The new Atlantis.* trans. H. Goitein. New York: E.P. Dutton & Co.

3. Barboza, D. 2003. You asked for it; you got it: the pint-size watermelon. *New York Times,* Sec. 1, p. 1 (Business) (June 15).

4. Cohen, S.N., Chang, A.C.Y., Boyer, H.W., and Helling, R.B. 1973. Construction of biologically functional bacterial plasmids *in vitro. Proceedings of the National Academy of Sciences USA* 70: 3240-3244.

5. Morrow, J.F., Cohen, S.N, Chang, A.C.Y., Boyer, H., Goodman, H., and Helling, R. 1974. Replication and transcription of eukaryotic DNA in *Escherichia coli. Proceedings of the National Academy of Sciences USA* 71: 1743-1747.

6. Bud, R. 1992. The zymotechnic roots of biotechnology. *British Journal of the History of Science* 25: 127-144.

7. Bud, R. 1993. *The uses of life: A history of biotechnology.* Cambridge, UK: Cambridge University Press, p. 34.

8. Krimsky, S. 1997. Revolution or evolution. *Future* II, Hoechst. p. 15.

9. Office of Technology Assessment, U.S. Congress. 1984. *Commercial biotechnology.* Washington, DC: USGPO, OTA-BA-218 (January), p. 1.

10. Judson, H.F. 1992. A history of gene mapping and sequencing. In *The code of codes: Scientific and social issues in the human genome project.* Edited by D.J. Kevles and L. Hood. Cambridge, MA: Harvard University Press, p. 62.

11. Gilbert, W. 1992. A vision of the grail. In *The code of codes: Scientific and social issues in the human genome project.* Edited by D.J. Kevles and L. Hood. Cambridge, MA: Harvard University Press, p. 92.

12. Krimsky, S. 1982. *Genetic alchemy.* Cambridge, MA: The MIT Press.

13. Commoner, B. 2002. Unraveling the DNA myth. *Harpers* 34: 44 (February).

14. Newmark, P. 1980. The origins of biotechnology. *Nature* 283: 124.

15. Morrow, J.F. et al. 1974. See note 5.

16. Johnson, I.S. 1977. Pharmaceutical applications: microbial production of insulin. In *Research with recombinant DNA: An academy forum,* March 7-9, 1977. National Academy of Sciences. Washington, DC: National Academy of Sciences, p. 162.

17. Valentine, R.C. 1977. Genetic engineering in agriculture with emphasis on biological nitrogen fixation. In *Research with recombinant DNA.* Washington, DC: National Academy of Sciences, pp. 224-236.

18. Krimsky, S., and Wrubel, R. 1996. *Agricultural biotechnology and the environment.* Urbana, IL: University of Illinois Press.

19. Weiner, C. 1982. Relations of science, government, and industry: The case of recombinant DNA. In *Science, technology and the issues of the eighties.* Edited by A. Teich and R. Thornton. Boulder, CO: Westview Press, pp. 71-97.

20. Kenney, M. 1998. Biotechnology and a new economic space. In *Private science.* Edited by A. Thackray. Philadelphia: University of Pennsylvania Press, p. 135.

21. Office of Technology Assessment, U.S. Congress. 1988. *New developments in biotechnology: 4 U.S., investment in biotechnology.* OTA-BA-360. Washington, DC: OTA. p. 79.

22. Office of Technology Assessment, U.S. Congress. 1991. *Biotechnology in a global economy.* OTA-BA-494. Washington, DC: OTA (October), p. 45.

23. Bud, R. 1993, see note 7.

24. Krimsky, S., and Wrubel, R. 1996. p. 15. See note 18.

25. World Bank. 1991. *Agricultural biotechnology: The next "green revolution"?* Technical Paper #133. Washington, DC: The World Bank, p. vii.

26. Krimsky, S., and Wrubel, R. 1996, Chapter 8. Frost inhibiting bacteria. See note 18.

27. Ibid., p. 235.

28. Yoon, C.K. 1999. Squash with altered genes raises fears of "superweeds." *New York Times* Sec. A, p. 22 (November 3).

29. Goldburg, R.J. 1992. Environmental concerns with the development of herbicide-tolerant plants. *Weed Technology* 6: 647-652.

30. "Because of the concentration of seed production in the hands of a few companies aiming to recoup large investments in GM seed development only a small number of highly bred species will be placed on the market." Berkhout, F. 2002. Novel foods, traditional debates: Framing biotechnology sustainability. *New Genetics and Society* 21:137.

31. Gould, F., Anderson, A., Jones, A., Sumerford, D., Heckel, D.G., Lopez, J., Micinski, S., Leonard, R., and Laster, M. 1997. Initial frequency of alleles for resistance to *Bacillus thuringiensis* toxins in field populations of *Heliothis virescens*. *Proceedings of the National Academy of Sciences USA* 94: 3519-3523.

32. Leary, W.E. 1996. Gene inserted in crop plant is shown to spread to wild. *New York Times* Sec. B, p. 14 (March 7).

33. Yoon, C.K. 2000. New study links biotech corn to butterfly deaths. *New York Times* Sec. F, p. 2 (August 22).

34. Hansen Jesse, L.C., and Obrycki, J. J. 2000. Field deposition of *Bt* transgenic corn pollen: Lethal effects on the monarch butterfly. *Oecologia* 125: 241-248.

35. Losey, J.E., Rayor, L.S., and Carter, M.E. 1999. Transgenic pollen harms monarch larvae. *Nature* 399: 214.

36. Editorial. 1999. Health risks of genetically modified foods. *The Lancet* 353: 1811.

37. Ewen, S.W.B., and Pusztai, A. 1999. Effects of diets containing genetically modified potatoes expressing *Galanthus nivalis* lectin on rat small intestine. *The Lancet* 354: 1353-1354.

38. Benbrook, C.M. 2003. Saving seeds of destruction. Op. Ed. *New York Times,* Sec. A, p. 19 (July 11).

39. Leonhardt, D. 2003. Talks collapse on U.S. efforts to open Europe to biotech food. *New York Times,* Sec. A, p. 1 (June 20).

40. Alvarez, L. 2003. Europe acts to require labeling of genetically altered food. *New York Times,* Sec. A, p. 3 (July 3).

41. Pollack, A. 2003. Trade pact on gene-altered goods to take effect in 90 days. *New York Times,* Sec. C, p. 3 (June 14).

42. Commoner, B. 2002. p. 39. See note 13.

43. A Canadian company called SemBioSys Genetics Inc. developed a transgenic canola variety that contains an anti-coagulant originally found in the saliva of leeches.

44. Kenney, M. 1986. *Biotechnology: The University-Industry Complex.* New Haven, CT: Yale University Press, p. 240.

45. Biotechnology Industry Organization. *Biotechnology industry statistics.* http://www.bio.org/er/statistics.asp (Accessed December 29, 2003).

Chapter 2

Controversy Around Terminology and Novelty: Engineered, Modified Biotechnology, and Transgenics

Brian Ellis

The highly polarized controversy over genetic engineering (GE) of crops has been complicated by two intertwined challenges. The first of these has been the need to establish a dialogue about complex biological concepts between parties whose scientific vocabulary and sophistication covers the full spectrum from the deep to the nonexistent. At the same time, the GE debate involves bridging the "two cultures" identified many years ago by C. P. Snow, since the issues at the core of the controversy are not purely scientific questions but also tap into ethical and sociopolitical concerns. It is probably fair to say that the conjunction of these mutually reinforcing challenges has contributed to the suspicion and frustration that typify the present scene, traits that are the hallmark of ineffective communication. However, resolution of this situation will require more than a communication effort focused on "better informing the public about the science," a model often promoted by the scientific community. Rather, the goal must be to develop a pattern of two-way communication that acknowledges and respects the interests and concerns of all those involved. It is hoped that this book makes a useful contribution to that global dialogue.

Genetically Engineered Crops
© 2007 by The Haworth Press, Inc. All rights reserved.
doi:10.1300/5880_02

THE CLASSIC PERIOD

If we are to place the GE controversy in its appropriate context, we need to appreciate both the origins and current state of this highly penetrant technology. Plant GE is the newest approach to one of the oldest agricultural challenges: How can we develop a better food source? For millennia, the human response to this challenge has been the selection and propagation of a desirable individual (or subpopulation) from a population of plants that displays variation in the trait(s) of interest. In the early days of plant domestication this selection process was probably not very systematic, but over time it did result in the establishment of distinct subpopulations of the species of interest, Within these subpopulations were concentrated gene variants (mutations) whose characteristics added tangible value (e.g., large seeds, uniform ripening, greater carbohydrate content) to the harvested product. Random mating between these selected genotypes typically exposed additional variation in their progeny, which provided ongoing opportunities to repeat the selection and create even more desirable forms of the crop plant.

The rediscovery of Gregor Mendel's laws of genetics in the early twentieth century finally created an intellectual framework for predicting the outcome of crossing, and this event is generally considered to mark the beginning of scientific breeding. Over the next eighty years, our knowledge of genetics allowed the systematic exploitation of the genetic potential in our major food crop species, resulting in unprecedented gains in crop yield and quality. It has also generated unique genotypes tailored to address specific agronomic needs such as tolerance to frost, resistance to new races of pathogens, and reduced levels of toxic metabolites. With few exceptions, these modifications have been the result of sexual crossing of selected parental genotypes, followed by additional cycles of selection within their progeny for new gene combinations. Plant breeding has thus become a highly sophisticated enterprise, creating thousands of new varieties of food, fiber, and ornamental crops each year.

Extrapolation from this success into the future, however, is not straightforward. The narrow genetic base of some of our major crops, a consequence of historical patterns of commercial introduction and improvement, has made it increasingly difficult to sustain the pattern of yield and quality gains to which growers have become accustomed. Changing agronomic practices and market demands, as well as evolution of new races of pests and pathogens capable of overcoming existing genetic resistance mechanisms, constantly force breeders to seek out novel variation. This may involve introducing less highly selected germplasm sources such as locally adapted varieties (landraces) or wild relatives of crop species, or using special treatments to force sexual crossing between parents that would normally be sexually incompatible.

An alternative approach to the creation of additional variability is the direct mutagenesis of existing germplasm. This, in effect, drastically speeds up the normal rate of spontaneous mutation by treating seeds or seedlings with chemical or physical agents that create changes in the DNA sequence (point mutations) or structure (DNA insertions/deletions and rearrangements). The location of such induced changes within the genome is essentially random, and mutagenized plants typically may carry dozens, if not thousands, of mutagenically introduced DNA changes. While mutagenesis did enjoy a period of popularity as a tool for induction of novel traits in plants in the mid-twentieth century, it generally proved to be an inefficient method for creating useful improvements in traits. Nevertheless, many advanced breeding lines in crop programs have one or more progenitors that arose from an earlier mutagenesis program, which means that a significant fraction of modern plant varieties probably contain partially mutagenized genomes whose full range of induced DNA changes has never been characterized or even quantified. At present in North America, new varieties whose phenotypic novelty may have been derived (in part) from induced mutation, are subject to the same regulatory testing and evaluation that is applied to genotypes derived from other modes of classical breeding. The underlying rationale is presumably that mutational change in genomes is a normal part of evolution, and artificial mutagenesis simply speeds this process up manyfold.

While these reservoirs of novel genetic material do expand the palette of variation with which breeders can work, their introduction typically comes at a price. Integrating a novel trait from an undeveloped germplasm source into an agronomically acceptable genetic background adds substantially to the time required to regenerate a new, commercially desirable variety, and time is a continuous challenge for a plant breeder. Depending on the species and the available germ-plasm, development of a new crop variety may require anywhere from seven to thirty or more years. Extensive collections of advanced breeding lines and innovations such as winter nurseries (shuttle breeding) can help shorten the cycle, but classical plant breeding is intrinsically a very slow process. At the same time, the pace of agribusiness and the demands for new and diversified products are steadily increasing. The increased speed of novel germplasm development offered by GE technology thus holds understandable attractions for those who must deal with the pressure to generate more novelty more quickly.

MOTHER NATURE'S GENETIC ENGINEER

As plant breeding continued to push the envelope of crop productivity, research into a relatively minor plant disease (crown gall disease) in the latter part of the twentieth century was quietly setting the stage for a genetic revolution. The painstaking studies that revealed how the crown gall disease agent, a soil bacterium called *Agrobacterium tumefaciens,* is able to attack and modify its host plant's cells have been well described elsewhere.[1] A key discovery was the observation that this bacterium *permanently* modified its host plant's genetic blueprint by transferring a small set of bacterial genes directly into the plant cell's DNA. However, the discovery that opened the door to GE crop development was the finding that the *Agrobacterium* gene transfer mechanism was blind to the nature of the genes being moved from the bacterial cell to the plant cell. In other words, it was possible to use routine DNA manipulation methods in the laboratory to replace the usual *Agrobacterium* genes with other pieces of DNA, of any origin, and the bacterium would proceed to transfer the new DNA efficiently into the plant cell genome. Several research groups immediately recognized the potential of this observation and the ensuing race led to the rapid development of practical methods for routinely inserting tracts of novel DNA into plant genomes to create transgenic plants.[2]

WHEN IN DOUBT, REACH FOR YOUR GUN

Agrobacterium-mediated gene transfer had been exploited to create an extraordinarily powerful technology, but this method suffered one major practical drawback—the biological host range of *A. tumefaciens*. While *Agrobacterium* can efficiently infect many plant species, there are far more species that are not susceptible to crown gall disease, and the list includes most crops of major commercial importance, including wheat, rice, corn, and soybeans.

This barrier was eventually overcome by the discovery in the mid-1980s that pieces of foreign DNA forcibly injected into living cells would also become incorporated into the cell's DNA. The early demonstrations of this phenomenon literally used gunpowder to drive microscopic "bullets" loaded with DNA into plant cells,[3] but more sophisticated tools were soon developed, largely based on high-pressure gas propulsion (e.g., Gene Gun®). The incorporation obtained via these "biolistic" approaches was far less efficient than that resulting from the *Agrobacterium*-mediated process, and the incoming DNA was often damaged. However, with sufficient screening and assessment of an array of biolistically modified plants, some of the incoming DNA that carried a functional version of the desired DNA insert could generally be recovered as a permanent part of the plant genome. The advantage of the biolistic gene transfer method is its universal applicability, which allows researchers to generate transgenic plants from virtually any desired species. However, the *Agrobacterium*-mediated gene transfer method is still preferred wherever possible because of its higher transfer efficiency and reduced tendency to generate aberrant DNA inserts.

THE BASIC TOOLKIT

No matter whether the new DNA has been inserted using *Agrobacterium* or a biolistic technique, the same basic DNA elements are being added to the genetic complement of the resulting transgenic plant. These consist of (1) a primary gene function unit, which would typically be a new gene, or a DNA element designed to modify specifically the action of an existing plant gene; (2) a promoter region, which controls the level and timing of expression and activity of the primary gene function unit; and (3) a selectable marker, which is a second gene function unit encoding a protein whose properties ensure that cells carrying this gene will survive treatments that kill all other cells, thus allowing selection of cells also carrying the primary gene or trait of interest.

The most commonly used selectable marker systems have been derived from early recombinant DNA studies in microbial species, and, therefore, consist of genes encoding various types of resistance to antibiotics. More recently, however, other selectable marker functions have been developed specifically for use in plants, including genes encoding various types of herbicide tolerance,[4] or the ability to metabolize unusual or toxic substrates.[5] The selectable marker gene function unit itself must be accompanied by another promoter region that ensures strong expression of the selection function in all the transformed cells.

The basic plant transformation "cassette" or "construct" thus consists of at least two functional gene units with their respective promoters. Within that basic model, there is virtually unlimited flexibility to create new gene combinations in plants. The transformation construct can include genes from any species, whether plant, animal, or microbe—the only constraint is that the structure and composition of the new DNA be compatible with the functional requirements of the cellular gene expression machinery in plants. For genes from plant and animal sources this generally poses no problems, but microbial gene sequences sometimes require modification to allow efficient processing within a plant cell milieu. Thus, in order to obtain high levels of expression in transgenic plants of the insecticidal Cry protein from the bacterium, *Bacillus thuringiensis,* which is used to create the so-called *Bt* crops now on the market, the nucleotide sequence encoding the protein was altered to reflect the preferences of the plant cell's protein synthesizing machinery.[6]

CHOOSING TARGETS

Beyond these operational constraints, the choice of functional gene units for use in plant GE is driven largely by the desired outcomes in the resulting transgenic plants. Interestingly, despite the availability of a vast global gene pool, the transgenic crops developed for commercialization over the past decade have focused on a very limited set of gene function units. These encode either (1) various classes of herbicide resistance, (2) a range of variations on the family of *B. thuringiensis* Cry proteins, or (3) constructs that trigger heightened resistance to viral diseases.

Several factors combined to create this pattern. First, these gene function units are relatively easy to develop and manipulate, since they all control discrete, positively acting genetic traits. This simplifies the design of the transformation constructs and the selection of the transformed progeny. Second, from a commercialization perspective, the traits in question also directly address operational needs of some field crop producers, who represent the target market for new seed varieties. Herbicide tolerance, for instance, greatly simplifies weed control in cropping systems, while Cry protein-containing plants are less susceptible to damage from insect pests. Pest and weed management is a challenge faced by growers every year, and the ability of the first-generation GE crops to make that management easier led to their rapid adoption, particularly within the industrial agriculture sector in North America.

The economic profile of that sector has, in turn, influenced the pattern of GE crop development. The research and development costs associated with creation and introduction of the first GE varieties have been very high, but the incremental added value for a GE seed relative to a non-GE seed is very small. These factors dictated a commercialization strategy that focused on crops whose scale of production would, given sufficient market penetration, provide an acceptable return on that investment. GE variety development has, therefore, concentrated on major commodities such as corn, soybeans, and cotton. Less attention has been devoted to smaller-volume crops such as potatoes, tomatoes, and canola, although GE canola has become a major component of oilseed production in Canada.

ROUND TWO

The lists of GE crop varieties currently approved for commercial release in Canada and the United States at first appear much longer than this narrow focus might suggest. However, this reflects the development of various versions of herbicide-tolerant germplasm by different companies, and deployment of different classes of the Cry family of insecticidal proteins to address particular pest problems. More recently, GE varieties carrying combinations of the same traits have been released but, in general, there have been very few novel traits added to the array of first-generation traits developed almost ten years ago. In part, this pattern reflects the general utility of these early traits, which are still attractive for the grower community. In addition, however, the paucity of novel traits is indicative of the challenges facing the next generation of trait developers.

The existing herbicide-tolerant genes, and *Bt* constructs, display high specificity for their targets without significantly perturbing plant metabolism and agronomic performance. The proposed targets of the next-generation traits, on the other hand, are complex biological phenomena such as photosynthetic efficiency, cold tolerance, or resistance to fungal disease. While single genes have been discovered that can positively affect these traits in transgenic plants, their mode of action often remains obscure. In addition, the ability of a gene isolated from one plant species to function efficiently in another species' genetic background, or to provide long-term field-level effectiveness, will need careful examination.

Thus, while many different transgenic crop improvements are currently under early or mid-stage development in the product pipeline, relatively few have made it to the marketplace. The increased development costs associated with engineering more complex traits, the lack of clarity and consistency in the GE crop regulatory process, and the uncertainty in the global market created by widespread rejection of GE crops in the international food supply system, have all combined to deter agbiotech companies from attempting to bring many second-generation varieties into production.

It is difficult to see how any of these pressures will be relieved in the near future. Our understanding of plant biology is rapidly growing, thanks to the availability of new genome-scale tools such as DNA microarrays that allow us to monitor the expression of every gene in the plant at a given time. As a result, our ability to predict, with much less uncertainty, the outcomes of specific modifications of a crop genome will steadily improve. This should shorten the development cycle and allow unsuitable modifications to be detected and rejected early on in the research and development cycle, a trend that will eventually reduce product development costs. Similarly, as the national regulatory agencies gradually accumulate more experience with performance and impacts of transgenic crops, we can anticipate movement toward an approval system that is able to draw upon a much richer scientific knowledge base.

The third challenge pertaining to further extensive deployment of GE technology may be less tractable. Public perceptions of this innovative technology currently cover the spectrum from enthusiastic acceptance to intense antipathy. The end of this spectrum that dominates can vary dramatically, depending on the continent or country concerned. While many factors contribute to this widespread unease, the two issues that dominate all others are (1) concern about potential human health impacts, and (2) uncertainty about environmental problems that might arise from large-scale deployment of GE varieties.

At first glance, the scale of this unhappiness with GE crop technology seems to be out of proportion with any identified risk. After multiple years of large-scale production and integration into the North American food supply system there have been no verified adverse health effects attributable to consumption of any approved commercial GE crop variety, nor have irremedial or catastrophic environmental impacts been reported. The continuing unease thus seems to stem more from the uncertainty that still attends the potential for negative impacts. As pointed out earlier, this uncertainty may be resolved over time by improvements in our understanding of the relevant science, and by accumulated experience. However, to place this in context, and to gauge the scale of the necessary scientific knowledge expansion it may be useful to consider the nature and scope of the gaps that need filling.

SO MANY GENES, SO MUCH TIME

The point has often, and correctly, been made that humans have been manipulating the genetic structure of their food sources for thousands of years. The entire process of crop species domestication has been essentially one of selection of certain gene combinations in preference to others. The consequence has been an unprecedented shift in the terrestrial ecological landscape in almost every corner of the planet. On the one hand, a tiny fraction of the world's 300,000+

plant species have been converted into modified genetic forms that now cover large parts of the landbase. At the same time, consumption of these modified genetic forms has allowed humans to multiply to the point where they now dominate the planet's ecology. Quantification of this phenomenon is challenging but a recent study concluded that humans now appropriate between 10 and 55 percent of the total terrestrial photosynthetic output on Earth,[7] and it would appear that this can only increase in the future. This represents a truly extraordinary concentration of the planet's resources on a single species, and emphasizes how much the future health of the biosphere depends on humans making wise choices in their collective activities.

The plant genotype manipulation that brought us to this point has been remarkably effective at rearranging existing gene combinations for specific human purposes, but it is important to recognize that the process has also operated almost exclusively upon discrete gene pools. Broadly speaking, these are the groups of organisms that we characterize as "species," since one of the classical criteria for species definition is genetic isolation from related genotypes. While not an absolute threshold, a species generally defines a gene pool (i.e., a group of related organisms) within which gene exchange and recombination will freely occur in conjunction with sexual breeding.

Given the large number of genes (perhaps 25,000 to 50,000) in a typical plant genome, the potential genetic variation at every DNA base position (estimated at 2 to 4 percent across the genome), and the constant "stirring" of this pool through spontaneous mutation, rearrangement, and recombination, there would appear to be an almost limitless range of genetic diversity available within any one species. However, we know that species generally remain recognizably distinct for thousands, or even millions, of years. The unavoidable conclusion is that there must be major constraints within an organism's genetic architecture on how much of the potential variation can be realized at the level of phenotype. These constraints are thought to reside in the gene interaction and regulation networks that ultimately control the expression of the genetic blueprint and the articulation of the final phenotype. Acting as a "systems level" integrator, the networks monitor the internal and external environment, and operate to maximize the likelihood of survival of the organism (and its genes). It is within this context of "biological outcomes" that new genetic variants are tested for their contribution to fitness and survival.

In the narrower context of agriculture, however, the definition of "fitness" is heavily biased toward the goals set by plant breeders and agronomists. Thus, the genetic variants created in crop species have tended to be those whose biology is focused on yield (maximal conversion of photosynthate to edible parts of the mature plant) when the plants are grown under highly managed conditions. It is a matter of current debate whether this drive toward ever greater agronomic yield can be sustained, but the available data suggest that physiological limits on yield gain are being approached in at least some species. What is not clear is whether this reflects an absolute yield ceiling based on the attributes of the underlying biochemical reactions, or whether the putative limits derive from the specific genetic architecture that has evolved within that species. If the former is correct, prospects for further progress will presumably be limited. However, if the apparent ceiling derives from a particular genetic configuration that represents the best that can be extracted from the available gene pool within that species, the potential exists, at least in principle, to reconfigure that genetic architecture by introducing additional genetic elements. Such a reconfiguration might, for instance, allow increased efficiency of energy capture and carbon allocation to seed production.

Introduction of novel genetic elements is, of course, the basic paradigm of crop GE. However, the first-generation GE crops in commercial production have only indirectly addressed yield increases by using gene function units that increase crop productivity through reduced weed competition and suppression of pest and disease losses. The addition of a gene conferring insensitivity to a synthetic herbicide, or driving accumulation of an insecticidal protein such as *Bt* toxin, is generally felt to be unlikely to affect the remainder of the transgenic plant's genetic and physiological functions, and the field performance of the first-generation GE varieties is consistent with this expectation, at least within an agricultural context. While it is difficult to distinguish between yield effects derived from the presence of a transgene and effects resulting from breeding-based genetic enhancement within the same germplasm collection, there does not appear to be any pattern of yield reduction associated with the first-generation GE varieties.

Less attention has been directed toward the impacts of these early transgenes on long-term reproductive success in plants growing outside of managed agroecosystems. One recent study demonstrated that four different GE crops carrying herbicide-tolerance or insect pest-tolerance functions competed relatively poorly with nontransgenic genotypes over a period of three years,[8] suggesting that expression of the transgene was imposing a significant fitness cost. On the other hand, transgenic plants of *Arabidopsis* (currently the model plant for plant biology studies) that carry another herbicide-tolerance gene construct have been reported to become more promiscuous (i.e., display higher rates of outcrossing) than wild-type plants.[9] Unintended effects such as these emphasize the need to thoroughly examine each transgenic genotype on a case-by-case basis if the full impact of the genetic modification is to be understood.

Some of the second-generation GE varieties under consideration are likely to attempt to address crop yield potential directly by introducing elements intended to modify carbon allocation, enhance photosynthetic rates, and stimulate growth under conditions of suboptimal nutrient supply. These changes will require direct intervention in the physiological systems of the plant, and thus raise obvious questions about the ecological consequences of such "fitness enhancement" genes. For example, acquisition of higher photosynthetic activity could conceivably provide a selective advantage for a wild relative, thus improving its reproductive success and allowing it to extend its normal range or local abundance, that is, to become more "weedy." However, it is important to emphasize that such novel GE traits will be deployed within the "genetic envelope" of highly selected and agronomically optimized parental genotypes, and the efficacy of the GE trait is evaluated within a managed agroecosystem context. What performs well within such a system and genetic background may be distinctly less functional when operating within a nonselected ("wild") genetic background, and growing in a much more diverse and challenging nonmanaged ecosystem. Nevertheless, the potential for unintentionally creating new pest and weed problems through dispersal of transgenes into native populations cannot be discounted, which means that analysis of the ecological outcomes of transgene introgression into wild relative populations remains a high-priority research area.

CONCLUSIONS

Technology development and deployment in the arena of transgenic crops has been extraordinarily rapid, leaving both regulators and plant scientists scrambling to understand the full implications both for agriculture and for ecosystem management. Fortunately, the same tools that led to creation of these novel products are also enabling researchers to probe their performance and impact in a much more comprehensive and informative fashion than could have been envisioned even ten years ago. The challenge will be to ensure that the knowledge thereby gained effectively guides society's future choices.

NOTES

1. Lurquin, P. 2001. *The green phoenix: A history of genetically modified plants.* New York: Columbia University Press.

2. Fraley, R.T., Rogers, S.G., Horsch, R.B., Sanders, P.R., Flick, J.S., Adams, S.P., Bittner. M.L., Brand, L.A., Fink, C.L., Fry, J.S., et al. 1983. Expression of bacterial genes in plants. *Proceedings of the National Academy of Sciences* USA 80: 4803-4807.

3. Klein, T.M.. Wolf, E.D., Wu, R., and Sanford, J.C. 1987. High-velocity microprojectiles for delivering nucleic acids into living cells. *Nature* 327: 70-73.

4. Wilmink, A. and Dons, J.J.M. 1993. Selective agents and marker genes for use in transformation of monocotyledonous plants. *Plant Molecular Biology Reporter* 11: 165-185.

5. Todd, R. and Tague, B.W. 2001. Phosphomannose isomerase: A versatile selectable marker for *Arabidopsis thaliana* germ-line transformation. *Plant Molecular Biology Reporter* 19: 307-319.

6. Perlak, F.J., Fuchs, R.L., Dean, D.A., McPherson, S.L., and Fischoff, D.A. 1991. Modification of the coding sequence enhances plant expression of insect control protein genes. *Proceedings of the National Academy of Sciences USA* 88: 3324-3328.

7. Rojstaczer, S., Sterling, S.M., and Moore, N.J. 2002. Human appropriation of photosynthesis products. *Science* 294: 2549-2552.

8. Crawley, M.J., Brown, S.L., Hails, R.S., Kohn, D.D., and Rees, M. 2001. Biotechnology: Transgenic crops in natural habitats. *Nature* 409: 682-683.

9. Bergelson, J., Purrington, C.B., and Wichmann, G. 1998. Male promiscuity is increased in transgenic plants. *Nature* 395: 25.

Chapter 3

Transgenic Crops, Agrobiodiversity, and Agroecosystem Function

Miguel A. Altieri

Transgenic or "genetically engineered" (GE) crops, the main products of agricultural biotechnology, are becoming a dominant feature of the agricultural landscapes of the United States and other countries. Worldwide, the areas planted with transgenic crops increased more than thirtyfold, from 3 million hectares in 1996 to nearly 58.7 million hectares in 2002.[1] The increase in area between 2001 and 2002 was 12 percent, equivalent to 6.1 million hectares. Despite expectations that transgenic crops will benefit third world agriculture, 99 percent of the total global GE crop area remains concentrated in four countries: United States—55 percent of global total, Argentina—27 percent, Canada—6 percent, and China—4 percent. Globally the main GE crops are soybean, occupying 36.5 million hectares, and maize with 12.4 million hectares, followed by cotton and canola. In the United States, Argentina, and Canada, over half of the average for major crops such as soybean, corn, and canola are planted in transgenic varieties. Herbicide-resistant crops (HRC) and *Bacillus thuringiensis* protein-engineered insect-resistant crops (*Bt* crops) have been consistently the dominant traits. Transnational corporations such as Monsanto, DuPont, Novartis, and so on, which are the main proponents of biotechnology, argue that carefully planned introduction of these crops should reduce or even eliminate the enormous crop losses due to weeds, insect pests, and pathogens. They argue further that the use of such crops will have added beneficial effects on the environ-

Genetically Engineered Crops
© 2007 by The Haworth Press, Inc. All rights reserved.
doi:10.1300/5880_03

ment by significantly reducing the use of agrochemicals.[2] Several scientists have argued that HRCs and *Bt* crops have been a poor choice of traits to feature this new technology given predicted environmental problems and the issue of resistance evolution. There is considerable evidence to suggest that both of these types of crops are not really needed to address the problems they were designed to solve. On the contrary, they tend to reduce the pest-management options available to farmers[3] and there are many effective alternative approaches (i.e., rotations, polycultures, cover crops, biological control, etc.) that farmers can use to regulate the insect and weed populations that are being targeted by the biotechnology industry.[4] For example, in Africa, scientists at the International Center of Insect Physiology and Ecology (ICIPE) developed a habitat-management system to control Lepidopteran stemborers, potential primary targets to be controlled via *Bt* crops. The push–pull system uses *Penissetum purpureum* (Napier grass) and *Sorhgum vulgare* (Sudan grass) planted in the borders of maize fields to act as trap crops. These draw stemborer colonization away from maize (the push) and two plants intercropped with maize, *Melinis minutifolia* (molasses grass) and *Desmodium uncinatum* (silverleaf) that repel the stemborers (the pull).[5] Border grasses also enhance the parasitization of stemborers by the wasp *Cotesia semamiae,* and are important fodder plants. The leguminous silverleaf suppresses the parasitic weed *Striga* by a factor of forty when compared with maize monocrop. *Desmodium*'s N-fixing ability increases soil fertility; and it is excellent forage. As an added bonus, sale of *Desmodium* seed is proving to be a new income-generating opportunity for women in the project areas. The push–pull system has been tested on over 450 farms in 2 districts of Kenya and has now been released for use by the national extension systems in East Africa. Participating farmers in the breadbasket of Trans Nzoia are reporting a 15 to 20 percent increase in maize yield. In the semiarid Suba district, which is plagued by both stemborers and striga, a substantial increase in milk yield has occurred in the last four years, with farmers now being able to support increased numbers of dairy cows on the fodder produced. When farmers plant maize together with the push–pull plants, there is a return of US$2.30 for every dollar invested, compared with only $1.40 obtained by planting maize as a monocrop.[6] To the extent that transgenic crops further entrench the current monocultural system, they impede farmers from using a plethora of alternative methods such as the push–pull system.[7]

I contend that GE crops will lead to further agricultural intensification and ecological theory predicts that as long as transgenic crops follow closely the pesticide paradigm, such biotechnological products will do nothing but reinforce the pesticide treadmill in agroecosystems, thus legitimizing the concerns that many environmentalists and some scientists have expressed regarding the possible environmental risks of genetically engineered organisms. In fact, there are several widely accepted environmental drawbacks associated with the rapid deployment and widespread commercialization of such crops in large monocultures,[8] including the following:

1. The spread of transgenes to related weeds or conspecifics via crop-weed hybridization.
2. Reduction of the fitness of nontarget organisms (especially weeds or local varieties) through the acquisition of transgenic traits via hybridization.
3. The rapid evolution of resistance of insect pests to *Bt*.
4. Accumulation of the insecticidal *Bt* toxin, which remains active in the soil after the crop is plowed under and binds tightly to clays and humic acids.[9]
5. Disruption of natural control of insect pests through intertrophic-level effects of the *Bt* toxin on natural enemies.[10]
6. Unanticipated effects on nontarget herbivorous insects (e.g., Monarch butterflies) through deposition of transgenic pollen on foliage of surrounding wild vegetation.[11]
7. Vector-mediated horizontal gene transfer and recombination to create new pathogenic organisms.

As a new form of industrial agriculture, the rapid spread of transgenic crops threatens crop diversity by promoting large monocultures on a rapidly expanding scale leading to further environmental simplification and genetic uniformity. Such simplification and the associated environmental impacts of GE crops can lead to reductions in agroecosystem biodiversity. Direct benefits of biodiversity to agriculture lie in the range of environmental services provided by the different biodiversity components such as nutrient cycling, pest regulation, and productivity. Disruptions in biodiversity levels prompted by GE crops are bound to affect such services and thus affect agroecosystem function. This chapter focuses on the known and potential ecological effects of the two dominant types of GE crops: HRCs and *Bt* crops on agroecosystem performance.

BIOTECHNOLOGY
AND THE LOSS OF AGROBIODIVERSITY

Worldwide, 91 percent of the 1.5 billion hectares of cropland are under annual crops, mostly monocultures of wheat, rice, maize, cotton, and soybeans.[12] This represents an extreme form of simplification of nature's biodiversity, because monocultures in addition to being genetically uniform and species-poor systems advance at the expense of natural vegetation, a key landscape component that provides important ecological services to agriculture such as natural mechanisms of crop protection.[13] Since the onset of agricultural modernization, farmers and researchers have been faced with a major ecological dilemma arising from the homogenization of agricultural systems: an increased vulnerability of crops to insect pests and diseases, which can be devastating when infesting uniform, large-scale monocultures.[14] Monocultures may have short-term economic advantages for farmers, but in the long term they do not represent an ecological optimum. Rather, the drastic narrowing of cultivated plant diversity has put the world's food production in greater peril.[15]

History has repeatedly shown that the uniformity, which characterizes agricultural areas sown with a smaller number of varieties as in the case of GE crops, is a source of increased risk for farmers, because the genetically homogeneous fields tend to be more vulnerable to disease and pest attack.[16] Documented examples of disease epidemics associated with homogeneous crops abound, including the $1 billion loss of maize in the United States in 1970 and the 18 million citrus trees destroyed by pathogens in Florida in 1984.[17]

Many proponents of the biotech revolution also promoted the Green Revolution in the developing world. These people assume that progress and the development of traditional agriculture inevitably require the replacement of local crop varieties with improved ones. They also presume that the economic and technological integration of traditional farming systems into the global system is a positive step that enables increased production, income, and community well-being.[18] However as evinced by the Green Revolution, modernization and integration entailed several negative impacts:[19]

- The Green Revolution involved the promotion of a package that included modern varieties, fertilizers, and irrigation, and margin alized a great number of resource-poor farmers who could not afford the technology.
- In areas where farmers adopted the package stimulated by government extension and credit programs, the spread of modern varieties led to greatly increased use of pesticides, often with serious health and environmental consequences.
- Enhanced uniformity caused by sowing large areas with a few modern varieties increased risk for farmers. Genetically uniform crops proved more susceptible to pests and diseases, and modern varieties did not perform well in marginal environments where the poor live.
- Crop diversity is an important nutritional resource of poor communities, but the spread of modern varieties was accompanied by a simplification of traditional agroecosystems and a trend toward monoculture that affected dietary diversity, thus raising considerable nutritional concerns.
- The replacement of folk varieties also represents a loss of cultural diversity, as many varieties are integral to religious or community ceremonies. Given this, several authors have argued that the conservation and management of agrobiodiversity may not be possible without the preservation of cultural diversity.[20]

Concerns have been raised about whether the introduction of transgenic crops may replicate or further aggravate the effects of modern varieties on the genetic diversity of landraces and wild relatives in areas of crop origin and diversification and, therefore, affect the cultural thread of communities. The debate was prompted by a controversial article in the journal *Nature,* reporting the presence of introgressed transgenic DNA constructs in native maize landraces grown in remote mountains in Oaxaca, Mexico.[21] Although there is a high probability that the introduction of transgenic crops will further accelerate the loss of genetic diversity and of indigenous knowledge and culture through mechanisms similar to those of the Green Revolution, there are some fundamental differences in the magnitude of the impacts. The Green Revolution increased the rate at which modern varieties replaced folk varieties, without necessarily changing the genetic integrity of local varieties. Genetic erosion involves a loss of local varieties but it can be slowed down and even reversed through in situ conservation efforts, which conserve not only landraces and wild-weedy relatives, but also agroecological and cultural relationships of crop evolution and management in specific localities. Examples of successful in situ conservation that preserve native crop diversity while respecting local cultures have been widely documented.[22]

The problem with introductions of transgenic crops into regions of diversity is that the spread of genetically altered grain characteristics to the local varieties favored by small farmers could dilute the natural sustainability of these races. Many proponents of biotechnology believe that unwanted gene flow from GE maize may not compromise maize biodiversity (and, therefore, the associated systems of agricultural knowledge and practice along with the ecological and evolutionary processes involved) and may pose no worse a threat than cross-pollination from conventional (non-GE) seed. In fact some industry researchers believe that DNA from engineered maize is unlikely to have an evolutionary advantage, rather that if transgenes do persist they may actually prove advantageous to Mexican farmers and crop diversity. Opponents to biotechnology pose that traits important to indigenous farmers (resistance to drought, food or fodder quality, maturity, competitive ability, performance on intercrops, storage quality, taste or cooking properties, compatibility with household labor conditions, etc.) could be traded for transgenic qualities that may not be important to those same farmers.[23] Under this scenario, risk will increase and farmers will lose their ability to adapt to changing biophysical environments and to obtain relatively stable yields with a minimum of external inputs while supporting their communities' food security.[24]

Most scientists agree that teosintle and maize interbreed. One problematic result from a transgenic maize–teosintle cross would be if the crop–wild relative hybrids became more successful by acquiring tolerance to pests.[25] Such hybrids could become problem weeds upsetting farmers' management but also out-competing wild relatives. Another potential problem derived from transgenic crop-to-wild gene flow is that it can lead to extinction of wild plants through swamping and outbreeding depression.[26]

ECOLOGICAL EFFECTS OF HRCS

Gene Flow: Super Weeds, and Herbicide Resistance

Just as it occurs between traditionally improved crops and wild relatives, pollen-mediated gene flow occurs between GE crops and wild relatives or conspecifics despite all possible efforts to reduce it. The main concern with trangenes that confer significant biological advantages is that they may transform wild/weed plants into new or worse weeds. Hybridization of HRCs with populations of free living relatives will make these plants increasingly difficult to control, especially if they are already recognized as agricultural weeds and if they acquire resistance to widely used herbicides. Snow and Palma[27] argue that widespread cultivation of HRCs could exacerbate the problem of gene flow from cultivated plants enhancing the fitness of sexually compatible wild relatives. In fact, the flow of herbicide-resistant transgenes has already become a problem in Canadian farmers' fields where volunteer canola resistant to three herbicides (glyphosate, imidazolinone, and glufosinate) has been detected, a case of "stacked" resistance or resistance to multiple herbicides.[28] The Royal Society of Canada[29] reported that herbicide-resistant volunteer canola plants are beginning to develop into a major weed problem in some parts of the Prairie Provinces of Canada. Transgenic resistance to glufosinate can introgress from cultivated *Brassica napus* into weedy populations of *B. napus,* and persist under natural conditions.[30] In Europe there is a major concern about the possibility of pollen transfer of herbicide-tolerant genes from *Brassica* oilseeds to *Brassica nigra* and *Sinapis arvensis*.[31]

Transgenic herbicide resistance in crop plants simplifies chemically based weed management because it typically involves compounds that are active on a very broad spectrum of weed species. Postemergence application timing for these materials fits well with reduced or zero-tillage production methods, which can conserve soil and reduce fuel and tillage costs.[32] Reliance on HRCs perpetuates and accelerates the weed-resistance problems and species shifts that are common to conventional herbicide-based approaches. Herbicide resistance becomes more of a problem as the number of herbicide modes of action to which weeds are exposed becomes fewer and fewer, a trend that HRCs may exacerbate due to market forces that encourage their use. Given industry pressures to increase herbicide sales, acreage treated with broad-spectrum herbicides will expand, further exacerbating the resistance problem. For example, it has been projected that the acreage treated with glyphosate will increase to nearly 150 million acres by 2005. Although glyphosate resistance is considered less likely to develop in weeds, the increased use of the herbicide will eventually result in weed resistance, as has been already documented with Australian populations of *Lolium multiflorum* (annual ryegrass), *Agropyron repens* (quackgrass), *Lotus corniculatus* (birdsfoot trefoil), and *Cirsium arvense*.[33] In Iowa, *Amaranthus rudis* populations showed delayed germination thus "avoiding" planned glyphosate applications and velvetleaf demonstrated greater tolerance to glyphosate.[34] *Conyza canadensis* (horseweed) that is resistant to glyphosate has been found in Delaware.[35] Even in areas where resistant weeds have not been reported, scientists are seeing shifts in dominant weed species that may be due to heavy use of glyphosate in engineered crops. For example, University of Illinois specialists suggest that increases in *Solanum ptycanthum* (eastern black nightshade) in Illinois soybean fields may be a result of widespread adoption of the glyphosate-resistant crop and the concomitant use of the herbicide in the state. Similarly, weed scientists in Iowa are finding populations of *Crotalaria juncea* (water hemp) that survive spraying in fields of glyphosate-resistant soybean.

Perhaps the greatest problem of using HRCs to solve weed problems is that they steer efforts away from alternatives such as crop rotation or use of cover crops and help to maintain cropping systems dominated by one or two annual species. Crop rotation not only reduces the need for herbicides, but also improves soil and water quality, minimizes requirements for synthetic nitrogen fertilizer, regulates insect pest and pathogen populations, increases crop yields, and reduces yield variance. Thus, to the extent that transgenic HRCs inhibit the adoption of rotational crops and cover crops, they hinder the development of sustainable agriculture.

One additional threat resulting from HRCs involves the increased used of glyphosate by farmers adopting no-till agriculture. In 2000, about 19.7 million hectares in the United States and 13.5 million hectares in Brazil were under no-till systems. Promoters of no-till methods affirm that the technology is sustainable because it conserves soils, enhances soil quality by improving structure, water infiltration, and biological activity, saves energy and labor, captures CO_2, and so on. Such benefits, however, are threatened by the widespread use of glyphosate. As farmers encounter more herbicide-tolerant volunteers in their fields, they increasingly resort to tilling.

HRCs and the Consequences of Total Weed Removal

The presence of weeds within or around crop fields influences the dynamics of the crop and associated biotic communities. Studies over the past thirty years have produced a great deal of evidence that the manipulation of a specific weed species, a particular weed control practice, or a cropping system can affect the ecology of insect pests and associated natural enemies.[36]

Many weeds are important components of agroecosystems because they positively affect the biology and dynamics of beneficial insects. Weeds offer many important requisites for natural enemies such as alternative prey/hosts, pollen, or nectar as well as microhabitats that are not available in weed-free monocultures. Many insect pests are not continuously present in annual crops, and their predators and parasitoids must survive elsewhere during their absence. Weeds usually provide such resources (alternative host or pollen and nectar), thus aiding in the survival of viable natural enemy populations. In the last twenty years, research has shown that outbreaks of certain types of crop pests are less likely to occur in weed-diversified crop systems than in weed-free fields, mainly due to increased mortality imposed by natural enemies. Crop fields with a dense weed cover and high diversity usually have more predaceous arthropods than do weed-free fields. The successful establishment of several parasitoids usually depends on the presence of weeds that provide nectar for the adult female wasps. Relevant examples of cropping systems in which the presence of specific weeds has enhanced the biological control of particular pests have been reviewed by Altieri and Nicholls.[37] A literature survey by Baliddawa[38] showed that population densities of twenty-seven insect pest species were reduced in weedy crops compared to weed-free crops. Obviously, total elimination of weeds as commonly occurs under HRC crops can have major ecological implications for insect pest management.

Recent studies conducted in the United Kingdom showed that reduction of weed biomass, flowering, and seeding of plants within and in margins of HR beet and spring oilseed rape crops involved changes in resource availability with knock-on effects at higher trophic levels, reducing abundance of relatively sedentary and host-specific herbivores including Heteroptera, butterflies, and bees. Counts of predaceous carabid beetles that feed on weed seeds were lower in HRC fields.[39] Data showed that weed densities were lower in the HRCs compared to their conventional forms while the biomass in GE beet and oilseed rape was, respectively, one-sixth and about one-third of that in conventional plots. Many weed species among these two HR crops produced lower biomass, which led to the conclusions that these "differences compounded over time would result in large decreases in population densities of arable weeds" and, "with a few exceptions, weed species in beet and spring oilseed rape were negatively affected by the HRC treatment." The abundance of invertebrates, which are food for mammals, birds, and other invertebrates, and important for controlling pests or recycling nutrients within the soil, was also found to be generally lower in HR beet and oilseed rape.

These reductions may be an underestimate because comparisons were made only between conventional and biotech plots. Organic systems were not included in the comparisons, hence the full spectrum of impacts on biodiversity was not captured. The studies did not address effects, if any, of biodiversity reductions on agroecosystem processes such as nutrient cycling or pest regulation.

ECOLOGICAL RISKS OF Bt CROPS

Pest Resistance

More than 500 species of pests have already evolved resistance to conventional insecticides; hence it is likely that these and other pests can also evolve resistance to *Bt* toxins present in transgenic crops.[40]

No one questions that *Bt* resistance will develop, the question is now how fast it will develop. Susceptibility to *Bt* toxins can, therefore, be viewed as a natural resource that could be quickly depleted by inappropriate use of *Bt* crops.[41] However, cautiously restricted use of these crops should substantially delay the evolution of resistance. The question is whether cautious use of *Bt* crops is possible given commercial pressures that have resulted in their rapid rollout, reaching 7.6 million hectares worldwide in 2002.

Like conventional pesticides, transgenic technologies represent a single-intervention approach. Pest populations are selected, typically resulting in resistance to the control, and pest predators are harmed either directly or through deprivation of their prey. To move beyond the "pesticide paradigm" followed by *Bt* crops, technologies should be designed to induce pest-damage tolerance rather than resistance to pests. Tolerance does not rely on toxicity to kill pests and, therefore, does not negatively impact nontarget organisms or promote resistance development.[42]

To delay the inevitable development of resistance by insects to *Bt* crops, bioengineers are preparing resistance-management plans, which include the use of patchworks of transgenic and non-transgenic crops (called refuges) to delay the evolution of resistance by providing susceptible insects for mating with resistant insects. According to a report of the Union of Concerned Scientists, refuges should cover at least 30 percent of the crop area, but Monsanto's new plan calls for only 20 percent refuges even when insecticides are to be used. Moreover, the plan offers no details regarding whether the refuges must be planted alongside the transgenic crops, or some distance away, where studies suggest they would be less effective.[43] In addition to refuges requiring difficult regional coordination among farmers, it is unrealistic to expect most small- and medium-sized farmers to devote up to 30 to 40 percent of their crop area to refuges, especially if crops in these areas may sustain heavy pest damage. In one of the few field studies assessing resistance development to *Bt* crops, Tabashnik et al.[44] found in 1997 that approximately 3.2 percent of pink bollworm larvae collected from Arizona *Bt* cotton fields exhibited resistance.

The farmers who face the greatest risk from the development of insect resistance to *Bt* are neighboring organic farmers who grow corn and soybeans without agrochemicals. Once resistance appears in insect populations, organic farmers will not be able to use *B. thuringiensis* in its microbial insecticide form to control the lepidopteran pests that move in from adjacent neighboring transgenic fields. Only about 8 percent of the organic farmers use *Bt* microbial sprays; therefore, this problem may not be widespread. In addition, genetic pollution of organic crops resulting from gene flow (pollen) from transgenic crops can jeopardize the certification of organic crops causing organic farmers to lose premium markets. Who will compensate the farmers for such losses? This may well become a major issue in the United States as commercial seeds of several major crops are "pervasively contaminated" with DNA from engineered varieties of those crops according to a recent study of the Union of Concerned Scientists. Commercial testing of eighteen non-transgenic varieties (six each of corn, soybean, and canola) showed low foreign DNA levels in at least half of the corn and soybean varieties and at least 83 percent of the canola tested.[45]

Bt Crops and Beneficial Insects

B. thuringiensis proteins are becoming ubiquitous, highly bioactive substances that persist for many months in agroecosystems. Most, if not all, nontarget herbivores that colonize *Bt* crops in the field, although not lethally affected, ingest plant tissue containing *Bt* protein, which they can pass on to their natural enemies in a more or less processed form. Polyphagous natural enemies that move between crop cultures are frequently found to encounter *Bt* containing nontarget herbivorous prey in more than one crop during the entire season. According to Groot and Dicke[46] natural enemies may come in contact more often with *Bt* toxins via nontarget herbivores, because the toxins do not bind to receptors on the midgut membrane in the nontarget herbivores. This is a major ecological concern given previous studies that documented the adverse effects of Cry1 Ab on the predaceous lacewing *Chrysoperla carnea* reared on *Bt* corn-fed prey larvae.[47] In another study, feeding three different herbivore species exposed to *Bt* maize to *C. carnea* showed a significant increase in mortality and a delay in development when predators were fed *Spodoptora littoralis* reared on *Bt*-maize. A combined interaction of poor prey quality and Cry 1Ab toxin may account for the negative effects on *C. carnea*.

Apparently the fitness of parasitoids and predators is indirectly affected by *Bt* toxins exposed in GE crops by feeding from suboptimal food or because of host death and scarcity.[48] Due to the development of a new generation of *Bt* crops with much higher expression levels, the effects on natural enemies reported so far are likely to be an underestimate.[49]

These findings are problematic for small farmers in developing countries who rely on the rich complex of predators and parasites associated with their mixed cropping systems for insect pest control.[50] Research results showing that natural enemies can be affected directly through inter-trophic level effects of the toxin present in *Bt* crops raise serious concerns about the potential disruption of natural pest control, as polyphagous predators that move within and between crop cultivars will encounter *Bt*-containing, nontarget prey throughout the crop season. Disrupted biocontrol mechanisms will likely result in increased crop losses due to pests or due to increased use of pesticides by farmers with consequent health and environmental hazards.

Effects on the Soil Ecosystem

The possibilities for soil biota to be exposed to transgenic products are very high. The little research conducted in this area has already demonstrated long-term persistence of insecticidal products (*Bt* and proteinase inhibitors) in soil after exposure to decomposing microbes.[51] The insecticidal toxin produced by *B. thuringiensis* subsp. *kurskatki* remains active in the soil, where it binds rapidly and tightly to clays and humic acids. The bound toxin retains its insecticidal properties and is protected against microbial degradation by being bound to soil particles, persisting in various soils for at least 234 days.[52] Furthermore, 25 to 30 percent of the Cry1A proteins produced by *Bt* cotton leaves remained bound in the soil even after 140 days. In another study researchers confirmed the presence of the toxin in exudates from *Bt* corn and verified that it was active in an insecticidal bioassay using larvae of the tobacco hornworm.[53] In a recent study, after 200 days of exposure, *Lumbricus terrestris* adults experienced a significant weight loss when fed *Bt* corn litter when compared to earthworms fed on non-*Bt* corn litter.[54] Potentially these earthworms may serve as intermediaries through which *Bt* toxins may be passed on to organisms feeding on these earthworms. Given the persistence and the possible presence of exudates, there is potential for prolonged exposure of the microbial and invertebrate community to such toxins, and, therefore, studies should evaluate the effects of transgenic plants on both microbial and invertebrate communities and the ecological processes they mediate.[55]

If transgenic crops substantially alter soil biota and affect processes such as soil organic matter decomposition and mineralization, this would be of serious concern to organic farmers and most poor farmers in the developing world. These people cannot purchase or do not want to use expensive chemical fertilizers, and rely instead on local residues, organic matter, and especially soil organisms for soil fertility (i.e., key invertebrate, fungal, or bacterial species), which can be affected by the soil-bound toxin. Soil fertility could be dramatically reduced if crop leachates inhibit the activity of the soil biota and slow down natural rates of decomposition and nutrient release. Accumulation of toxins over time during degradation of plant biomass could have worse and longer-term impacts on soil biology because the doses of *Bt* toxin to which these soil organisms are exposed may increase with time.[56]

HRCs can also act indirectly on soil biota through effects of glyphosate, which appears to act as a soil antibiotic inhibiting mycorrhizae, antagonists, and nitrogen-fixing bacteria. Root development, nodulation, and nitrogen fixation are impaired in some HR soybean varieties, which then exhibit lower yields with effects being worse under drought stress or infertile soils.[57]

GENERAL CONCLUSIONS AND RECOMMENDATIONS

The available, independently generated, scientific information suggests that the massive use of transgenic crops poses substantial potential risks from an ecological point of view. The environmental effects are not limited to pest resistance and creation of new weeds or virus strains.[58] As I have argued, transgenic crops can produce environmental toxins that move through the food chain and also end up in the soil where they bind to colloids and retain toxicity that affects invertebrates and possibly nutrient cycling.[59] No one can really predict the long-term impacts on agrobiodiversity and the processes they mediate from the massive deployment of such crops.

Not enough research has been done to evaluate the environmental and health risks of transgenic crops, an unfortunate trend, as many scientists feel that such knowledge was crucial before biotechnological innovations were upscaled to current commercial levels. There is a clear need to assess further the severity, magnitude, and scope of risks associated with the massive field release of transgenic crops.[60] Risk evaluation must now move beyond comparing GE crop fields with conventionally managed systems to include alternative cropping systems that feature crop diversity and low-external input approaches. These systems express higher levels of biological diversity and thus allow rigorous studies to capture the full range of impacts on biodiversity and agroecosystem processes.

Moreover, the large-scale landscape homogenization by transgenic crops will exacerbate the ecological problems already associated with monocultural agriculture.[61] Unquestioned expansion of this technology into developing countries may not be wise or desirable. There is strength in the agricultural diversity of many of these countries, and it should not be inhibited or reduced by extensive monoculture, especially when consequences of doing so may be serious social and environmental problems.[62]

The repeated use of transgenic crops in an area may result in cumulative effects such as those resulting from the buildup of toxins in soils. For this reason, risk assessment studies not only have to be of an ecological nature in order to capture effects on ecosystem processes, but also of sufficient duration so that probable accumulative effects can be detected. Decades of careful ecologically monitored field work and larger-scale results are necessary to assess the full potential for risks from GE crops to the environment. Decreases in pesticide use are not acceptable as proxies for environmental benefits, because even if pesticide use is reduced with GE crops the ecological impacts of such crops remain unchanged. The application of multiple diagnostic methods to assess multitrophic effects and impacts on agroecosystem function will provide the most sensitive and comprehensive assessment of the potential ecological impact of transgenic crops.

Until these studies are completed, a moratorium on transgenic crops based on the precautionary principle should be imposed in the

United States and other regions. This principle advises that instead of using the criterion of "absence of evidence" of serious environmental damage, the proper decision criterion should be the "evidence of absence," in other words avoiding "type II" statistical error: the error of assuming that no significant environmental risk is present when in fact risk exists.

Although biotechnology could be considered an important tool to manage agroecosystems, at this point alternative solutions exist to address the problems that current GE crops, developed mostly by profit motives, are designed to solve. The recent study conducted by scientists at ICIPE in Africa highlighted the dramatic positive effects of rotations, multiple cropping, and biological control on crop health, environmental quality, and agricultural productivity, also confirmed by scientific research in many parts of the world. The evidence is conclusive: new approaches and technologies spearheaded by farmers, NGOs, and some local governments around the world are already making a sufficient contribution to food security at the household, national, and regional levels. A variety of agroecological and participatory approaches in many countries shows very positive outcomes even under adverse conditions. Potentials include raising cereal yields from 50 to 200 percent, increasing stability of production through diversification, improving diets and income, contributing to national food security and even to exports and conservation of the natural resource base and agrobiodiversity.[63]

In its present form, biotechnology is incompatible with more agroecological approaches because of its cascading effects on agroecosystem function. Moreover, this technology is under corporate control, excluding it from the realm of the international public goods, a major barrier when it comes to promoting socially equitable and accessible agricultural technologies.

NOTES

1. James, C. 2002. Global review of commercialized transgenic crops: 2000. *International service for the acquisition of agri-biotech application.* ISSA Briefs No. 23-2002, Ithaca, NY.

2. Krimsky, S. and Wrubel, R.P. 1996. *Agricultural biotechnology and the environment: Science, policy and social issues.* Urbana: University of Illinois Press.

3. Altieri, M.A. 2000. The ecological impacts of transgenic crops on agroecosystem health. *Ecosystem Health* 6: 13-23.

4. Altieri, M.A. 1995. *Agroecology: The science of sustainable agriculture.* Boulder, CO: Westview Press.

5. Kahn, Z.R., Ampong-Nyarko, K., Hassanali, A., and Kimani, S. 1998. Intercropping increases parasitism of pests. *Nature* 388: 631-632.

6. Ibid.

7. Altieri, M.A. 1995. See note 4.

8. Rissler, J. and Mellon, M. 1996. *The ecological risks of engineered crops.* Cambridge, MA: MIT Press; Snow, A.A., and Moran, P. 1997. Commercialization of transgenic plants: Potential ecological risks. *BioScience* 47: 86-96; Kendall, H.W., Beachy, R., Eismer, T., Gould, F., Herdt, R., Raven, P.H., Schell, J., and Swaminathan, M.S. 1997. Bioengineering of crops. *Report of the World Bank Panel on Transgenic Crops.* Washington, DC: World Bank, p. 30; Altieri, M.A. 2000. See note 3.

9. Saxena, D., Flores, S., and Stotzky, G. 1999. Insecticidal toxin in root exudates from *Bt* corn. *Nature* 401: 480.

10. Hilbeck, A., Baumgartner, M., Fried, P.M., and Bigler F. 1998. Effects of transgenic *Bacillus thuringiensis* corn fed prey on mortality and development time of immature *Chysoperla carnea* (Neuroptera: Chysopidae). *Environmental Entomology* 27: 460-487.

11. Losey, J.J.E., Rayor L.S., and Cater, M.E. 1999. Transgenic pollen harms monarch larvae. *Nature* 399: 241.

12. Soule, J.D. and Piper, J.K. 1992. *Farming in nature's image.* Washington, DC: Island Press.

13. Altieri, M.A. and Nicholls, C.I. 2004. *Biodiversity and pest management in agroecosystems.* Binghamton, NY: The Haworth Press.

14. Ibid.

15. Robinson, R.A. 1996. *Return to resistance: Breeding crops to reduce pesticide resistance.* Davis, CA: AgAccess.

16. Ibid.

17. Thrupp, L.A. 1998. *Cultivating diversity: Agrobiodiversity for food security.* Washington, DC: World Resources Institute.

18. Netting, R.M. 1993. *Smallholders, householders.* Stanford, CA: Stanford University Press.

19. Tripp, R. 1996. Biodiversity and modern crop varieties: Sharpening the debate. *Agriculture and Human Values* 13: 48-62; Lappe, F.M., Collins, J., and Rosset, P. 1998. *World hunger: Twelve myths.* New York: Grove Press.

20. Brush, S.B. 2000. *Genes in the field: On farm conservation of crop diversity.* Boca Raton, FL: Lewis Publishers.

21. Quist, D. and Chapela, I.H. 2001. Transgenic DNA introgressed into traditional maize landraces in Oaxaca, Mexico. *Nature* 414: 541-543.

22. Brush, S.B. 2000. See note 20.

23. Jordan, C.F. 2001 Genetic engineering, the farm crisis and world hunger. *BioScience* 52: 523-529.

24. Altieri, M.A. 2004. *Genetic engineering in agriculture: The myths, environmental risks and alternatives.* Oakland, CA: Food First Books.

25. Ellstrand, N.C. 1988. Pollen as a vehicle for the escape of engineered genes? In *Planned release of genetically engineered organisms.* Edited by J. Hodgson and A.M. Sugden. Cambridge, UK: Elsevier, pp. S30-S32.

26. Stabinski, D. and Sarno, N. 2001. Mexico, centre of diversity for maize, has been contaminated. *LEISA Magazine* 17: 25-26.

27. Snow, A.A. and Moran, P. 1997. See note 8.

28. Hall, L.M., Huffman, J., and Topinka, K. 2000. Pollen flow between herbicide tolerant canola is the cause of multiple resistant canola volunteers. *Abstracts of the Annual Meeting of the Weed Science Society of America* 40: 56-64 .

29. Barrett, S., Beare-Rogers, J.L., Brunk, C.G., Caufield, T.A., Ellis, B.E., Forton, M.G., Ham-Pong, A.J., and McNeil, J.N. 2001. *Elements of precaution: Recommendations for regulation of food biotechnology in Canada.* The Royal Society of Canada. Ottawa, Ontario.

30. Snow, A.A. and Moran, P. 1997. See note 8.

31. Goldberg, R.J. 1992. Environmental concerns with the development of herbicide-tolerant plants. *Weed Technology* 6: 647-652.

32. Duke, S.O. 1996. *Herbicide resistant crops: Agricultural, environmental, economic regulatory, and technical aspects.* Boca Raton: Lewis Publishers, p. 420.

33. Gill, D.S. 1995. Development of herbicide resistance in annual ryegrass populations in the cropping belt of Western Australia. *Australian Journal of Experimental Agriculture* 3: 67-72.

34. Owen, M. 1997. North American development in herbicide resistant crops. *Proceedings of British Crop Protection Conference.* Brighton, England. http://www.weeds.iastate.edu. November 14, 2005.

35. VanGessel, M.J. and Glasgow, J.L. 2001. *Conyza canadensis* insensitivity to glyphosate. *Proceedings of the Northeastern Weed Science Society* 55 (January 2-5): 32.

36. Altieri, M.A. and Nicholls, C.I. 2004. See note 13.

37. Ibid.

38. Baliddawa, C.W. 1985. Plant species diversity and crop pest control: An analytical review. *Insect Science and Applications* 6: 479-487.

39. Hawes, C. Haughton, A.J., Osborne, J.L., Roy, D.B., Clark, S.J., Perry, J.N. Rothery, P., Bohan, D.A., Brooks, D.J., Champion, G.T., et al. 2003. Responses of plants and invertebrate trophic groups to contrasting herbicide regimes in the farm scale evaluations of genetically modified herbicide-tolerant crops. *Philosophical Transactions of the Royal Society of London* B 358: 1899-1913.

40. Paoletti, M.G. and Pimentel, D. 1996. Genetic engineering in agriculture and the environment: Assessing risks and benefits. *BioScience* 46: 665-671.

41. Mellon, M. and Rissler, J. 1998. Now or never: Serious plans to save a natural pest control. Washington, DC: Union of Concerned Scientists.

42. Welsh, R., Hubbell, B., Ervin D.E., and Jahn, M. 2002. GM crops and the pesticide paradigm. *Nature Biotechnology* 20: 548-549.

43. Mellon, M. and Rissler, J. 1998. See note 41.

44. Tabashnik, B.E., Oatin, A.L., Dennehy, T.J., Liu, Y., and Antilla, L. 2000. Frequency of resistance to *Bacillus thuringiensis* in field populations of pink bollworm. *Proceedings of the National Academy of Sciences USA* 97: 12980-12984.

45. Mellon, M. and Rissler, J. 2004. *Gone to seed: Transgenic contaminants in the traditional food supply*. Washington, DC: Union of Concerned Scientists.

46. Groot, A.T. and Dicke, M. 2002. Insect-resistance transgenic plants in a multi-trophic context. *The Plant Journal* 31: 387-406.

47. Hilbeck, A. et al. 1998. See note 10.

48. Groot, A.T. and Dicke, M. 2002. See note 46.

49. Dutton, A., Klein, H., Romeis, J., and Bigler, F. 2002. Uptake of *Bt*-toxin by herbivores feeding on transgenic maize and consequences for the predator *Chrysoperla carnea*. *Ecological Entomology* 27: 441-446.

50. Altieri, M.A. 1995. See note 4.

51. Donegan, K.K. and Seidler, R.J. 1999. Effects of transgenic plants on soil and plant microorganisms. *Recent Research and Development in Microbiology* 3: 415-424 .

52. Palm, C.J., Schaller, D.L., Donegan, K.K., and Seidler, R.J. 1996. Persistence in soil of transgenic plant produced *Bacillus thurigiensis* var. *Kustaki* endotoxin. *Canadian Journal of Microbiology* 42: 1258-1262.

53. Saxena, D. 1999, see note 9.

54. Zwahlen,C., Hillbeck, A., Howland, R., and Nentwig, W. 2003. Effects of transgenic *Bt* corn litter on the earthworm *Lumbricus terrestris. Molecular Ecology* 12: 1077-1082.

55. Altieri, M.A. 2000. See note 3.

56. Donegan, K.K., Palm, C.J., Fieland, V.J., Porteous, L.A., Ganis, L.M., Scheller, D.L., and Seidler, R.J. 1995. Changes in levels, species, and DNA fingerprints of soil micro organisms associated with cotton expressing the *Bacillus thuringiensis* var. *Kurstaki* endotoxin. *Applied Soil Ecology* 2: 111-124.

57. Benbrook, C. 2001. Troubled times amid commercial success for roundup ready soybeans. *Ag Bio Tech InfoNet Technical Paper No. 4.* http://www.biotech-info.net. November 14, 2005.

58. Kendall, H.W., et al.,1997. See note 8.

59. Altieri, M.A. 2000. See note 3.

60. Kjellsson, G. and Simonson, V. 1994. *Methods for risk assessment of transgenic plants*. Basel: .Birkhauser Verlag, p. 214.

61. Altieri, M.A. 2000. See note 3.

62. Altieri, M.A. 1995. See note 4.

63. Altieri, M.A. 2002. Agroecology: The science of natural resource management for poor farmers in marginal environments. *Agriculture, Ecosystems and Environment* 93: 1-24.

Chapter 4

Ecological Risk Assessment of GE Crops: Getting the Science Fundamentals Right

Michelle Marvier*
Sabrina West

Many human activities that confer enormous benefits also entail risk. Consider, for example, the use of pharmaceuticals. Although modern drugs have saved untold numbers of human lives, the use of new drugs or vaccines always has some potential for unexpected consequences. But quantifying risks is seldom a simple task. The side effects of a drug can manifest themselves in myriad ways, and researchers typically test the allergenicity and toxicity of new drugs, screen for effects on different organs and organ systems, and even look for long-term delayed effects. To be safe, drug trials should also examine subjects who vary in age, sex, and physical condition, while guarding against interactions with other drugs. Even after well-replicated drug trials have been completed, some level of scientific uncertainty regarding risks will always persist, because some serious consequences may only manifest themselves under extremely rare conditions or only after long exposures.

Although the assessment of health risks associated with drugs is clearly complex, it is relatively simple when compared to the assessment of environmental risks associated with the release of a novel

*Michelle Marvier was supported by EPA grant CR-832147-01 during preparation of this chapter.

organism or, of more concern to us here, an organism with novel genetically engineered traits. The added complexity of environmental risk assessment stems from a less developed understanding of how ecosystems function and the fact that, unlike a drug, a novel organism can evolve over time. Furthermore, there are a large number of routes through which a novel organism could affect other species. For example, a crop that is genetically engineered (GE) to resist insect pests might also reduce populations of an important pollinator species, a soil-dwelling detritivore, or a parasitoid that perhaps also controls pest populations. Moreover, the potential for collateral impacts is not limited to species that interact directly with the novel organism. The direct effects on the abundance of either the pest or nontarget populations can potentially cascade into indirect effects on seemingly unconnected species in ways that can be quite difficult to anticipate. For instance, the deliberate (although unauthorized) introduction of *Myxoma* virus to control rabbit populations in England led to the local extinction of a widespread butterfly species, *Maculina arion*. This extinction occurred not because the virus affected the butterfly directly—it did not—but because successful reduction of rabbit numbers led to changes in grassland vegetation that caused the elimination of an ant species in whose nests the caterpillars developed (references in Simberloff and Stiling 1996).[1]

Not only are the risks of releasing a novel organism more difficult to assess than are the risks of drugs, but the field of environmental risk assessment is far younger and much less funded than the field of human health risk assessment. Consequently, we should not be surprised by the public confusion and heated debate regarding the possible understatement or overstatement of environmental risks as they pertain to GE organisms (GEOs); in fact, the development of risk-assessment protocols and analyses relevant to the release of GEOs is in its infancy.

In this chapter, we take a look at the environmental risk-assessment practices currently implemented in the process of moving GE crops down the regulatory pathway toward unregulated commercial production within the United States. In particular, we review the quality of evidence used by petitioners (typically industry) to argue that gene flow and weediness are not major concerns for those GE crop varieties that have already been approved for commercial production. We pay particular attention to the two most fundamental ingredients of science: What types of data are used and how are these data synthesized and analyzed? In addition, we ask what we might learn from the much older field of classical biological control regarding environmental "side effects" of novel organisms. Finally, we return briefly to the analogy between environmental risk assessment and the assessment of human health risks. Despite the added complexity of assessing the risks associated with release of a novel organism, many of the same general principles apply.

ENVIRONMENTAL RISK ASSESSMENT FOR GE CROPS

Current production of GE crops is dominated by herbicide-resistant varieties of soybean, cotton, and corn and insect-resistant varieties of corn and cotton.[2] To date, the major ecological concerns regarding GE crops have revolved around two issues: toxic effects on nontarget organisms, especially those that are beneficial to humans, and changes in the weediness of either crop or non-crop plants, in the latter case as a result of interspecific gene flow.[3] Additional environmental issues revolve around changes in farm-management practices and the evolution of resistance to plant-incorporated pesticides.[4]

Earlier reviews have found that, for the majority of U.S. petitions for deregulation, the approaches used to support the environmental safety of GE crops rely primarily on the following:

1. Conducting small (meaning poorly replicated) trials to test for effects and
2. Citing published and unpublished studies or using letters from expert scientists to establish an absence of risk.[5]

These approaches generally provide feeble evidence of safety.[6] In fact, it is typically much more difficult to satisfactorily demonstrate the absence, rather than the presence, of an effect (by effect, we mean a true difference among treatment groups such as GE versus control plants). For example, an effect may exist, but may be small relative to the variability among replicates. In such a case, it is unlikely that the effect will be detected in a poorly replicated experiment. Alternatively, an effect may be large, but rare. Again, it is unlikely that the effect will be detected by a study that includes only three or four replicates or that is performed over a short time period.

Unfortunately, risk-assessment studies applied to GE crops have typically been weak, resulting in data that, by themselves, shed little light on the issue of GE safety.[7] The lack of scientific rigor that characterizes most GE risk-assessment trials was made clear by a recent review of U.S. Department of Agriculture (USDA) petitions for deregulation that documented the typically small sample sizes used to test the acute toxicity of insect-resistant GE crops to nontarget invertebrates under laboratory conditions.[8] Sample size is a key aspect of a study's quality: the more replicates that are included in a study, the more credible are the study's results. However, the majority of the studies that Marvier reviewed[9] included four or fewer replicates per treatment (thirteen out of nineteen studies; Figure 4.1). Studies with so few replicates are unlikely to detect even large effects on nontarget organisms, yet the authors of the reviewed petitions for deregulation repeatedly pointed to the absence of statistically significant effects as evidence of safety.

Earlier reviews of petitions for deregulation have found that petitions rarely include quantitative data, frequently rely upon poorly

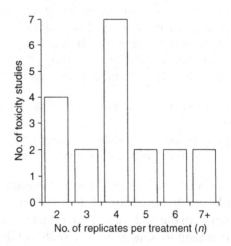

FIGURE 4.1. Sample sizes for nineteen experiments that tested the acute toxicity of *Bt* protein on nontarget invertebrate species. *Source:* Adapted from Marvier (2002) Improving risk assessment for nontarget safety of transgenic crops. *Ecological Applications* 12: 1119-1124.

designed experiments, and tend to use heuristic arguments.[10] We revisit some of these same issues to examine whether the quality of evidence presented in petitions for deregulation has improved over time. In particular, we assess the types and quality of evidence that have been used to gauge the environmental risks associated with GE crops by again examining USDA petitions for deregulation for a variety of GE traits and species. Our focus here is on gene flow from GE crops, changes in the weediness of the GE crop itself, and changes in the weediness of non-crop plants following hybridization. Gene flow and its possible consequences for weed management have been a central concern in the debate over GE crops.[11] Indeed, most crop species hybridize, often quite readily, with weedy relatives in some part of their distribution.[12] Moreover, Snow et al.[13] have recently documented that hybridization with a GE crop can confer a fitness advantage to a wild relative. Thus gene flow from at least one GE crop (*Bt* sunflower) has the potential to make a related weedy species even more weedy. Table 4.1 summarizes information regarding weediness and tendency to

TABLE 4.1. Weediness of selected crop species and their tendency to hybridize with wild species.

Examples of GE crops grown commercially in the United States	Related wild species in the United States with which GE crop can hybridize	Can the crop itself become a weed in the United States?
Beta vulgaris (beet)	*B. vulgaris* var. *Maritima* (hybrid is a weed)	Present and weedy but importance unclear
Brassica napus (oilseed rape, canola)	*Brassica rapa* (field mustard) *Brassica juncea* (Indian mustard)	Troublesome in some locales
Cucurbita pepo (squash)	*Cucurbita texana* (wild squash)	Occasionally
Gossypium hirsutum (cotton)	Hybridizes with wild congeners that are not weedy but that may be threatened by hybridization	No
Oryza sativa (rice)	*Oryza sativa* f. *spontanea* (red rice)	Serious weed

Source: Summary of information from Keeler, K.H., Turner, C.E., and Bolick, M.R. (1996). Movement of crop transgenes into wild plants. In *Herbicide-resistant crops: Agricultural, environmental, economic, regulatory, and technical aspects.* Edited by S.O. Duke. Boca Raton, FL: Lewis Publishers. pp. 303-330; Snow, A.A. and Palma, P.M. (1997). Commercialization of transgenic plants: potential ecological risks. *BioScience* 47: 86-96; Ellstrand, N.C., Prentice, H.C., and Hancock J.F. (1999). Gene flow and introgression from domesticated plants into their wild relatives. *Annual Review of Ecology and Systematics* 30: 539-563.

hybridize for a handful of crop species for which GE varieties have been developed and approved for commercial production in the United States.

For the current study, we examined seventeen petitions for deregulation, three of which were updates on earlier petitions (Table 4.2). These seventeen petitions covered ten different crop species. All seventeen petitions were approved by the USDA, resulting in deregulation (approval for unregulated commercial planting) of the particular GE crop varieties. From our pool of seventeen petitions, we identified

TABLE 4.2. Petitions for deregulation examined in this review.

Petition #	Crop	Phenotype	Petitioner	Update?
92-201-01	Squash	VR	Asgrow	No
93-196-01	Cotton	HT	Calgene	No
93-258-01	Soybean	HT	Monsanto	No
94-257-01	Potato	IR	Monsanto	No
94-319-01	Corn	IR	Ciba	No
95-352-01	Squash	VR	Asgrow	No
96-051-01	Papaya	VR	Cornell & UH	No
97-205-01	Rapeseed	HT	AgrEvo	No
97-287-01	Tomato	IR	Monsanto	No
97-336-01	Sugarbeet	HT	AgrEvo	No
98-173-01	Sugarbeet	HT	Novartis & Monsanto	No
98-238-01	Soybean	HT	AgrEvo	Yes
98-329-01	Rice	HT	Liberty Link	No
99-173-01	Potato	IR and VR	Monsanto	Yes
00-136-01	Corn	IR and HT	DowAgroSciences	No
01-206-01	Rapeseed	HT	Aventis	Yes
02-042-01	Cotton	HT	Aventis	No

Note: Petitions for crops modified to be insect resistant (IR), viral resistant (VR), or herbicide tolerant (HT) were chosen. Petitions for crops with altered agronomic properties were excluded. The newest and oldest approved petition for each crop were selected. In some cases, there was only one approved petition for a crop. "Update?" refers to whether the petition is or is not an update of an earlier petition.

three that provided reasonably detailed descriptions of new (not previously published) field trials designed specifically to test whether GE crops or GE-wild hybrids possess a competitive advantage over non-transgenic or wild controls. We only considered field trials for which at least some information regarding either the number of replicates, plot sizes, or density of plants per pot was included. In all, we found nine qualifying field studies in this collection of petitions. For each field study, we recorded the number of replicate plots used and other details of experimental design. For three of the nine field studies there was no indication of the sample size, an oversight that may seem small, but in fact makes it impossible to gauge the credibility of the studies' findings (Table 4.3). Of the six studies that did report

TABLE 4.3. Details of experimental design for nine field studies submitted to the USDA as parts of petitions for deregulation.

Petition number	Crop	No. of replicates	Plot size	Plot density
98-173-01p	Sugar beet	?	?	5 to 43 plants for transgenic treatment and 75 plants in the control treatment
98-173-01p	Sugar beet	?	?	100-103 plants per treatment
98-173-01p	Sugar beet	2 (1 rep in each of 2 seasons)	"Small"	8 roots/plot
98-173-01p	Sugar beet	?	?	64 plants/m^2
98-173-01p	Sugar beet	12 (3 reps × 4 harvest dates)	?	16-64 plants/m^2 for monocultures and 32-128 plants/m^2 for binary mixtures
97-205-01p	Canola	4	1.5 m × 1.5 m	?
97-205-01p	Canola	4	1.25 m × 7 m	100 seeds/m^2
94-257-01p	Potato	3	single 20'row	20 tubers per plot
94-257-01p	Potato	3	single 20'row	20 tubers per plot
94-257-01p	Potato	3	single 20'row	20 tubers per plot

sample size, only one used more than four replicate plots per treatment (actually three replicates but at four harvest times; Table 4.3). Given an estimate of the variation among samples, it is straightforward to calculate the number of replicates needed for a desired type II error rate and a specified effect size.[14] Calculations of this sort previously showed that sample sizes for studies of nontarget effects should be bolstered.[15] The use of so few replicates in the studies reviewed here might be understandable if plots were very large, but this was not the case; plot sizes in these poorly replicated studies, if indicated at all, were typically small (Table 4.3).

Given the consistent weakness of the environmental-risk studies included in the petitions for deregulation reviewed here, one is left wondering how regulators ever decided to approve these particular GE crops. By analogy, if a pharmaceutical company were to claim that a new drug has no important side effects on the basis of a single clinical trial in which only four people were given the drug and four were given a placebo, no one would feel particularly confident in the claim. Part of the explanation may lie in the belief that a collection of several small studies, each finding no effect, must provide strong evidence of safety. In fact, it is not possible to discern whether this is the case without synthesizing the studies in a statistically appropriate manner. Simply tallying up the findings from a set of similar studies has been labeled "vote-counting" by statisticians and is widely recognized as biased in favor of concluding "no effect."[16] Rather than simply counting votes, the results of the various studies should be brought together by a meta-analysis. Meta-analysis provides a statistically powerful means of combining results and can detect the presence of an effect even when all of the individual component studies failed to do so.[17] However, the power of meta-analysis is limited when basic details such as sample size and within-group variation are not reported. These deficiencies are not limited to studies included in petitions to deregulation—in fact omissions of sample size and variance plague even the highest quality professional journals.[18] It is simply impossible to judge the results from a set of experiments unless sample sizes, means, and variances are clearly reported. At a minimum, regulatory agencies should enforce these basic reporting standards.[19]

In addition to recording the sample sizes used in field trials, we tallied the number of published references and expert opinions relevant to gene flow and weediness that were cited within each of our seventeen sample petitions (see Table 4.2 for list of petitions). Given that new research papers are constantly added to the literature, one might expect that the number of journal articles cited in a petition for deregulation would also increase over time. However, we found that the number of journal articles cited did not increase over time, even when we ignore updates of earlier petitions that tend to be shorter and offer relatively little new information (Figure 4.2). More troubling perhaps is the finding that eight of the seventeen petitions cited no field studies and no expert opinions regarding the potential for gene flow and enhanced weediness (of these eight, two petitions were for GE corn, two for GE squash, and one each for GE cotton, GE potato, GE soybean, and GE tomato). Of the remaining nine petitions, six cited more expert opinions than published or unpublished field studies with respect to the potential for gene flow and increased weediness (Figure 4.3). This heavy reliance upon expert opinion should be met with some degree of suspicion. The opinions offered may depend on whose experts are used (industry's expert or an environmental group's expert). Moreover, experts tend to deliver "yes" or "no" issuances about effects with no confidence intervals surrounding each expert's vote.[20]

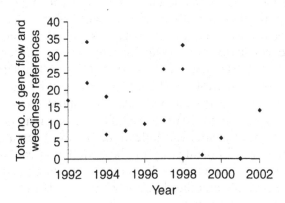

FIGURE 4.2. References cited in petitions versus the year of submission to the USDA. *Note:* References are defined as books, book chapters, workshop proceedings, grey literature, and journal articles cited in relation to gene flow and/or weediness of the GE crop, its wild relatives, or GE-wild hybrids. The three lowest values are for updates of earlier approved petitions.

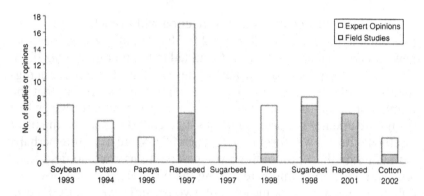

FIGURE 4.3. The number of field studies and expert opinions cited in relation to gene flow and the potential to exacerbate weed problems for our sample of seventeen successful petitions for deregulation.

WHAT CAN WE LEARN FROM THE LONGER HISTORY OF CLASSICAL BIOLOGICAL CONTROL?

Classical biological control is the importation and release of a nonnative species to reduce the numbers of some other (usually nonnative) species that is considered a pest. Whereas the nontarget effects of GE species have been considered for little more than a decade, over 100 years of information (albeit largely anecdotal) have accumulated regarding the release of nonnative species as biological control agents. There are several lessons we can learn from this longer experience with releases of novel organisms that are relevant for risk assessment applied to GE species. Biological control is thought by many agronomists to be a more "environmentally friendly" approach to pest control than the use of chemical pesticides, and the effects may be more targeted than those of pesticides. Indeed, there have been a number of dramatic successes in the history of classical biological control, and there will likely be some similarly dramatic successes in the deployment of insecticidal and other kinds of GE crops. However, attempts to control pests with nonnative species have not always been without serious environmental consequences.[21]

By taking a more detailed look at a single case study—the biological control of European thistles in North America—several additional lessons can be learned (reviewed by Louda et al.[22]). The flowerhead weevil, *Rhinocyllus conicus,* was imported from Europe and released in North America to control a suite of highly invasive, nonnative thistles. Feeding trials performed prior to the release of *R. conicus* indicated that this weevil could successfully feed on thistle species native to North America. However, its preference for the nonnative thistles seemed to be quite strong, and the weevil was, therefore, considered to be reasonably risk-free relative to its benefits.[23]

Unfortunately, infestation of native thistles by *R. conicus* has increased dramatically over time.[24] Today there is concern about the conservation of native thistles, many of which are quite beautiful and rare. The expansion of the host range of *R. conicus* is having strong direct effects on nontarget native thistle species. The production of seeds per seed head in native thistles was reduced by 86 percent in weevil-infested seed heads compared to weevil-free seed heads (see Figure 4.4a[25]). Earlier studies had already established that the population growth of this thistle is limited by seed production. The weevils are also exerting strong indirect effects, reducing the occurrence of native tephritid flies that feed in thistle flowerheads (Figure 4.4b). Admittedly, the effects of *R. conicus* on nontarget, native thistles should not be a complete surprise given the natural history of this insect; however, the extent of the impact and the indirect effects on native flies were unanticipated.

The flowerhead weevil provides a clear demonstration of nontarget effects stemming from the release of a biological control species. Note, however, the large numbers of samples that were used in this study (Figure 4.4). In contrast, most of the studies of nontarget effects for genetically modified crops that have been included in successful petitions for deregulation have employed fewer than five samples per group (Figure 4.1). If Louda et al.[26] had examined only five thistleheads per group, (assuming the same 86 percent reduction in seed production and the same pooled variance), they would have erroneously concluded that there was no difference in seed production between the two groups, with and without the introduced weevil.

Although releases of GEOs will likely have more predictable risks than releases of biological control agents,[27] our experience with nontarget effects of biological control organisms should cause us to consider the following factors that can complicate risk assessment applied to the release of any novel organism (GE or not):

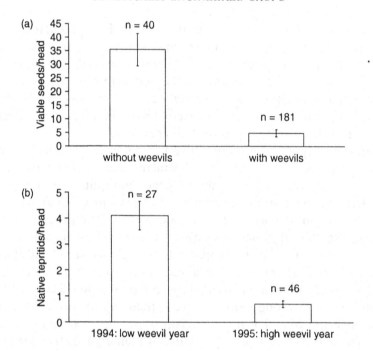

FIGURE 4.4. Effect of the biological control agent, *Rhinocyllus conicus*, on non-target species. Although this seed weevil was introduced to North America to control nonnative thistles, it is causing a severe reduction in (a) seed production by the native thistle *Cirsium canescens* and (b) the occurrence of native tephritid flies that also feed in flowerheads of native thistles. *Source:* Data are from Louda, S. M., Kendall, D., Connor, J., and Simberloff. D. (1997) Ecological effects of an insect introduced for the biological control of weeds. *Science* 277: 1088-1090.

Absence of evidence. One important point that the biocontrol literature makes abundantly clear is that "absence of evidence of harm" cannot and should not be equated with "evidence of absence of harm."[28] The number of ways by which one species may affect another is huge, while monitoring of populations is scanty or absent for most species. We should, therefore, expect that we will fail to detect most environmental impacts.

Evolutionary change. Releasing novel organisms into the environment to control pests is different from releasing chemicals because, if the released organisms establish self-sustaining populations, they can spread without dilution of their concentration and their behavior and impacts can evolve over time.

Time lags. The time between the introduction of *R. conicus* and detection of its host range expansion was approximately twenty years. Long lags between the time of introduction and the time when a problem is finally detected are a common feature of introductions. In fact, the probability of problems increases with time since release, because there is more time for evolutionary changes, gene flow, and dispersal to additional environments. The implication of this generality is that two, five, or even ten trouble-free years following the release of an organism cannot be interpreted as an "all-clear" signal. Long-term monitoring following a release will be essential.[29]

Indirect effects. The introduced flowerhead weevil appears to be competing with a native herbivore for the thistles, providing a relatively straightforward example of an indirect nontarget effect. Sometimes, however, indirect effects can be much harder to anticipate, as illustrated by the *Myxoma* example described earlier. In fact, effects can cascade from any species that is strongly affected by the novel organism to other, seemingly unconnected, species. Although it is not feasible to test for all possible indirect effects prior to an organism's release, familiarity with the natural history of a community can allow one to select some likely candidates for indirect effects *a priori.*

Importance of large sample sizes. Demonstrating the absence of an effect is often more difficult than demonstrating the presence of an effect.[30] Because scientific studies are generally designed to attempt to falsify a hypothesis of no effect, accepting a hypothesis of no effect usually requires a greater burden of proof.

Introductions are usually irreversible. Once an organism is released, it usually cannot be "recalled."[31] In many cases, such as the now infamous introduction of the cane toad to control beetle populations in Australia, the introduced species, itself, became an extremely noxious pest that will be extremely difficult, if not impossible, to eradicate.[32]

Cumulative effects. Currently, any given natural ecosystem is likely to be exposed to only a handful of GEOs. However, 100 years from now it may be hard to find an ecosystem without a prevalence of GEOs that have escaped into the wild. If this conjecture seems farfetched, one only needs to consider a remarkable study completed in

Hawaii. Henneman and Memmott[33] found that only 3 percent of the parasitoids collected from mostly native caterpillars in a relatively "pristine," high-elevation park were species native to Hawaii. In fact, 83 percent of all parasitoid individuals found at these high elevations represented species that had been introduced to Hawaii for biological control of pests on low-elevation farms.

CONCLUSION

The time for assessment of environmental risks is prior to the release of a GE crop, and studies of risk must be scientifically rigorous if they are to provide us with any useful information at all. Unfortunately, the blend of poorly replicated experiments, overreliance on expert opinions, and claims regarding an absence of statistically significant effects that characterizes the reviewed petitions for deregulation provides an extremely weak scientific approach to illuminating GE risks. It is imperative that regulators recognize that small studies (even a collection of small studies if they are not properly synthesized) finding no effect provide very weak evidence of safety. A recent, hard-hitting report by the National Research Council[34] recommended that the USDA should improve the rigor and transparency of its decision-making process by increasing peer review and explicitly presenting data, methods, analyses, and interpretations in its determination documents. Among other specifics, the NRC recommended that the USDA not use general traits of weeds to assess the potential weediness of GE crops because these traits have no predictive value.

Despite the lack of rigorous studies, the absence of hard evidence documenting environmental effects of GE crops may seem reassuring. Our experience with the environmental effects of biological control organisms has exposed the fallacy of this reasoning. In fact, evidence for environmental effects of biological control organisms often comes to light only under highly unlikely combinations of events—such as when a researcher already happened to be studying an affected non-target organism for completely unrelated purposes.[35] Such cases point to the need for long-term monitoring of GE crops even after commercial release has been approved.[36]

Even after well-replicated studies have been completed, some amount of scientific uncertainty will always remain. Scientific uncertainty can be reduced by (1) using larger trials (increasing sample size to decrease type II error), (2) repeating studies under similar conditions, and (3) repeating studies under different conditions to assess the generality of the findings. There will always be charges that a particular risk assessment failed to consider some indirect effect or some extremely low-probability surprise. But, at a minimum, the fundamentals of experimental design and data analysis for GE risks should be of sufficient quality to conclude that obvious and straightforward risks such as gene exchange with wild plants and direct impacts on nontarget species are well studied. The uncertainty regarding risks cannot be eliminated, but rigorous, well-replicated scientific studies that are synthesized appropriately can go a long way toward reducing the enormous uncertainties that exist today.

Some have argued that, particularly in light of the potentially immense benefits that some GE crops promise for developing nations, requirements for extensive and rigorous testing for environmental harms are overly restrictive to industry and "antitechnology."[37] Yet, it seems clear that serious consideration of the environmental risks associated with GEOs is no more "antitechnology" than requiring safety screening for new drugs is "anti-medicine." The fact that a particular drug may save millions of lives does not eliminate the need for extensive investigation of the drug's side effects. Just as many novel drugs are never brought to market, we anticipate that many proposed releases of GEOs will be considered too risky relative to the potential benefits. Certainly, we should demand, at a minimum, data quality and statistical rigor for GEO risk assessment on par with what we require of new drugs. After all, if a newly released drug turns out to be dangerous, we can simply take the new drug off the market. In contrast, if genes from problematic GEOs find their way into nature, they cannot be so easily recalled.

NOTES

1. Simberloff, D. and Stiling, P. 1996. How risky is biological control? *Ecology* 77: 1965-1974.

2. James, C. 2005. *Global status of commercialized biotech/GM crops.* ISAAA Brief No. 34-2005. Ithaca, NY: ISAAA.

3. Snow, A.A. and Palma, P.M. 1997. Commercialization of transgenic plants: Potential ecological risks. *BioScience* 47: 86-96; Wolfenbarger, L.L., and Phifer,

P.R. 2000. The ecological risks and benefits of genetically engineered plants. *Science* 290: 2088-2093; Poppy, G. 2000. GE crops: Environmental risks and non-target effects. *Trends in Plant Sciences* 5: 4-6; Connor, A.J., Glare, T.R., and Nap, J.-P. 2003. The release of genetically modified crops into the environment. Part II. Overview of ecological risk assessment. *Plant Journal* 33: 19-46; Snow, A.A., Andow, D.A., Gepts, P., Hallerman, E.M., Power, A., Tiedje, J.M., and Wolfenbarger, L.L. 2005. Genetically engineered organisms and the environment: Current status and recommendations. *Ecological Applications* 15:377-404.

    4. Snow, A.A. et al. 2004. See note 3; Wolfenbarger, L.L., and Phifer, P.R. 2000. See note 3.

    5. Parker, I.M. and Kareiva, P. 1996. Assessing the risks of invasion for genetically engineered plants: Acceptable evidence and reasonable doubt. *Biological Conservation* 78: 193-203; Marvier, M.A. 2002. Improving risk assessment for nontarget safety of transgenic crops. *Ecological Applications* 12: 1119-1124.

    6. Underwood, A. J. 1997. Environmental decision-making and the precautionary principle: What does this principle mean in environmental sampling practice? *Landscape and Urban Planning* 37: 137-146; Marvier, M.A. 2001. Ecology of transgenic crops. *American Scientist* 89: 160-167.

    7. Marvier, M.A. 2002. See note 5.

    8. Ibid.

    9. Ibid.

    10. Purrington, C.B. and Bergelson, J. 1995. Assessing weediness of transgenic crops: Industry plays plant ecologist. *Trends in Ecology and Evolution* 10: 340-342; Parker, I.M., and Kareiva, P. 1996. See note 5.

    11. Ellstrand, N.C. 2001. When transgenes wander, should we worry? *Plant Physiology* 125: 1543-1545; Snow, A.A. 2002. Transgenic crops—why gene flow matters. *Nature Biotechnology* 20: 542.

    12. Keeler, K.H., Turner, C.E., and Bolick, M.R. 1996. Movement of crop transgenes into wild plants. In *Herbicide-resistant crops: Agricultural, environmental, economic, regulatory, and technical aspects*. Edited by S.O. Duke. Boca Raton, FL: Lewis Publishers, pp. 303-330; Ellstrand, N.C., Prentice, H.C., and Hancock J.F. 1999. Gene flow and introgression from the domesticated plants into their wild relatives. *Annual Review of Ecology and Systematics* 30: 539-563.

    13. Snow, A.A., Pilson, D., Riesberg, L.H., Paulsen, M.J., Pleskac, N., Reagon, M.R., Wolf, D.E., and Selbo S.M. 2003. A *Bt* transgene reduces herbivory and enhances fecundity in wild sunflower. *Ecological Applications* 13: 279-286.

    14. Zar, J.H. 1999. *Biostatistical Analysis, 4th ed.* New Jersey: Prentice Hall.

    15. Marvier, M.A. 2002. See note 5.

    16. Hedges, L.V. and Olkin, I. 1985. *Statistical methods for meta-analysis.* Orlando, FL: Academic Press; Arnqvist, G., and Wooster, D. 1995. Meta-analysis: Synthesizing research findings in ecology and evolution. *Trends in Ecology and Evolution.* 10: 236-240.

    17. Hedges, L.V. and Olkin, I. 1985. See note 16; Gurevitch J., Morrow L.L., Wallace A., and Walsh J.S. 1992. A meta-analysis of competition in field experiments. *American Naturalist* 140: 539-572.

    18. Gurevitch, J. et al. See note 17.

19. NRC. 2002. *Environmental effects of transgenic plant: The scope and adequacy of regulation.* Washington, DC: National Academy Press.

20. Morgan, M.G. and Henrion, M. 1990. *Uncertainty: A guide to dealing with uncertainty in qualitative risk and policy analysis.* Cambridge, UK: Cambridge University Press.

21. Louda, S.M., Pemberton, R.W., Johnson, J.T., and Follet, P.A. 2003. Nontarget effects—The Achilles' heel of biological control? Retrospective analyses to reduce risk associated with biocontrol introductions. *Annual Review of Entomology* 48: 365-396.

22. Ibid.

23. Louda, S.M., Kendall, D., Connor, J., and Simberloff. D. 1997. Ecological effects of an insect introduced for the biological control of weeds. *Science* 277: 1088-1090.

24. Ibid.

25. Ibid.

26. Ibid.

27. Howarth, F.G. 1991. Environmental impacts of classical biological control. *Annual Review of Entomology* 36: 485-509.

28. Ibid.; Simberloff, D. and Stiling, P. 1996. See note 1.

29. Pool, R. and Esnayra, J. 2001. *Ecological monitoring of genetically modified crops: A workshop summary.* Washington, DC: National Academy Press; NRC. 2002. See note 19.

30. Underwood, A.J. 1997. See note 6.

31. Howarth, F.G. 1991. See note 27.

32. Lampo, M. and De Leo, G.A. 1998. The invasion ecology of the toad *Bufo marinus*: From South America to Australia. *Ecological Applications* 8: 388-396.

33. Henneman, M.L. and Memmott, J. 2001. Infiltration of a Hawaiian community by introduced biological control agents. *Science* 293: 1314-1316.

34. NRC. 2002. See note 19.

35. Simberloff. D. and Stiling, P. 1996. See note 1.

36. Pool, R. and Esnayra, J. 2001. See note 29.

37. Miller, H.I. 2001. The biotechnology industry's Frankensteinian creation. *Trends in Biotechnology* 19: 130-131; Paarlberg, R.L. 2003. Reinvigorating genetically modified crops. *Issues in Science and Technology* 19:86-92.

Chapter 5

Coping with Uncertainty: The Human Health Implications of GE Foods

Paul R. Billings
Peter Shorett

INTRODUCTION

Food has been a therapeutic since the time that earliest physicians altered diets to treat illnesses. Food (or its withholding) can cause and ameliorate illnesses, play a role in health-related rituals, and may interact with drugs to enhance or inhibit their effects. By law, the safety and efficacy of pharmaceuticals must be demonstrated through extensive documentation prior to public release. But foods, in particular novel or genetically engineered (GE) foods, are not subjected to similar scrutiny. Virtually no experimental data exist that could allow an adequate assessment of the human health effects of GE food consumption. Few, if any, private industry-sponsored studies have been published in peer-reviewed journals evaluating the toxicological and epidemiological effects of consuming GE foods. A survey in *Science* several years ago concluded that less than ten direct studies had been published on their potential health impacts.[1] And while industry scientists continue to conduct safety assessments of new GE crops for governmental review on a voluntary basis, the unwillingness of companies to submit their research for wider scientific review raises serious concerns. As Jose Domingo notes, "The general population and the scientific community cannot be expected to take it on faith that

Genetically Engineered Crops
© 2007 by The Haworth Press, Inc. All rights reserved.
doi:10.1300/5880_05

the results of such studies are favorable. Informed decisions are made on the basis of experimental data, not faith."[2]

While many observers have blamed this predicament on lax regulatory structures, others have pointed to the challenge of applying safety-assessment techniques traditionally used for additives and chemicals to novel whole foods. In Western Europe and North America, governments have taken the position that once a new food is shown to be similar in its chemical composition to conventional counterparts, it is safe for consumers. This concept, known as "substantial equivalence," is a policy construct that has not been tested by comparison with other standards to evaluate whether the public good or safety is served when it is applied.

We provide an overview of current safety concerns and challenges that must be confronted by policymakers and the informed public as they consider the risks and benefits that may arise from the use of genetically modified plants. We argue that there is an urgent need for rigorous, long-term, multi-methodological approaches to GE safety evaluation. Each of the major strategies that have been proposed—from animal feeding studies using whole foods to postmarket monitoring of human populations—may produce important data. However, we note several major drawbacks in current toxicological and epidemiological models for tracing the health impact of exposure to novel foods.

FOOD IS A DRUG

One of the triumphs of twentieth-century biomedical science has been the demonstration of the role of the constituents of food (its proteins, fats, sugars, vitamins, and minerals) in our metabolism. Basic processes such as movement, temperature regulation, and thinking depend on a consistent supply of key nutrients. The components of food constitute the basic prescription for all aspects of our normal life and health. When we are ill, eating can be effective treatment. Fevers, faints, high blood pressure, even heart diseases can find remedy in diet. Also, the actions of prescribed medications are often modified by what we eat. Antibiotics, ulcer treatments, and blood thinners are examples of prescription drugs, the effectiveness of which can be altered by food.[3]

We present evidence that genetically modified foods may not have the same medicinal properties as their conventional counterparts, including nutrient composition. They can introduce new food allergies and other immunological complications. Like swallowed pills, novel genes and proteins are broken down in our gastrointestinal systems and absorbed. Genes that confer antibiotic resistance, widely used to manufacture novel foods, can move, from what we have eaten, into us or into the bacteria that inhabit our intestines. Such transfers may alter our health directly or change the beneficial symbiosis that exists between bacteria that reside as parts of the normal human intestinal flora and the organs in which they live. Given the important medical and nutritional roles that food plays in everyday life, societies should take precautions when introducing novel proteins and other food components into the human diet.

USING TOXICOLOGY AND EPIDEMIOLOGY TO EVALUATE FOOD SAFETY

Toxicology is the study of the adverse effects of chemical agents on biological systems. Using databases of previous experiments and new tests, it can assist the evaluation of the safety of ingested or otherwise administered products, by assessing the impact of acute and chronic exposure to chemicals and toxicants on human or animal model systems. The U.S. Environmental Protection Agency, for example, uses toxicology as the primary framework for health studies on *Bt* proteins and other plant-incorporated pesticides produced by genetically modified crops. Toxicological evaluations of GE crops attempt to identify the behavior and impact of novel proteins and metabolites through high dose exposure to a small number of laboratory animals under controlled settings.

While critics have pointed to the limitations of a toxicological framework for identifying conventional environmental health threats, such as synthetic pollutants and other single chemical entities, GE foods could pose an even greater challenge. First, genetic modification can introduce a broad array of changes in gene expression. Both direct and episomal alterations may affect gene expression, result in insertional changes, promote gene conversion events, and alter important interactions among genes. While all of these events have occurred in experimental systems, their import and any associated health risk have not been adequately assessed. A priori, there is no reason to believe that the risks associated with these events are different from those arising from the natural "background" frequency of these events. The only reason that they should necessarily engender concern is because we lack the ability, insight, and methods to measure relevant outcomes with proper experiments. Thus, more nuanced toxicological methods need to be developed and applied, including long-term exposure studies on animals.[4] Companies have applied these methods with some success in evaluating food allergens.

The second framework for assessing GE food safety is the use of epidemiological studies in human populations. Given the central role of diet in both the maintenance of health and the development of acute or chronic diseases, epidemiology can play a useful role in evaluating GE food safety. For example, to understand the epidemiology of human immune responses to novel foods, interactions between food intake and measurements of metabolism, immune system factors, genetic background, lifestyle, and socioeconomic conditions are examined. These studies depend on the availability of unbiased population-wide data samples. An epidemiologist determines adverse effects through a controlled comparison of reactions between exposed and unexposed individuals.

Such a process, if applied, for instance, to postmarket surveillance of GE food consumption in human populations, would face a number of obstacles. First is the question of sensitivity: How can the adverse effects of GE foods be separated from a background of unhealthy conventional diets? Many toxicologists doubt that these effects can be distinguished.[5] Other factors including interacting variables and barriers to conducting multiple regression analyses will also limit this type of approach to evaluation and risk assessment.

SUBSTANTIAL EQUIVALENCE AND ITS CONSEQUENCES

In 1992, the U.S. Food and Drug Administration (FDA) issued a seminal ruling claiming that GE crops similar in composition to their conventional counterparts are "generally regarded as safe" and do not introduce unique health risks to consumers. The ruling also stated that genetic-engineering methods represent "extensions at the molecular level" of traditional plant breeding and should, therefore, be regulated in an equivalent manner.[6] Since this ruling, the FDA and other regulatory bodies in the United States, Canada, and Western Europe and have adopted a so-called substantial equivalence standard for the regulating GE crops. This standard states that if a GE plant is shown to be substantially equivalent to its conventional counterpart in its protein, carbohydrate, fat and nutrient content, no further safety evaluation should be required.[7]

In practice, regulators have used the concept of "substantial equivalence" to justify introducing GE foods into the market without long-term nutritional and toxicological testing on animals. Without such studies, we have few ways of assessing the full effects of foreign gene insertion. In response to concerns about the scientific inadequacy of "substantial equivalence," FDA standards have recently begun to evolve. In 2001, the FDA issued a proposed ruling that would require developers of GE foods to provide safety and nutritional data, including compositional analysis of whole GE food products, to the agency 120 days before their introduction to market. In contrast to its previous position, the FDA acknowledged that GE foods "are likely in some cases to present more complex safety and regulatory issues than seen to date."[8] The notice acknowledges that genetic modification can substantially alter levels of toxins and nutrients in host plants, and that position effects make these changes difficult to control or predict. Lastly, the FDA notes that "wide crossing," the introduction of genes from distantly related species, can alter plant composition in ways that could generate unique and unintended effects on food safety.[9] Critics point out, however, that the new rule has been poorly enforced,[10] has no clear data requirement standards, and does not require long-term premarket animal testing.

NUTRITION

GE crop developers and government regulators alike often speak of genetic engineering (GE) as a precise and predictable way to introduce desired traits into food. The FDA notes, for example, that traits "whose chromosomal location or molecular identity is known can be transferred to another organism . . . without simultaneously introduc-

ing undesirable traits."[11] These models fail to account for several uncertainties involved in genetic modification. First, because it is not
generally possible to predict the chromosomal location at which foreign genes will integrate within the recipient organism, transgene
insertion may result in unexpected silencing or activation of other
genes. Second, regulatory elements used to promote the expression
of the transgene can also effect "upstream" and "downstream" gene
expression within the chromosome. Third, the effect may alter the
physical structure of the chromosome, which in turn can alter even
distant interactive genetic events.

One of the results of such unintended effects is the expression of
endogenous proteins that alter levels of nutrients and other compounds that reduce nutrient utilization. Changes in the nutrient composition of a food can have serious human health consequences,
particularly for infants, pregnant women, and populations whose diets depend on a high intake of that food product.

Research points to several cases of such detrimental changes in nutrition. In 1999, Marc Lappé and colleagues published a study in the
Journal of Medicinal Food indicating that soybeans genetically modified for herbicide tolerance contained significantly lower levels of
phyto-estrogens than their conventional counterparts.[12] Industry studies of GE soybeans have shown heightened trypsin inhibitor levels in
defatted, non-toasted soybean meal.[13] Trypsin is a ubiquitous protein
in humans that participates in a variety of metabolic and inflammatory
phenomena. Limited experiments on herbicide-resistant GE maize
have also shown unexpected changes in fat and carbohydrate content.[14]
Although the results of these studies are controversial, they suggest
that genetic modification can, in some cases, cause unpredictable,
damaging changes in plant nutrients.

Much current attention, however, focuses on the use of GE to enhance the nutritional value of food. Although there are few GE foods
in commercial markets that target nutrition, many such products are
in the process of development and approval. DuPont has developed
genetically modified soybean and canola oils that exhibit increased
levels of oleic acid and reduced levels of saturated and polyunsatu-
rated fatty acids.[15] Genetically modified rice varieties have been
created with enhanced beta-carotene and iron content.[16] The Plant Biotechnology Institute in Canada has developed GE wheat varieties

containing only B granules (which reduce oil absorption) and higher levels of resistant starch (intended to reduce the caloric content of wheat).

A full review of the potential risks and benefits involved in the new generation of consumer-oriented GE food varieties is beyond the scope of this chapter. It should be noted, however, that arguments used to promote these crops, often promising a reduction in malnutrition and disease, have been widely disputed.[17] For example, Syngenta has heralded the introduction of GE rice varieties with increased beta-carotene content as a solution to Vitamin A deficiency in the developing world. Further analysis, however, suggests that the nutritional value of Golden Rice will largely depend on bioavailability.[18] Fat, which is required for efficient conversion of beta-carotene to Vitamin A, and protein, which provides carrier molecules that help deliver nutrients to cells, are both in short supply in the average diet of many poor countries. Without attention to the broader dietary context into which novel food products are introduced, consumer-oriented GE foods could bring little more than false promises. They may also divert our attention away from other solutions to disease and nutrient deficiency, such as conventional nutritional supplementation, diversification of diet, greater access to food, and basic public health improvements such as clean water and sanitation.

Both intended and unintended nutritional changes resulting from genetic modification can be evaluated through nutritional composition analyses and long-term animal feeding studies using whole GE food products.[19] These experiments can monitor physiological changes, such as altered metabolism and changes in tissue structure. They should be part of any prudent safety evaluation of GE foods.

ALLERGENICITY

The last few years have witnessed a substantial rise in the worldwide incidence of food allergies and food-borne anaphylactic shock.[20] This growing health dilemma has drawn added public and scientific attention to the question of whether the introduction of genetically modified products into the human food chain might cause unintended, and in some cases fatal, allergic reactions.

Food allergies are the result of adverse immunological reactions to proteins, and to a lesser extent other components, in food. The most common type of food allergies are immediate hypersensitivity reactions, which occur when immunoglobulin E (IgE) antibodies bind to an allergen, causing symptoms that range from mild itching and diarrhea to life-threatening anaphylactic shock.[21] Approximately 2 percent of people, including 8 percent of children, in industrialized countries are affected by food allergies.[22] Ninety percent of these moderate-to-severe food allergies result from exposure to a narrow range of nuts, cereal grains, seafood, soybeans, and dairy products.[23]

In general, food allergies are difficult to detect, measure objectively, and assess in terms of their impact on human health generally or as they interact with other factors that may modify health or disease. Few factors for predicting the severity of food allergies are known. Relevant animal models are not available. Despite a consensus that food allergies can be a significant problem, many scientific uncertainties persist and methods for addressing issues are in short supply.

Genetic modification of food crops, whether through the introduction of a foreign gene or an alteration in gene expression, could cause them to become allergenic. The process can introduce proteins into plants from known allergens, as well as from viral, bacterial, and plant sources for which allergenicity data is unavailable. As a recent Royal Society report noted, "There is at present no evidence that GM foods that are commercially available cause any clinical manifestations of allergenicity, and assertions to the contrary have not been supported by systematic analysis."[24] As we discuss later, however, two cases have demonstrated the risk that genes coding for allergens and immunogens may be inadvertently transferred into the human food supply.

Approaches to assessing potential allergenicity of GE foods containing genes from known allergens were developed and adopted by a joint consultation between the Food and Agriculture Organization (FAO) and the World Health Organization (WHO) in 2000.[25] The FAO/WHO consultation proposed a "decision-tree" framework to analyze the following:

1. amino acid sequence homology of novel proteins to known allergens;
2. immunoreactivity of proteins with IgE from blood samples of allergenic individuals;

3. digestive stability of the novel food source; and
4. stability of the novel protein to heat or processing.[26]

The methodology for carrying out such an assessment protocol is relatively straightforward, and has been outlined elsewhere.[27] Several organizations have proposed that such methods be amended to include analysis of reactions to pollen and dust inhalation. Attempts to assess the risk of significant food allergies arising from GE products need empirical data to gauge adequately the overall risks, likely severity, and possible risk interactions.

Other allergenicity risks from genetic modification may be more difficult to tackle. These include gene transfer from biological sources with unknown allergenicity and the unanticipated creation of novel allergens through gene inactivation or overexpression of genes that code for a minor allergen. FAO/WHO reports acknowledge that current assessment criteria are unreliable for predicting whether such crops will be allergenic.[28] As one observer notes, this gap in predictive capability is made all the more troubling by the fact that "biotechnology companies increasingly use microorganisms rather than food plants as gene donors, or are designing proteins themselves, even though the allergenic potential of these proteins is unpredictable and untestable."[29]

The most well-known study documenting inadvertent transfer of allergens from one plant species to another through genetic modification was published by Julie Nordlee and colleagues in the *New England Journal of Medicine* in 1996.[30] At that time, scientists at Pioneer Hi-Bred were in the process of developing a GE soybean containing a gene that coded for a methionine-rich protein from Brazil nuts. Through tests of IgE immunoreactivity on blood samples of individuals with relevant allergies, Nordlee et al. demonstrated that a major Brazil nut allergen had been transferred to Pioneer Hi-Bred soybeans.

A second case involved Starlink corn, which had been restricted to animal consumption by the USDA after tests indicated that cryc9c, the toxin encoded in the corn due to *Bt* gene insertion, was resistant to human digestion.[31] Despite the USDA ruling, ineffective segregation of corn intended for feed caused Starlink corn to contaminate hundreds of household food products. The controversy pointed to potential public health risks from inadvertent GE contamination of the food supply. Even industry studies have demonstrated allergenicity in premarket transgenic crop lines. In a report by Monsanto scientists in a 1996 issue of the *Journal of Nutrition,* glycophosate-tolerant soybeans showed a 28 percent increase in Kunitz trypsin inhibitor, a known antinutrient and allergen.

In cases where foreign genetic material is transferred from a known allergen, safety testing requires only a basic series of in vitro tests on blood samples from sensitized individuals. However, GE not only brings proteins into the food supply from known source of common allergens but many GE crops currently in development involve the introduction of transgenic elements from organisms never previously consumed as food. Population-wide data on allergenicity does not exist for most of the novel proteins from such sources. However, initiatives such as the University of Nebraska Food Allergy Research and Resource Program's Protein Allergen Database, a public database of all 1,191 known food- and environment-related allergenic proteins, will hopefully begin to address such uncertainties.

Comparisons between novel proteins and known food allergens, through identification of short-sequence homologies and common allergenic characteristics, can yield important predictive information. But the current so-called decision-tree approach has major limitations for assessing the risks of gene transfer from plants with unknown allergenic potential.[32] Efforts to conduct premarket allergenicity testing have also come up against the fact that, in contrast to nutritional and toxicological studies, no strong animal models exist for predicting human allergens.[33]

TOXICITY

Food toxicity is an enduring part of human dietary experience. Many plants, whether commonly consumed or considered dangerous, contain substances that can be toxic to humans when consumed in sufficient quantities. Spinach and rhubarb, for example, contain oxalic acid, an anti-nutritional compound that inhibits calcium and iron absorption and is poisonous in large amounts. Onions contain sulfuric acid that in sufficient quantities can act as an ingest corrosive in the upper gastrointestinal tract of humans. Many species of mushrooms are lethal in small doses and difficult to distinguish from their edible counterparts.

Genetic modification can alter both existing and unanticipated toxicological characteristics of foods. Although toxicity data on genetically modified foods is rare, some scientific literature indicates that gene insertion can generate unexpected increases in levels of naturally occurring toxins. A study involving recombinant yeast cells, where genes from yeast were cloned and then reintroduced through conventional GE techniques, showed a three-fold increase in the accumulation of an enzyme in the glycolytic pathway and a 40 to 200 fold increase in the production of methyl-glyoxal, a substance that is toxic and mutagenic in high concentrations.[34]

Uncertainty over the location, structure, and stability of foreign gene integration into the plant genome raises a persistent concern that cannot be addressed by evaluation of known toxic compounds. For example, one study involving the genetic modification of tobacco, intended to enable the production, of gamma-linolenic acid, activated a metabolic pathway that produces higher quantities of octadecatetraenic acid, a toxic compound not present in conventional tobacco plants.[35] While significant new toxicity may not be a likely outcome, particularly if previously unaltered products have no toxicity as assessed by reliable methods, the current data and methods do not allow either adequate data to be accumulated or risk estimates to be based on anything but sophisticated speculation. The possibility that GE foods will produce rare or previously unknown toxins has not yet led to serious changes in safety-assessment methods.

Most GE crops already released or currently undergoing field trials contain a promoter from the cauliflower mosaic virus, known as the CaMV 35S transcript. Precisely because it is so effective in promoting transgene expression, CaMV has raised significant safety concerns. As Michael Hansen notes, "the CaMV 35S promoter effectively puts the transgene(s) outside of virtually any regulatory control by the host genome as the natural plant promoters for each gene allow."[36] A study by Mae-Wan Ho, Angela Ryan, and Joe Cummins demonstrated that CaMV 35S can affect gene expression thousands of base pairs upstream and downstream from the insertion site on a given chromosome and even alter the behavior of genes on other chromosomes.[37] However, the CaMV promoter and other viral sequences are already widely present in the conventional food supply, with no clearly demonstrated impact on human health.

In order to determine the potential toxicity of GE foods, scientists rely on the chemical analysis of known nutrients and toxic compounds. Regulatory agencies, such as the U.S. Environmental Protection Agency (EPA), have also used high-dose oral testing of GE foods in mice as an indirect measurement of toxicity to humans.[38] This model of risk assessment has limited utility in its application to GE foods. As the Society of Toxicology notes, "Methods have not yet been developed with which whole foods (in contrast to single chemical components) can be fully evaluated for safety."[39] In assessing the risk posed by human exposure to most environmental pollutants, as well as conventional medicines, toxicologists usually study the role of an isolated chemical agent in biological systems. But foods are complex mixtures, and the interaction between their components may affect toxicity in ways that cannot be predicted by analysis of isolated proteins.

Furthermore, while conventional toxicology studies use high-dose exposure to animal models over a limited time period, potential exposures to toxic risks through GE foods are likely to be gradual and in low-dose quantities. As a result, acute oral toxicity tests may be poor predictive tools. Instead, many scientists recommend long-term feeding studies focused on immunology, organ development, and metabolism.

CONCLUSION

In this chapter, we have reviewed several issues surrounding the introduction into the human food supply of novel products, applying a health and disease framework to discuss the evaluation of these new foods. We have tried to point out where information is lacking and, of even greater concern, where no methods are currently available to begin the proper assessment of phenomena and impacts.

Unfortunately, the politics and rhetoric surrounding the introduction and dissemination of GE foods in human populations have obscured a new opportunity to review and conduct further studies on the medicinal role of foods. The varying role of food in the causation, amelioration, treatment, and culture that surround disease and illness are well known and clinically documented.[40] But modern assessment methods of increasing sensitivity have not been introduced or well accepted, nor have determinations of food-related pharmacogenomic or metabalogenomic effects helped understand, predict, or manipulate food impacts in predictable ways.

Simply noting the inadequacy of current data or methods is important because it argues against any assurances of safety. But since new foods are introduced to satisfy real markets and important public needs, there will be an appropriate pressure to act on improvements in currently available risk assessments. Experts who participate in these estimations in the absence of adequate empirical data need to define more clearly what information is absolutely necessary to proceed and then insist that pre-introduction testing be conducted to produce this data prior to human consumption. Only in unusual circumstances, and with prior consent, should Phase 4-like surveillance be the only safety assessment; humans should not be used as nonconsensual participants in GE product evaluations. While better labeling can inform the public of the content of a food and uncertainties about its impact, it does not relieve scientists and regulators of responsibility for defining a minimal level of knowledge required to protect the public from adverse consequences arising from the use of any food or drug.

Finally, by renewing interest in the role of foodstuffs (including GE products) in health, disease, and as medicinals, a more contemporary approach to these studies may occur. It is certainly possible that the alteration of individual foods or combinations in diets may produce beneficial outcomes for individuals coping with disease or populations burdened by illnesses. New methods may produce more nuanced knowledge and powerful modes to improve human nutrition generally and in the service of disease control or illness treatment. But honesty about our current state of ignorance, and the ad hoc nature of many risk evaluations, ought not to be denied nor obscure future possibilities.

NOTES

1. Domingo, J. 2000. Genetically modified foods: Many opinions but few data. *Science* 288: 1748-1749.

2. Ibid.

3. Harris, R.Z., Jang, G.R., and Tsunoda, S. 2003. Dietary effects on drug metabolism and transport. *Clinical Pharmacokinetics* 42(13): 1071-1088.

4. Taylor, S.L. 2003. Safety assessment of foods produced through agricultural biotechnology. *Nutrition Review* 61: 139-140.

5. Ritter, L. 2002. *Long-term monitoring of health effects related to genetically modified foods in Canada: Post-market surveillance and the role of labeling.* Report by the Canadian Network of Toxicology Centres to the Canadian Biotechnology Advisory Committee. Guelph, ON: University of Guelph.

6. US Food and Drug Administration. 1992. Statement of policy: Foods derived from new plant varieties. *Federal Register* 57: 22984.

7. See, for example, The Royal Society, 2002. *Genetically modified plants for food use and human health—an update* (http://www.royalsoc.ac.uk), p. 5. [Accessed April 6, 2004].

8. US Food and Drug Administration. 2001. Premarket notice concerning bioengineered foods: Proposed rule. *Federal Register* 66: 4706-4738. Quoted by Krimsky, S., and Murphy, N.K. 2002. Biotechnology at the dinner table: FDA's oversight of transgenic foods. *Annals of the American Association of Political and Social Science* 584: 80-96.

9. Krimsky, S. and Murphy, N.K. 2002. See note 10.

10. Gurian-Sherman, D. 2003. *Holes in the biotech safety net: FDA policy does not ensure the safety of genetically engineered foods.* Washington, DC: Center for Science in the Public Interest.

11. Quoted in Krimsky, S. 2000. Risk assessment and regulation of bioengineered food products. *International Journal of Biotechnology* 2: 231-238.

12. Lappe, M., Bailey, E.B., Childress, C., and Setchelle, K.D.R. 1999. Alterations in clinically important phytoestrogens in genetically modified, herbicide tolerant soybeans. *Journal of Medicinal Food* 1: 241-245.

13. Padgett, S.R., Taylor, N.B., Nida, D.L., Bailey, M.R., MacDonald, J., Holden, L.R., and Fuchs, R.L. 1996. The composition of glyphosate-tolerant soybean seeds is equivalent to that of conventional soybeans. *Journal of Nutrition* 126: 702-716.

14. Pusztai, A. et al. 2003. See note 9.

15. Kinney, A.J. 1997. Genetic engineering of oilseeds for desired traits. In *Genetic engineering.* Edited by J.K. Setlow. New York: Plenum Press, pp. 149-166.

16. Ye, X., Al-Babili, S., Kloti, A., Zhang, J., Lucca, P., Beyer, P., and Potrykus, I. 2000. Engineering the pro-vitamin a biosynthetic pathway into rice endosperm. *Science* 287: 303-305.

17. GeneWatch UK. 2000. *The next generation of GM foods: Good for whose health?* Briefing Paper Number 10. Derbyshire, UK: GeneWatch UK.

18. Potrykus, I. 2003. Nutritionally enhanced rice to combat malnutrition disorders of the poor. *Nutrition Reviews* 61: 101-104.

19. Pusztai, A. et al. 2003. See note 9.

20. Legall, D. 2003. Why is food allergy such a hard nut to crack? *Gastrointestinal Nursing* 1(10): 14-17.

21. Taylor, S.L. 2002. Protein allergenicity assessment of foods produced through biotechnology. *Annual Review of Pharmacology and Toxicology* 42: 99-112.

22. Sampson, H.A. 1997. Food allergy. *Journal of American Medical Association* 278: 1888-1894.

23. Food and Agricultural Organization and World Health Organization. 2000. *Safety aspects of genetically modified foods of plant origin.* Report of a Joint FAO/WHO Expert Consultation on Foods Derived from Biotechnology, Geneva, Switzerland.

24. The Royal Society. 2002. See note 8.

25. FAO and WHO. 2001. *Allergenicity of genetically modified foods.* Report of a Joint FAO/WHO Expert Consultation on Foods Derived from Biotechnology, Rome, Italy.

26. Ibid.

27. Taylor, S.L. 2003. See note 5.

28. FAO and WHO, 2000. See note 27.

29. Consumers Union, N.D. Comments to the FDA on biotech in the year 2000 and beyond. http://www.consumersunion.org/food/fdacpi100.htm. Accessed April 6, 2004.

30. Nordlee, J.A., Taylor, S.L., Townsend, J.A., Thomas, L.A., and Bush, R.K., 1996. Identification of a Brazil-nut allergen in transgenic soybeans. *New England Journal of Medicine* 334: 688-693.

31. Taylor, S.L. and Hefle, S.L. 2001. Will genetically modified foods be allergenic? *Journal of Allergy and Clinical Immunology* 107: 765-771.

32. Pusztai, A. et al. 2003. See note 9.

33. Helm, R.M. and Burks, A.W. 2000. Mechanisms of food allergy. *Current Opinion in Immunology* 12: 647-653.

34. Inose, T. and Murata, K. 1995. Enhanced accumulation of toxic compounds in yeast cells having high glycolytic activity: A case study on the safety of genetically engineered yeast. *International Journal of Food Science and Technology* 30: 141-146.

35. Reddy, S.A. and Thomas, T.L. 1996. Expression of a cyanobacterial delta 6-desaturase gene results in gamma-linolenic acid production in transgenic plants. *Nature Biotechnology* 14: 639-642.

36. Hansen, M. 2000. *Genetic engineering is not an extension of conventional plant breeding.* Yonkers, NY: Consumers Union.

37. Ho, M.-W., Ryan, A., and Cummins, J. 2000. The cauliflower mosaic viral promoter—a recipe for disaster? *Microbial Ecology in Health and Disease* 12: 6-11.

38. Murphy, N. and Krimsky, S. 2003. Implicit precaution, scientific inference and indirect evidence: The basis for the US Environmental Protection Agency's regulation of genetically modified crops. *New Genetics and Society* 22: 127-143.

39. Society of Toxicology. 2002. *The safety of genetically modified foods produced through biotechnology.* http://www.toxicology.org/Information/Government Media/GM_Food.html. Accessed April 6, 2004.

40. Harris, R.Z. et al. 2003, see note 4. See also Mintz, S.W., and Dubois, C.S. 2002. The anthropology of food and eating. *Annual Review of Anthropology* 31: 99-119; Katz S., and Weaver, W. eds.; 2003. *The encyclopedia of food and culture.* New York: Scribner.

Chapter 6

Future Research Tackling
the Technology Divide:
A Research Agenda
for Crop Biotechnology

Michiel Korthals

INTRODUCTION

Biotechnology will be on the research agenda for the next twenty-five years. In the Western world there is a huge need to restore trust, and research agendas must consider topics of future research incorporating strategies to enhance and rebuild consumer trust. This applies not only to market organizations, but for policy and civil societies as well. If biotechnology is to fulfill its promise in the next twenty-five years, consumers and citizens together must be included in planning the research agenda. The situation is often different in the developing world because in these countries both the private sector and governments are financially weak and are not able to invest in biotechnology research or biotechnological transfer. This is one reason for the existing technology divide.

Governments and civil society actors, such as NGOs and the media, are confronted by much uncertainty with regard to the possibilities of future biotechnologies, the societal trends vis-à-vis foods and agriculture, and broader trends with respect to lifestyles, trade policies, and power structures. However, this should not inhibit either reflection or action, because some problems will remain on the research agenda, like the demand for safe and healthy food, the rising meat demand, and its implications for animal feed[1] and food for aquaculture.

In this chapter, I address the following four major issues:

- *The state of affairs with respect to four major constituents in the field of biotechnology: governments, markets, civil societies, and (bio-)science.* These four categories are well known in contemporary sociology.[2] It is not clear how far they are applicable to non-Western societies and to global networks, but they serve to prevent us from thinking in terms of single-minded solutions such as either market or technology.[3]
- *Seven ethical problems, which will confront biotechnology in the next decades, and propose an overall ethical framework.*
- *The relationship between the four constituents and the seven problems and their possible diverging trajectories.* I connect these problems to future research agendas of Beta-Gamma cooperation in biotechnology and suggest ways to tackle the problems, in particular the technology divide between the developed and developing worlds.
- *The social embedding of research agendas.*

THE CURRENT STATE OF AFFAIRS: CRISIS IN GOVERNMENTS, MARKETS, CIVIL SOCIETIES, AND SCIENCE AND AGRICULTURE

Crisis in Governments

Governments in most countries are confronted with three general issues with respect to food: food security, food safety, and food quality. Decline of food production, zoönoses, trade conflicts, trade shifts, and crises in consumer confidence are constantly present and threatening an acceptable supply of good and healthy food stuffs. In the next decades, however, national government actions will be increasingly influenced by pressures from local, regional, and international interests.

On one hand, governments face two equally avoidable types of uncertainty. They may take actions that have uncertain, maybe bad, effects, or they may do nothing and face other uncertain, maybe bad, effects. On the other hand, their actions may lead to significant benefits from the applications of modern crop biotechnology. The essential cultural and pluralistic nature of food also means that governments often face strong reactions against substantive change from both consumers and farmers.

Large international and regional blocks are subject to severe internal and external pressures. The European Union (EU) is transformed by each new member, and at the same time it must defend its new Common Agricultural Policy (CAP) vis-à-vis the United States and other countries. The Cancun round of trade negotiations is still pending in 2006, and it remains to be seen what consensus can be reached. Controversies in and between trade blocks on GE food, intellectual property, and the formation of regimes to assist the poor will continue, as the Asian, Latin American, and African countries form associations. Furthermore, the large UN organizations, such as Food and Agriculture Organization (FAO), International Rice Research Institute (IRRI), and Consultative Group on International Agricultural Research (CGIAR), will strive for better food regimes, even though they are sometimes at odds with each other. Last but not least the World Trade Organization (WTO), and its international agreements like the General Agreement on Trade and Tariffs (GATT), can conflict with the Cartagena Protocol on Biosafety,[4] which took effect in 2003. The protocol requires GE food to be regulated differently and requires that international shipments of GE grains be labeled. This protocol focuses on the production method and is strongly linked to the precautionary principle. However, other international regulations and bodies, like GATT and WTO, favor the view of GE products as substantially equivalent to conventional food products, determined using Greedy Randomized Adaptive Search Procedures (GRASP) methods, and have little accommodation for public opinion on risks and food safety.

These differences lead to a major controversy between utilitarian ethics and connected strategies, such as cost benefit analyses (CBA), which are presumably culture neutral and science based, and deliberative ethics that emphasize serious deliberations between stakeholders and incorporate rights, consent, respect, and recognition of underprivileged positions and historical abuses. The protesters against globalization have put this problem on the agenda. However, their protests have often been violent, to which so-called decent societies can only react by practicing "decent" behavior, for example, by showing respect.

The rise of biotechnology has led to new cooperations between private and public organizations, referred to by Thackray[5] as Triple Helix, including an enormous increase in patent applications from which new private–public corporations hope to earn profits.[6] It is very difficult to define clear borders between governmental (public) organizations and the private sector and many tensions exist.

Crisis in Markets for Seeds and Plants

The actors in the marketplace vary a lot. Some large international companies respect ethical policies, such as People, Power, and Profit (PPP), while others are solely profit oriented. Furthermore, there are smallholders' cooperatives that are often driven by small-scale and crop-specific interests. Producers of seed and plants are mostly large, often competing companies, with or without a background in pharmaceutical research and development. In the life science industry, there is a clear trend toward formation of large-scale enterprises through horizontal (merging biochemical and seed companies) and vertical expansion. In 1998, ten large enterprises dominated 85 percent (US$31 billion) of the world market for pesticides. The world market for seed also shows remarkable monopolistic traits: here approximately ten enterprises share 30 percent of the total value (US$23 billion), and only four of them, Monsanto, Novartis, AstraZenica, and DuPont, dominate the markets for GE seed. Salta and Pine Land share 72 percent of the U.S. cotton seed market and Pioneer Hi-Bred (Dupont) alone controls 42 percent of the maize market.[7]

In most African, Latin American, and Asian countries experiments are directed toward modifying indigenous crops such as cassava and rice. In China, use of GE crops is particularly widespread,[8] and it could be that in twenty-five years Chinese biotech companies will outcompete those in the Western world. However, both small farmers, who form the worldwide majority of farmers, and most of the world population are seldom sustained by technology programs.[9]

Retailers in the Western world are influential stakeholders, because of their large share of the consumer market, and their constant desire to control their national and international supply lines. Two major European retailers, Carrefour and Ahold, have huge contracts with farmers in the developing world. They have strong, direct contact with consumers; nevertheless, they do not always reflect consumer interests. Even the official representatives of consumer organizations are not very effective in formulating and pursuing their constituents' interests.

Voluntary Organizations

In some parts of the world, voluntary organizations serve an important function by providing information and encouraging public debate. In other parts, these organizations are absent or have a strong bias toward their private or political associations.[10] When they have a larger role, they do not directly influence policymaking but, perhaps more important, they influence public opinion by exploiting the global networks provided by the mass media. Greenpeace, with its yearly budget of approximately US$100 million, has only a tiny financial capacity compared with the transnational corporations, but its influence on food and agricultural policies has proved to be enormous.[11]

Large international governmental institutions, like the UN Food and Agricultural Organization (FAO) and the International Livestock Research Institute (ILRI of CGIAR), support or stimulate research through national organizations. Many international organizations such as FAO, United Nations Development Program (UNDP), and Nuffield Council have subscribed to the view that research and production of GE crops are instruments for alleviating the world hunger and malnutrition problem. Some African, Asian, and Latin American organizations, for example, the Zimbabwe Community Technology Development Trust, have also supported this view. Opposing organizations, including ActionAid Brazil and Greenpeace, have been joined by the prominent American, Indian, and English philosophers Mark Lappé, Vandana Shiva, and Ben Mepham. They all subscribe to the view that biotechnology will push more people into the hunger trap.

PLANT BIOTECHNOLOGY AND AGRICULTURE

The history of science and plant biotechnology bubbles has become tiresomely repetitive. The current messages claim that new plants will save us from hunger or healthy plants will give us a longer life.

During the latter half of the twentieth century we heard that nuclear power was a panacea. Currently it is biotechnology. Certainly the improvements in plant biotechnology are enormous and the next decades will surprise us with new possibilities, as genetic modification becomes more sophisticated and oriented toward consumer-friendly products, such as edible vaccines[12] and protein-enriched crops, and research in medical diagnostics and genetic markers (genomics) points to potentially enormous benefits for humanity. Some advantages of genomics will be in the field of food safety: new models inspired by a predictive microbiology will provide early warning systems against harmful food microbes. It is likely that improved food security and quality, nutrigenomics, and other predictive approaches will produce crops that result in functional foods, that is, food with additional health effects. Edible vaccines and food enriched with vitamins and minerals (like Golden Rice) that were predicted a few years ago[13] are now in late development stages or have become commercially available.

Contemporary agriculture has at least two global components. On the one hand, there is a system of monocultural crop unification, where high-tech solutions are developed and tested, as, for example, in large parts of the United States, Canada, Argentina, and China.[14] In 1996 there were few GE crops in the fields, but by 2002 GE plantings worldwide exceeded 60 million hectares. The other farming system is enormously differentiated, multifunctional, and complex, and embraces the different cultural infrastructures of mostly small farmers. Both systems are clearly connected by the type of plants that are produced, and the diversity of the plantings creates different ways of life.

SEVEN PERSISTENT ETHICAL PROBLEMS OF BIOTECHNOLOGY

An Inventory

Globally, there are at least seven large uncertain issues and controversies around crop biotechnologies that form an ongoing battleground for industries, NGOs such as Greenpeace, research scientists, and governments.

1. GE crops may pose a risk to the environment by causing a further decline in biodiversity. The long-term effects of GE crops

on other organisms are still unclear. Does biotechnology affect the variety of wild life (plants, animals)? Does cross pollination of GE crops with wild relatives or conventional varieties pose severe dangers? I have previously discussed the risks of GEOs in detail,[15] so here I stress the need for long-term monitoring, early warning systems, and evaluatory agencies, and for some kind of harmonization of risk definition.

2. Should GE and unaltered crops be separated at the retail level and should they be labeled? As many consumers do not accept and/or agree with expert scientific views or because of the scientific controversies themselves, there is demand to grow non-GE crops and to have access to guaranteed non-GE foods. The need to respect consumer opinions is still very controversial. The United States and its large GE industry seem adamantly opposed to labeling GEOs, both as grown and after food production processes.[16] Voluntary labeling is opposed, because it may give the impression that GE crops are less safe. The EU does respect consumer opinion and has now stringent rules, using terms such as "GE free" and "non-GE" (meaning content of less than 0.9 percent). It remains to be seen how this trade conflict will be solved. The problem of labeling or not is also linked to the costs of ensuring strict separation of crops (who pays?) and to industry's willingness to comply with separation (Nestle 2003).[17]

3. Biotechnology seems unconcerned to manage the microbial resistance spiral; indeed it may even have accelerated this spiral over time through the use of antibiotic markers. There is an urgent need for biotechnology to contribute to reducing the speed of this spiral. Microbial evolution is clearly faster than new drug development but there may be novel methods to dampen the spiral, by more careful use of antibiotics, investing more research in the efficiency of disease-reducing measures, and the development of sustainable fungus resistance[18]. For example, some virulent forms of the fungus *Phytophthora infestans,* the causal agent of potato late rot, have genetically based mechanisms to stop expression of those proteins, for which the potato has built up resistance.[19] The puzzle is to bypass this difficulty.

4. Whether GE crops are really acceptable for the developing world is a hotly debated issue. As possible solutions for chronic

hunger and frequent bad harvests, the often large economic and environmental advantages of GE crops seem to make them socially advantageous.[20] For several years the larger agricultural and biochemical corporations as well as the U.S. government have used the argument that for hungry people, protests against possible risks of GE crops are a Western luxury. Although GE crops could alleviate the world hunger problem, it is not at all clear that organizations and governments that have cared so little for poor people in the past have a sudden, newfound commitment to them. Furthermore, it is patronizing to presume that poor people cannot develop an opinion about the things they eat. Do hungry people not have an opinion on GE food? Surely so, because everywhere food is embedded in local culture and lifestyles.

5. There is considerable controversy surrounding the methods of risk assessment (e.g., based upon substantial equivalence, methods implied by the precautionary principle, and CBA) and their relationship with risk management and risk communication. Risk definitions, which are only science based (like those of the USA FDA or EPA), often conflict with risk definitions that accommodate the essentially cultural and ethical aspects of risks and food safety. North American tests start with the idea of "substantial equivalence," which presupposes that a GE crop only differs from the same non-modified crop by possession of manipulated gene or genes that code for one or more proteins. Tests of such a GE crop involve the use of rats (or other test animals), which are fed with the particular proteins that are encoded by the added genes. If no negative effects are detected, it is supposed that all of the non-modified proteins in the modified plant will act as they would in the non-GE plant.[21] This is an uncertain assumption, because it is well known that many proteins (and genes) act in several regulatory ways. The emphasis of European test procedures is to assess the whole modified crop, not just the added gene products, and until now no significant health or ecosystem risks have been found. The U.S. National Academy of Sciences has recently advocated this more comprehensive procedure.[22]

6. The very complex issue of intellectual property, which can be fully or partly regulated in both private and public ways,[23] has

serious national and international implications. Private owner-ship of knowledge may not be ethically acceptable, is not easily regulated, and may be a denial of justice. The germplasm resources of developing countries have often been taken and used, some would say pillaged, by the richer countries. The WTO's agreement on Trade-Related Intellectual Property Rights (TRIPS) includes possibilities to protect farmers' traditional practices. The new International Treaty on Plant Genetic Resources for Food and Agriculture recognizes the contributions of farmers to the conservation and use of plant genetic resources over time and for future generations. Strategies against biopiracy, as it is called, and of benefit sharing will be discussed later.

7. The final, but by no means the least, important ethical issue is the handling of social risks, especially the emergence of bio-technology monopolies and the relegation of small farmers to the role of indentured laborers. Socially, the biotechnology di-vide may cause nightmarish scenarios because access to these technologies is often restricted to the wealthy few, while the poverty of the large population of poor farmers in the develop-ing world gives them little choice but to comply with the indus-trial interests. The lack of national resources for technology transfer and innovation in developing nations can lead to new feudal-like bonds between technological societies and the re-source-producer nations.[24] The rapid introduction of biotech-nology, with economics dictated by world powers, often disrupts established cultural traditions and does not allow sufficient time for the less developed supply countries to adapt themselves to new technological and social circumstances. These issues all connect very strongly with this most obvious problem. These may not be intrinsic to biotechnology, but to contemporary cha-racteristics of biotechnology and the aforementioned issues 1, 2, and 6 can contribute to the biotechnology divide.

Ethics in Response to These Persistent Problems

The global community has reached consent (apart from the mani-fest on human rights) on some very general ethical principles, such as those directed to sustainability, which are intended to ensure our respect for future generations. Nevertheless, it is unclear how we should handle these principles when there are obvious conflicts between the needs of contemporary and future generations or when new technologies like biotechnology have potentially severe impacts on agriculture and the cultural foundations of societies.

The need for just access to and distribution of resources seems to be a very important principle.[25] A utilitarian perspective of justice is often used in deciding the deployment of a technology, based upon calculation of future costs and benefits. This perspective does not account for the global obligations to respect cultural diversity, farming styles, and food styles.[26] A deontological position is often used to discount technology and its possibilities,[27] but it does have the advantage of taking into account respect for rights. I, therefore, introduce a deliberative perspective, by emphasizing the importance of deliberations with stakeholders, and incorporation of the principles of justice stated by Rawls and Habermas.[28] They both proposed rather universalistic versions of a principle of justice that presupposes a sharp distinction between universal norms that require reasons that are valid for everyone and contingent values that are dependent on particular social contexts and reasons. In the international context of biotechnologies, there are many different principles of justice, and it is often unclear what reasons are universal and what are contingent. Hence the principle requiring the equal distribution of food and the respect of individual autonomy is not seen a priori as valid but as being contextually impregnated. One cannot be distracted from historical contexts; indeed this is not desirable in ethical deliberations. The principle of justice according to this interpretation indicates that we should respect people as well as their involvement in coping with life and its conflicts. Anticipation of consensus or discrimination between rational and universal norms and nonrational and contingent values is not necessary. When trying to cooperate peacefully, it is more urgent to coordinate and fine-tune the different value and norm interpretations.[29]

As the utilitarian and deontological approaches cannot take into account these technological diversities and their sociocultural contexts, I connect technology development with a deliberative approach of crop biotechnology, that is, an ethic that emphasizes ethical sensibilities of stakeholders participating in ethical deliberations. For crop biotechnology (GE food), neither hell nor paradise is the outcome, but one moves toward respect for metaphysical and religious opponents, and gives serious consideration to risks and the sociocultural implications of these technologies. Clearly these technologies have some advantages, like the possibilities of using less pesticides, and producing food that is enriched with minerals, less allergenic nutrients, and more healthy ingredients for certain groups, such as the elderly, young children, allergic patients.

Role of the Government, Industry, and Civil Society in These Problems

Governments as well as industry and its representatives and supporters have often aggravated the ethical problems of biotechnology. The perceived aggressive actions of the transnational Monsanto Corporation have given it a notoriety that has contributed to a loss of consumer and government trust.

Western governments are often responsive to their citizens' concerns, although they very often act according to short-term, short-sighted interest. Some have tried to build nationally defined controls for food safety and food security, although microorganisms do not recognize national borders. Food supply, being essentially a cultural and pluralistic issue, requires regulatory progress to depend on a co-evolution of biotechnologies and societal values. Hence, other governments, such as those of the EU member states, have chosen an inclusive strategy and face two tasks in rebuilding public confidence. One is to facilitate public debates between the main stakeholders (consumers, producers, scientists, and government) on the future course of the food system, and the other is to direct research that includes monitoring and assessing outcomes, by encouraging responsible cooperation between corporations and scientists. International agencies also waver between these two alternative trajectories (authoritarian excluding rule versus democratic including rule). For Codex Alimentarius, Sanitary and Phytosanitary Committee (SPS of WTO), and WTO, these trajectories are complicated by more or less science-based orientations. However, in developing countries governments often do not have the resources to engage in debate or research.

Industry often demands as little regulation and (national or international) government intervention as possible. Many biotech companies see consumer confidence only as a device to enhance technology sales, or to gain a competitive advantage (see www.whybiotech.com). In the long run, however, critical and rational consumers are more important than a public that only reacts fearfully to new foods, particularly to modification of staple commodities. A second strategy that seems more favorable for industry is to open and sustain communication with the public, not only from producers to consumers but also vice versa. Furthermore, it is doubtful that smaller private companies, in particular small farmers, will be fostered by the first strategy. In many developing countries the private sector is very small and works under many disadvantages. So it may be economically overwhelmed by transnational organizations, often resulting in national and cultural chaos. Some industrial organizations are conscious of this, and their corporate responsibility strategy includes local interests. However, trust must be built and transparency about different interests is necessary. It sounds suspicious when private corporations with little history of altruism suddenly announce that they will use their new technology to end world hunger.

Societal and voluntary organizations in the Western world are often deliberately single-minded and not favorably disposed toward trade-offs with other considerations. Sometimes this strategy succeeds because it provides the mass media with a clear and strong position that is more attractive and easy to communicate, but it remains to be seen what happens to public trust in the long run. This strategy can be distinguished from a more argumentative strategy, in which non-entrenched positions are taken to provide enlightenment through communication. However, in the developing world, civil society is structured differently (often along religious groups), and the single-minded approach is often preferred.

RESEARCH AGENDAS OF NATURAL AND SOCIAL SCIENCES (BETA-GAMMA INTERACTION): TACKLING THE PROBLEMS

From a deliberative perspective, the seven ethical problems mentioned earlier can be addressed by integrating natural and social sciences to search for new, multilevel participation models, both upstream of the innovation process (research priorities) and downstream (social embedding of the research agenda). Some of the issues can be reframed, while others may have to be decided in the context of a particular ethical framework and scientific reasoning, as long as this choice remains within the framework of a deliberative ethic.

Coevolution of Biotechnology and Society

The deliberative approach removes the conflict between technology push and societal pull, and takes into account societal and identity changes in response to new technologies. The technological pathways chosen depend on societal processes. Coevolution of technology and the extent to which it is socially embedded mean that the different aspects coevolve, for example, relationships between institutions and technologies; social norms and values and technical standards; and persons acting outside and inside technological institutions.

The rising demand for meat makes it necessary to develop new types of animal (including fish) feed and new types of enhanced crops. Biotechnology can contribute to these new directions,[30] and is well placed to study the possibilities for edible vaccines against zoönoses, new technologies to improve yields, and changing elemental composition of animal waste. Biotechnological methods for diagnosing and intervening in living organisms can be developed within the framework of well-accepted norms such as sustainability, the respect for variety and justice; thus entrenched but often unclear perceptions of both biology and chemistry can be bypassed.[31] In the case of vitamin A deficiency, the International Vitamin A Consultative Group (IVACG) of USAID and the International Life Science Institute (ILSI) have made it clear that correction of widespread vitamin A deficiency in developing countries by diet alone is probably impossible where populations remain dependent on conventional plant-based foods. These realities point to the need for independent research institutions like CGIAR and IRRI to increase support for their traditional and biotechnological programs for crop improvement.

Development of sustainable diagnostic genomics instruments to survey large areas of crops may allow farm advisors to predict which crops are vulnerable to which diseases and then prevent use of unnecessary chemicals. These techniques can help to detect host responses of plants toward bacteria and fungi. Research is required to develop cheap and convenient technical instruments to undertake this kind of survey. Biotechnology solutions are not the only answer to these problems, and social improvements are equally or more important. There is no one technological solution to hunger or food quality problems; therefore, technological citizenship should be probed, balancing local/national and global considerations and interests.

A Global Network of Food Safety and Quality Agencies

Although many would argue that global policy institutions should be shaped in the same way as their national counterparts, the deliberative approach favors a network system in which the various agencies check and balance each other. Risk-assessment and risk-management institutions can fulfill their tasks better when there is a network that urges them to compete and cooperate in ways that allow consumers to appeal for help or redress. The many definitions of risks and food safety (all permeated with visions of good life and the quality of food) point to the need for a network rather than a single monolithic organization that is responsible for both food safety and food quality. The desired shape of this network is not yet clear and much theoretical and empirical research on the social and technological remains to be done.

International Benchmarking and Standards to Measure Progress of CSR

In the better managed parts of the private sector it is now more or less established that private corporations must ensure Corporate Social Responsibility (CSR).[32] Currently most of the mission statements of the "People, Planet and Profits" communities remain little more than statements and lack both long-term and short-term goals, including measurements and mid-term reviews. Few global, independent institutions guard, evaluate, and advise those companies that are indexed as CSR corporations (CSR 2002). In fishery, there is the London-based Marine Stewardship Council, but it only includes a very small number of fishery companies, has a naïve system of monitoring, and lacks any compulsory devices. Clearly, we need independent international organizations like FAO, CGIAR, and IRRI, to develop models that encourage setting of international benchmarking and ethical accounting.

Public Participation Models: Multilevel Deliberation

Most participation models are at a grassroots level, but they must be coordinated locally, nationally, and internationally to ensure that regulations are supported by effective means to deal with transgression at all levels. This is necessary not only to narrow the global technology divide, but also to develop new deliberative strategies that are established after organized debates on top priorities for research topics between experts, stakeholders, small crop holders, and representatives of consumer organizations.

Models of Biotech Benefit Sharing

The International Treaty on Plant Genetic Resources for Food and Agriculture recognizes the contributions of farmers to the conservation of plant genetic resources over time and for use by future generations. It also provides an international framework to regulate access to plant genetic resources and establishes a mechanism to share the benefits derived from their use.[33] Intellectual property models are now required that give equal access to new technologies and allow benefit sharing of biological material. Exemptions for poorer countries from strict patent and trade rules with respect to important crops and easy access for poorer countries (and farmers) to new patented crops can be regulated, just as they are for generic medicine.[34] Although generic quality crops should be as affordable as drugs, the essentially plural and cultural definitions of food do require participation by consumers and farmers.

In the literature, benefit sharing is often exclusively conceptualized as a process downstream from the innovation (sharing results of successful innovations in genomics agriculture and health projects), and problems of property rights emerge.[35] There is also the prospect of benefit sharing upstream, in advance of doing research through public consultation and participation in setting the research agenda. Until now the EU ethics committee has not made recommendations on this issue (http://europa.eu.int/comm/european_group_ethics/docs/cp17_en .pdf) but only on ethical issues of clinical trials, germplasm research, and stem cells. Some forms of benefit sharing seem rather silly, such as organizing PhD grants for nomads,[36] and serve only to satisfy the consciences of the Western researchers as they enjoy the benefits of their patents. Here again, social research is essential to identify local stakeholders' needs.

Social Embedding of Research Agendas

Coevolution of biotechnology and society means not only combinations of technological and social solutions upstream from the innovation process, through the direction of research priorities, but also downstream to meet the needs of the end users.

National governments, private companies, and civil societal organizations face different possibilities to embed food biotechnologies in local contexts. Promising in my view are several tailor-made research programs, such as the Andhra Pradesh The Netherlands Biotechnology Programme (APNLBP), the Kenya Agricultural Biotechnology Platform (KABP) and Biotechnology Trust Africa, Science Technology Policy Research Institute (STEPRI, Ghana), INIFAT of Cuba and the Biotechnology Development Trust Zimbabwe (BTZ).[37] Successful programs take into account the daily requirements of farmers, both men and women, as well as their short-term and long-term future expectations and preferences. The goals should be to meet the more humble needs of their daily routines rather than far-reaching, ambitious, and often unrealistic national or international goals. Ethical principles are at work here. There should be strong condemnation when the daily routines are unjust or do not respect human rights or when there is female suppression or child abuse. Perhaps more important, there should be policies that assist victims.

CONCLUSION

Biotechnology to improve food safety, food security, and food quality requires something more than laboratory work at the frontiers of science. The twenty-first century will certainly see increased cultural, social, and ethical pluriformity with respect to the definitions of food safety, food security, and food quality. Ethical issues of respect and participation should belong at the core of food biotechnology.

This does not mean that ethics must overwhelm, but it does mean that ethical standards should be developed and made explicit in prioritizing and doing research. Food biotechnologies have enormous potential when these ethical concerns have been addressed, and issues both upstream and downstream of the innovation process have been articulated. I have argued specifically for a combination of agricultural and social technologies that will decrease the poor/rich divide. Upstream, coevolution of technologies and societal norms, stakeholder participation, networks of assessment, and benefit sharing must be elaborated and explicitly understood, whereas downstream, models of embedding technology in local contexts are essential. The future of food biotechnology will then look bright.

NOTES

1. Hodges, J. and Han, K., eds. 2000. *Livestock, ethics, and quality of life*. New York: CABI; Keyzer, M., Merbis, M.D., Pavel, I.F.P.W., and vanWesenbeeck, C.F.A. 2003. *Can we feed the animals? Origins and implications of rising meat demand*. SWOW Working Paper. Amsterdam: VU. www.whybiotech.com. Accessed March 1.

2. Busch, L. 2000. *The eclipse of morality: Science, state and market*. New York: Gruyter.

3. Korthals, M. 2001. Ethical dilemmas in sustainable agriculture. *International Journal for Food Science and Biotechnology* 36, 813-820; Korthals, M. 2003. Gruene Gentechnik. In *Bioethik*. Edited by K. Steigleder and M. Duwell. Frankfurt a/M: Suhrkamp, pp. 354-362.

4. Cartagena Protocol on Biosafety. 2000. United Nations Environment Program. http://www.biodiv.org/biosafety/protocol.asp. Accessed March 10, 2006.

5. Thackray, A., ed. 1998. *Private science. Biotechnology and the rise of the molecular sciences*. Philadelphia: University of Philadelphia Press.

6. Nestle, M. 2003. *Safe food*. Berkeley, CA: University of California Press.

7. United Nations Development Program. 2000. Annual Report.

8. Huang, J., Rozelle, S., Pray, C., and Wang, Q. 2002. Plant biotechnology in China. *Science* 295: 674-676.

9. Fresco, L. 2000. Scientific and ethical challenges in agriculture to meet human needs. *Food, Nutrition and Agriculture* 27: 4-11.

10. Keane, D. 2001. *Global civil society?* Cambridge: Cambridge University Press.

11. Jepson, W.E. 2002. Globalization and Brazilian biosafety: The politics of scale over biotechnology governance, *Political Geography* 21: 905-925; Kareiva, P., and Laurance, W.F. 2002. In brief. *Trends in Ecology & Evolution* 17: 63.

12. McNamee, D. 1999. Transgenic potatoes produce oral HBV vaccine. *Lancet* 354: 1707.

13. Pinstrup-Andersen, P. 2001. *Seeds of contention.* Baltimore: Johns Hopkins University Press.

14. Arvanitoyannis, I. 2003. Genetically engineered/modified organisms in foods. *Applied Biotechnology, Food Science and Policy* 1: 3-13; Huang, J. et al. 2002. See note 8.

15. Clark, E.A. and Lehman, H. 2000. Assessment of GM crops in commercial agriculture. *Journal of Agricultural and Environmental Ethics* 14: 3-28; Korthals, M. 2001 and 2003. See note 3.

16. Castle, D. and Ruse, M. 2002. *Genetically modified foods: Debating biotechnology.* Amherst: Prometheus Books; Sherlock, R.. and Morrey, J. 2002. *Ethical issues in biotechnology.* Lanham: Rowman & Littlefield.

17. Nestle, M. 2003. See note 6.

18. Castle, D. and Ruse, M. 2002. See note 16.

19. Lee, T., Robold, A., Testa, A., van 't Klooster, J.W., and Govers, F. 2001. Mapping of avirulence genes in *Phytophthora infestans* with amplified fragment length polymorphism markers selected by bulked segregant analysis. *Genetics* 157: 949-956.

20. Sherlock, R. and Morrey, J. 2002. See note 16; Keyzer, M., et al. 2003. See note 1.

21. Ibid.

22. National Academy of Sciences. 2002. *Environmental effects of transgenic plants: The scope and adequacy of regulation.* National Research Council Board on Agriculture and Natural Resources, Committee on Environmental Impacts Associated with Commercialization of Transgenic Plants. Washington, DC: National Research Council.

23. van den Belt, H. 2003. Enclosing the genetic commons: Biopatenting on a global scale. In *Patente am Leben? Ethische, rechtliche und politische Aspekte der Biopatentierung.* Edited by D. Mieth and C. Baumgartner. Paderborn: Mentis-Verlag, pp. 229-243.

24. Byerlee, D. and Fischer, K. 2002. Accessing modern science: Policy and institutional options for agricultural biotechnology in developing countries. *World Development* 30: 931-948.

25. Burley, J. and Harris J., eds. 2002. *A companion to genethics.* Cambridge: Blackwell; Farrelly, C. 2002. Genetic intervention and the new frontiers of justice. *Dialogue* 41: 139-154.

26. Korthals, M. 2001. See note 3.

27. Mepham, B. 1999. *Novel foods.* London: Ethical Council.

28. Rawls, J. 1972. *A theory of justice.* Cambridge MA: Harvard University Press; Habermas, J. 1992. *Faktizität and Geltung. Beiträge zur Diskurstheorie des Rechts und des demokratischen Rechtsstaats.* Frankfurt a/M: Suhrkamp.

29. Korthals, M. 2002. Functional foods. *Journal of Agricultural and Environmental Ethics* 16: 35-42.

30. Arnzen, C.J. 1997. Edible vaccines. *Public Health Report* 112: 190-197.

31. Fresco, L. 2002. See note 9; Singer, P.A., and Daar, A. S. 2001, Harnessing genomics and biotechnology to improve global health equity. *Science* 294: 87-89.

32. CSR: Corporate Social Responsibility. 2002. Food for Thought: www. iblf.org/csr. Accessed April 11, 2006.

33. Knoppers, B.M., Daar, A. S. a.o., 2000. Genetic benefit sharing. *Science* 290: 49.

34. van den Belt, H. 2003. See note 23.

35. Ibid.; Conference of the Parties to the Convention of Biological Diversity. 2002. Bonn Guidelines on Access to Genetic Resources and Fair and Equitable Sharing of the Benefits Arising out of their Utilization. *UNEP/CBD/COP/6/20.* www.biodiv.org/decisions/default.asp?m=cop-06&d=24&print=1. Accessed April 11, 2006.

36. WIPO/UNEP. 2000. The role of intellectual property rights in the sharing of benefits from the use of biological resources and associated traditional knowledge: selected case studies. A joint submission by the World Intellectual Property Organization (WIPO) and the United Nations Environment Programme (UNEP). pp. 5-30. www.wipo.int/globalissues/documents/pdf/cs_oct_2000.pdf. Accessed April 11, 2006.

37. Sithole-Niang. I. 2001. Future of plant science in Zimbabwe. *Trends in Plant Science* 6: 493-494.

Chapter 7

Next Challenges for Crop GE:
Maturing of Governance
and Moves Beyond Food Issues

Iain E. P. Taylor

INTRODUCTION

It is clear that genetic engineering (GE) methods are here to stay, but it is much less clear whether they will have the impact on crop production that has been predicted by the advocates of this technology. Their raison d'etre from the beginning has been to use the techniques of molecular biology to improve the effectiveness of crop production by adding desirable traits that are not accessible through traditional plant breeding, and to reduce the time to bring a new crop variety to marketplace production. The successful complete descriptions of whole genomes[1] have paved the way to a level of genetic precision that few predicted in the decade after the first reliable inventions. As with any new inventions, the justifications have been often hyperbolic, which is to be expected from inventors. For the stated goals to fulfill the worldwide needs for food, fiber, medicinal, and other resources, especially in the developing world, success will have impact, but as with most new inventions, the real benefits will appear more slowly and many will be unanticipated.

GE crops are grown on an enormous scale. Just five years ago, in 2000, there were 40 million hectares planted in Canada, the United

Genetically Engineered Crops
doi:10.1300/5880_07

States, and Argentina alone.[2] The area planted with smaller-scale introductions is not well documented. Past experiences and public perceptions of previous so-called agricultural revolutions have left many members of the public and an increasing number of scientists with doubts about hyperbolic claims of "miracle cures" for food shortages. The objectors cite divine prohibitions against "messing with nature": dangers to organic ways of farming, scientific doubts about long-term ecological impacts such as gene escape during natural reproductive processes, possible gene escape into soil microflora, and claims that GE foods are in some way dangerous. While there are no verifiable records of human deaths attributed directly to GE crop consumption, there are unaddressed concerns about long-term human health and environmental impacts as well as fears of further erosion of crop diversity. Some governments prohibited cultivation and/or import of GE crops.[3] On the other hand, the Brazilian government has just legalized the production of GE crops[4] and the U.S. government continues to support multinational corporations as they try to persuade "developing" world governments to devote substantial parts of their farmland to the production of GE crops for export.

Most of us know that scientists have not been particularly prescient about the uses and applications of new knowledge and technology. It seems that aquaculture, forestry, and environmental remediation are ready for large-scale GE effort, but forecasting specific new directions is unlikely to be accurate. We can probably engineer any trait into any crop that we choose. I propose to address four complex matters that should concern us most as GE crop science and technology passes from infancy through adolescence to an adulthood that includes manipulation of enormously complex systems associated with aquatic plants, perennial life histories in the forests, and the uncertainties of creating plants for life under man-made hazardous conditions.

First, science and society must surely ask what plants can do for us and what the most appropriate subjects for GE technology are. Priority must surely go to plants that are uniquely able to meet needs and for which traditional breeding is either too slow or not an option. Second, we should avoid "knee-jerk" dismissal of the rhetoric about environmental impact, especially if we engineer trees that are wind-pollinated and hence will produce and disperse untold quantities of pollen and seed each year. The negative impact of traditional and modern agriculture on biodiversity is well documented, so there is some biological obligation to find the positive uses for all agricultural practices including uses for GE crop plants and to undertake the rigorous precautionary work from the earliest stages of GE crop development. Third, we should ask whether the incentives for industry can be adjusted to direct GE crop developments toward more reliable, long-term benefits for producers, particularly in local cultures and economies that require food production without resort to mega-farming and international trade and traffic. The capacity to meet local needs may include developments such as growth and harvest of heavy metal accumulating plants that can be contained at sites of planting. Fourth, as agricultural trade is increasingly dependent on international politics, we will need to understand the diverse contexts in which new inventions are controlled by regulations. If governments are to regulate GE crops, or indeed any other invention, society has a right to demand transparent separation of promotion policies, often driven by inventor lobby groups, from the regulatory arm that must give priority to society's interests.

WHAT CAN PLANTS DO FOR US?

Can We Redirect Ge Technology to Where it is Uniquely Able to Meet Our Needs? Future research will certainly continue to find more effective methods of meeting agricultural needs or addressing the continuing problems, such as improving yield and pest management. In the first instance, priority will go to desired traits that are the most economically viable and are amenable to specific selection or synthesis of pieces of DNA (genes) for addition, deletion, up-regulation, or down-regulation of particular pest resistance and nutritional traits. As attempts to engineer more complex systems proceed, we can reasonably expect more attention directed toward the control of plant chemistry, such as the regulatory steps in seaweed mucilage production, wood formation,[5] and nutrient uptake. Progress in studies to engineer plants for production of non-plant molecules, such as vaccines, fuels, and bioplastic materials will probably be determined as much by research ingenuity as commercial opportunity. This research will require specific understanding of one or more facets of plant structure and function and will certainly have unpredicted applications. A major criterion for success is that the engineered gene(s) in plants are expressed in the appropriate plant organs or tissues. Studies in the recently defined subdisciplines of genomics, proteomics, and metabolomics are expected to provide information that will improve engineering precision and lead to more stable and reliable gene expression.

Herbicide- and pest resistance are currently the most widely used traits. While resistance to herbicides has potentially great value to the large-scale farmer and to the forester, long-term resistance to disease and pests is more important for engineering of perennials because of longer time available for the pest mutations to overcome the engineered host resistance. The continuing biological warfare between host and parasite/pest may be tilted in favor of the host by use of GE to provide resistant strains more quickly than can be done through traditional plant breeding. This tipping of the balance can only be achieved if there is enough genetic diversity available to provide new resistance traits. Clearly, the acceleration of GE technology must not swamp the relatively low technology of nurturing the wild forms and the "heritage" strains that are no longer in mainstream production. Reports of new virulent pest strains and pesticide-resistant forms can be expected to increase as more land is brought into cultivation. GE may well be a major contributor to the production of resistant forms, but the technology alone clearly will be inadequate.

Global warming, increased resource extraction, agricultural intensification, and the increasing use of so-called marginal land for agriculture point to needs for drought, pollutant, and salt-tolerant strains. There must be major efforts to resolve these environmental problems and to understand the fundamental responses of plants to these pressures before engineered solutions will be feasible. Recorded and oral histories are replete with stories of catastrophic social damage due to drought-caused crop failure. Farm crops have never developed immunity from climatic disasters, so the drought, heavy metal, and salt-tolerant varieties developed by selection or GE will offer real benefit. However, unlike herbicide resistance and to some extent pest resistance, the genetic control of drought tolerance appears to be far more complex than single gene systems. The plant breeder will continue the search for naturally resistant strains and try to breed that resistance into the currently available crop stock in hopes of finding a new combination of acceptable productivity and drought resistance.

Salt marsh and brackish water-tolerant species as well as those that grow on mine tailings and other highly toxic sites provide a natural source of tolerant forms in which researchers can search for usable mutants.

While there have been few efforts to cultivate such natural forms, intensive irrigation of deserts has led to increased salinity in these cultivated soils for which new salt-tolerant forms are required. The value of grain crop parental species is well understood by breeders and genetic engineers, but opportunities also exist within the gene pools of salt marsh and brackish species, such as *Salicornia, Sueda,* and mangroves, to locate genes for transfer to develop new salt-tolerant GE crops. In the first instance, the well-studied systems in wheat, barley, maize, and rice seem to provide a clearer market opportunity for GE salt-tolerant forms and they may prove useful in developing novel methods for soil remediation. While some low-fertility land is in fact nutrient-deficient or may even support plants that are toxic to grazing farm animals (CSIRO Australia),[6] the removal of fertile land for nonagricultural uses continues to force farmers to crop more marginal land and to look for the most stress-resilient forms.

Improved nutritional value has always been a goal for plant breeders but the notion of a GE crop that meets all needs for essential nutrients seems somewhat far-fetched. If GE has a function in improving nutrient availability, it seems that enhancement of already useful strains, such as lycopene levels in tomato, vitamin A precursors in rice, or essential amino acid content in protein foods,[7] will continue to be an effective use of GE technology.

Recent research suggests that GE will have considerable impact on the development of high-tech niche crops, such as those engineered to synthesize vaccines, human antibodies or replacements for nonrenewable resources, such as fossil fuels.[8] Certainly, we can expect many opportunities in microbial and animal biotechnology, in particular to move microbial processes, such as atmospheric nitrogen fixation, into plants or into mycorrhizal fungi.[9] The pioneer work to engineer microbes, such as bacteria to produce human insulin and yeast in the wine industry, remains central to technology that can provide large quantities of high-quality single products. The use of mammalian milk as a vehicle for production of human gene products, such as immuno-globulins, has raised ethical, moral, and safety concerns that led re-searchers to turn to plants as possible substitutes. Patents have been filed on the use of food crops, such as bananas, as vehicles for deliv-ery of vaccines. We can expect many very valuable additions to the list of GE niche crop plants in addition to the thirty to forty so far de-veloped, and expect that some will be expanded to mainstream agri-culture, but also that some will prove to have singular disadvantages. We can expect many to hear concerns about both short- and long-term efficacy as well as health and environmental safety, which will require inventors to test and report potential and real hazards in addi-tion to promoting the obvious benefits of their new GE product. The fact that we never know what we do not know does not excuse us from keeping an open mind when new scientific and societal con-cerns emerge as a result of growing experience with new GE forms.

Although the genetic basis of plants, animals, and microbes is sim-ilar, there are some aspects of plants that make them uniquely suitable for GE. There are few, if any, concerns that engineering plants is mor-ally reprehensible, as is the case for mammals, particularly primates. More technically important is the fact that many plants can be ren-dered obligately self-compatible and hence genetic leakage through pollen from a plant engineered to produce a non-plant substance, such as an antibody or an animal hormone, can be eliminated. In ad-dition, plants have substantial developmental plasticity that allows genetically identical individuals to survive in very different habitats, long enough for rare and random mutations to occur that improve survival over the original individual(s). This characteristic is cur-rently being investigated by researchers who look for plants that grow and thrive on what are considered to be very toxic soils. Such plants have potential for use in remediation of soil polluted by mining waste, oil, or man-made toxins. Initially they can be collected, propa-gated, and grown in bulk using standard horticulture methods, but high throughput genetic screens and tissue culture that are basic tools of GE suggest that this technology can be used to produce the millions of plants that may be needed for a large soil remediation program.

THE ENVIRONMENTAL FUTURE

There was little surprise in the plant ecology community when the first peer-reviewed reports appeared[10] that herbicide-resistant genes in GE rapeseed/canola were detected in the wild forms and relatives of *Brassica napus* from which these crops were developed. The species is an outbreeder and the flowers attract pollinating insects. The case was similar to the report[11] that pollen from corn engineered to resist the corn borer (*Bt*-corn) caused negative effects on the caterpillars of the *Danaus plexippus* (Monarch butterfly) when they were fed their normal food leaves coated with *Bt*-corn pollen. A more recent report[12] claimed that GE genes had contaminated traditional maize landraces in Mexico. These peer-reviewed reports were adequately well founded to be taken seriously, even if perceived conflicts of interest may have slanted details of experimental design. There are many short- and long-term uncertainties about genetic "leakage" within and between natural populations and GE crops are no better understood. Promoters of GE crops interpreted the early reports of "leakage" as inflated and alarmist impediments to essential crop improvement. The contrary view saw the observations as "canary in the mine" reminders that genetic pollution of wild forms will cause irreversible long-term damage to critical genetic resources. There is a widespread belief that authors of peer-reviewed research reports, which contain results that are unfavorable to the industry, are subject to various types of harassment and efforts to discredit their work. There is also no reason to believe that such practices will cease so long as massive industrial investment continues to require unquestioning promotion of GE and any other new products.

The introduction of synthetic genes or genes that would never have moved naturally from their biological origin into the target species raises fundamental questions about the uncertainties facing current regulatory scientists. Future success will rely on minimizing unpredicted developmental side effects in both the gene's expression and the target organism as a result of its new chromosomal environment. The use of vector viruses or other methods to stabilize insertion of the new DNA is still a work in progress. Increased knowledge of genomics will certainly be valuable and already short-term tests exist to assess genetic stability of the new forms, and we can hope to rely less on

success defined as "the plants grew, were healthy and expressed the desired trait." While the short-term results submitted to support regulatory approval for GE forms seem positive, the aforementioned studies
point to the need for long-term, albeit expensive and time-consuming,
studies that are designed with the full rigor of high-quality, statistically robust ecological experiments. It is increasingly clear that environmental risk from GE crop introduction is nontrivial.

The policy decision by legislators and their scientific advisors who
regulate crop introductions in Canada, the United States, and some
other countries to consider GE crops as "substantially equivalent" has
become a serious inhibitor of vital research efforts that can only provide long-term benefit to farmers, consumers, the GE industry, and
the environment. In the case of species introductions, there are many
records of plants, animals, and microbes that were well adapted and
integrated into one ecologically balanced community becoming invasive and destructive of biological diversity when introduced into a
new environment. Although many crop species do not survive well
under fallow conditions, those individuals that do can continue to
contribute to the natural genetic and reproductive dynamic of the new
environment. The same technological systems that are used to develop
GE crops are being used to show that gene transfers have occurred
much more frequently during evolution than was detected using older
methods of genetic analysis. Clearly, there is some precautionary ur-
gency to address these problems with large-scale, interdisciplinary,
rigorous, and long-term studies that will allow better understanding
of the impact of GE and provide a scientific basis to include both biological and regulatory controls for GE crop introductions.

THE INDUSTRIAL INCENTIVE

The increased development costs associated with engineering more
complex traits, the lack of clarity and consistency in the GE crop regulatory process, and the global market uncertainty created by widespread rejection of GE crops in the international food supply system,
have all combined to deter Agbiotech companies from attempting to
bring many second generation varieties into production.[13]

Biotechnology is flourishing in the so-called industrialized world,
but it is growing largely under the control of multinational corporations

and industrial interests are inextricably intertwined with national and international governance. As with many earlier inventions, the capitalization of new products has relied on a source of venture capital, which is attracted by an inventor's active strategy to identify the benefits from which the venture investor can expect to reap profit. Industrialized world governments begin commitment to new technology by pursuing perceived benefits to their own citizens. In the case of agriculture, an assured national food supply requires support of current farming practices as well as support of any new technology that may increase the farmers' economic health. In the case of GE crops, the industrial strategy has been to persuade legislators, farmers, and academic plant researchers of the benefits that accrue by developing methods to use this new technology to improve crops. The GE approach promises benefits such as pest resistance, greater productivity by elimination of competitive weeds, and more high-quality product for the export market. In addition, it is argued that these improvements will strengthen the national economy by increasing the strength of the agricultural sector. The perceived market need in the "industrialized" world will drive a GE commercial niche. The academic community invented many of the methods used in GE crop development. The methods are the foundation of major advances in our understanding of plant biology. The relatively recent formation of the industry–university partnerships in plant molecular biology has led to recruitment of many academics into the conflicting world of industry-planned and industry-directed research conducted with public funds. The notion that industry sponsorship of academic research is inherently unethical has little theoretical foundation. The long-standing, but only recently exposed, pharmaceutical industry practices of obtaining regulatory approval based only on positive drug trial reports[14] has led to major policy changes in research journal publication practices[15] whereby the results of trials must be fully disclosed before manuscripts will be considered for publication. While economic interest should always be a major concern for industrial enterprises, inappropriate manipulations of impact and biological safety are becoming increasingly unacceptable to the scientific community. Industry will surely see the benefits of following both the letter and the spirit of the GE crop regulatory processes.

As industry is advocating the new technology, governments are seeing manufacturing jobs being lost to the developing world where farm labor costs are lower. The new "industrialized" world direction is toward the "knowledge based" (a.k.a. high-tech) economy and governments have instructed the appropriate ministries to promote such technology. Companies seeking to market their products, mostly in food and medicine, have made massive lobbying efforts to accelerate the regulatory approvals. GE crops, especially those that grow well in countries where farm labor is cheap, provide an opportunity for economic expansion by the manufacturers as well as to expand the crop "menu" to improve general food and fiber supply for developing world citizens. They also may allow for faster, locally inspired development of crop forms that are suited to local climate or soil conditions or may even meet local medical needs such as the production of vaccine that cannot be easily transferred from an industrialized-world manufacturing plant. These and other opportunities provide excellent incentives for industrial investment, but concerns expressed by the government of Zambia[16] regarding import of GE corn and by citizen groups within the EU indicate the reemergence of a problem associated with old-style colonial government. Historically, the economies of many less-developed countries were built on their role as primary producers of export crops. While the benefits were promoted as providing local jobs and hence making a contribution to the local economy, the downside effects of mono-cropping and environmental degradation by displacement of local biodiversity were not perceived as important. The emergence of GE crops happens to have occurred in parallel to a new form of political independence that includes local economic and social control and clearly international industry can only succeed with local and regional cooperation.

Local biodiversity clearly has much more than tourism value. It is a national treasure that must be held and maintained for the local good and for the heritage of humanity. The introduction of GE crops, many of which originated from the natural resources of second- and third world countries, is now seen as a major hazard to the natural biodiversity. Political and economic pressures such as that exerted on Zambia through the actions of the USDA, and the threat by Monsanto to downsize its research efforts in Europe, point to the need for both international industry and the developed countries to attend to their own future by a much broader consideration of the GE crop impact. It seems reasonable to predict that pressures from industrialized-world countries will require international regulation that is honored both in the spirit and the letter of the law.

THE REGULATORY FUTURE

The already polarized opinions about GE crops are obviously founded in the personal, scientific, and political history of the parties. The pressures of public opinion in some parts of Europe and the intensive lobbying of members of the U.S. Congress by industrial interests have had serious repercussions for the sound governance of this new technology. It is clear from other chapters in this book[17] that few national governments have successfully separated their promotion of GE crop development from the regulatory mandates that are given to the same agency that is charged with promotion. While the incentives to biotechnology are often provided by tax incentives, formal and informal pressures on government and their advisers are growing to lower the regulatory barriers and to accelerate the approval processes that allow release of new products.

One indirect means of achieving faster approval has been through the current political agenda of reducing the size of government and departmental budgetary allocations. These practices, often referred to as "rationalization," effectively limit the conduct, application, and enforcement of the regulatory processes by ensuring inadequate numbers of regulatory staff. It seems that regulators who speak out in support of stronger regulatory action may be either reassigned to nonregulatory tasks or even dismissed.[18] Further pressures are rumored when the industrial applicants complain to the political leadership, often with well-placed leaks to the news media, that approval of their new product is being obstructed by the bureaucracy.[19] This is occasionally accompanied by insinuations of scientific bias and portrayal of regulatory scientists as irresponsible "whistleblowers." Speedier government approval shortens the time available for scientifically responsible scrutiny, but it also improves prospects of profitability, which increases industry attractiveness to the entrepreneur and the investor.

Recent controversy over supply of generic anti-HIV drugs to Africa has shown that some governments, acting on behalf of their own industrial organizations, are less than enthusiastic in making biotech-nological benefits available at less than industrialized-world prices. In sharp contrast, there have been attempts, driven in part by researchers who have not sought patents for their inventions, to make seemingly valuable products available to third world countries. The "Golden Rice" strain, in which gene expression for production of vitamin A precursors is engineered to occur in the rice seed (normally restricted to the leaves) to provide supplemental vitamin A, is perhaps the best-known case. Organizations such as CAMBIA work to apply GE tech-nology to resolve crop problems on a local level. They rely on locally cultivated strains as resources and try to develop crops that meet local climatic and environmental challenges. Production may be regionally based, with more consideration for specified local needs. Clearly, the technology must be available within or close to the target communities and it may be less attractive for the industrialized-world balance sheet, and responsibility for legislation and regulation must move to the na-tional governments in the region and be protected from what some see as a new form of old-style colonialism where industrialized-world interests predominate.

Issues of health and environmental risk should not be far from the decision-making processes that support regulatory decisions. Many of the public concerns raised concerning GE crops can be interpreted as part of normal fear-driven risk patterns that are associated with many new inventions. This is particularly the case when the new tech-nology is directed to improve human health. Since the publication of Rachel Carson's *Silent Spring*,[20] fears of environmental hazards have been added to the list of public concerns.

The future will bring more demands for rigorous and open assess-ment of the impact of any new technology, and biotechnology is al-ready being challenged to address real and perceived harms. We can expect that "the precautionary principle"[21] will have much more impact on the granting of regulatory approval. While every invention is pro-moted on the basis of clearly stated benefits, these statements often reach remarkable hyperbole. GE is the latest case in point. The initial benefit statements rarely separated real, relatively short-term advan-tages such as creation of a herbicide-resistant crop plant, from the more speculative, long-term possibilities such as large-scale reduction of third world malnutrition, a target that requires major and undefined contributions from the political and economic arenas. Even if the benefits are achieved, there are enormous uncertainties concerning any negative impacts both on individuals and groups within human society.

The often hyperbolic statements of benefit for the common good lead to the presumption that the risks of negative outcomes are of much lower concern, because the benefits to human disease or agricultural production are so overwhelmingly positive. Plant biotechnology certainly provides a faster and possibly more certain way of introducing desired traits into crop species, but the cost of the technology and the expectation that a crop will be ready for commercial production sooner than traditionally produced forms raises expectation of fast regulatory approval. As mentioned earlier, this often means that less attention can be given to assess the longer-term environmental impacts of the GE form of the crop. Many claimed benefits or harms require much more basic research, while the regulatory bodies are constantly under pressure to accept the results of short-term experimental work as valid for the longer term.

It is reasonable to predict that regulatory policy for GE crops will begin to change from the dominance of "substantial equivalence" when new GE products emerge that have medical/pharmaceutical functions. Clearly a banana plant engineered to provide insulin to a diabetic patient cannot be regulated as if it were the unengineered plant: an overdose would have serious if not fatal consequences. The Canadian policies on substantial equivalence are extremely ambiguous but in most cases, GE crops are not regulated as if they are not engineered. They are approved if they are determined—through the regulatory process—to be substantially equivalent to unengineered crops. Still, the newer applications will obviously require more scrutiny. The case for oral vaccines was made successfully with the development of the polio-protecting Salk vaccine. The notion of using plants to produce vaccine seems an easy and effective way to deal with losses of vaccine effectiveness during storage.[22]

While the proponents of GE crops tend to see the questioning of apparently obvious, scientifically supported benefits as Luddite alarmism, the fact remains that historically, regulatory approval has almost always given relatively little attention to the potential for harm that is inherent in most new technology. The attitude seems to be that objectors need not be taken seriously because they are misinformed about the robust scientific evaluations and that they are not asking the correct questions (from the scientific viewpoint). Most governments who have the resources are actively encouraging development of GE crops as part of their drive toward a "knowledge-based economy." It is clear that the many biological, economic, and political concerns raised within and outside the scientific community require regulatory scientists to be freed of the scientific, political, and bureaucratic conflicts of interest. The obvious step is for all GE crop regulatory approval to depend in part on submission of peer-reviewed publications that report sound and complete experiments and observations that assess both benefit and harm. The current willingness of regulatory agencies to accept industry-supported research results in the same light as independently obtained information ignores the obvious, and to some extent perfectly human, conflicts of interest associated with proponent-sponsored research.

AQUACULTURE

GE of Atlantic, and, more recently, Pacific salmon[23] is relatively far advanced, although the scale of public response to GE salmon has been somewhat less than that for plant GE crops. As with plants, the technology has proved relatively easy to apply[24] and I presume that regulatory approval will begin from the presumption of substantial equivalence with wild salmon. The ecological impacts will be rather easier to predict given that GE salmon, carp, tilapia, and any other fish, which should also be so altered, will probably be held in confined spaces from which some escape is likely.[25] It is also clear that engineering sterility in any hybridization to wild stock will be somewhat easier than is the case in plants. However, hybridization between Atlantic salmon and their Pacific cousins is now documented, so the physiological or genetic barriers to wild GE hybridization must be strongly understood to avoid contamination.

Other aquacultural GE is somewhat more difficult to predict. There are increasing needs to improve production of marine bioproducts, such as carrageenans, agar, and agarose used in food processing, which have led to improved and more reliable production through selection of desirable types. However, the genetic complexity of many useful macroalgae currently suggests that this selection of high-quality strains is more likely to lead to improvement than the GE of the type used in land plant breeding. Recently the Hawaii State Board of Agriculture has approved a field trial, the first of its kind in the United States, of *Chlamydomonas reinhardtii* that has been engineered by addition of seven different monoclonal human antibodies, hormones, and interleukins. The risk of escape is high, given the difficulties that most aquarium facilities have in managing their wastewater. In spite of the fact that *C. reinhardtii* is one of the better-known microalgae, its genetics may require much more study to assess the risks of genetic leakage of the type that is reported from several land crops.

FORESTRY

Wood quality and long-term protection against pests have long been identified as desirable goals for GE of trees. The metabolic pathways that lead to production of lignin and related polymers have been widely studied and it is apparent that successful GE to either increase or (usually the case) to decrease lignin content depends on modification of the rate-limiting biochemical steps at the beginning of the processes. An interesting and potentially valuable observation was that engineering decrease in lignin formation led to increase in cellulose production,[26] a desirable trait at least in the paper industry.

The lessons learned from the widespread use of crop plants engineered to include the *Bacillus thuringiensis* (*Bt*) insecticidal protein also seem applicable to trees. The most valuable tropical species may currently prove more difficult to alter but the generally rapid growth in tropical regions may eventually make GE an option for forest management. In the meantime, progress is being made in the description of the genome of fast-growing poplar clones and there are major efforts to determine the genomes of several conifer species. Efforts to control forest insect pests face two long-term problems. The control must be viable for several decades and many natural controls, such as an annual period of extreme cold or drought, are increasingly unreliable in these times of global climate change. One intriguing possibility is to try to either breed, wean, or engineer the pest away from the food source on which it is a pest.[27] Another possibility is to engineer the secondary metabolism in such a way that the plant attracts a new biological partner that is a predator of an insect pest.[28] Were such engineering to be successful, it would widen the genetic gap that a pest would have to cross to attain protection from the controlling agent.

ENVIRONMENTAL REMEDIATION

Many compromised habitats and ecosystems are the result of dumping or storage of industrial or human waste. Remediation thus becomes almost a case-by-case situation. Work in several countries, including that by A. D. Bradshaw and collaborators in the 1950s and later,[29] set the scene to search for mutants that arose naturally under various conditions of heavy metal stress. Bradshaw's work showed that while the heavy metal tolerant forms thrived where levels were high, they were relatively poor competitors on uncontaminated soil. There have been several large-scale revegetation projects, especially in German and South Wales, all of which have used forms (ecotypes) obtained by field collection. It seems unlikely that molecular biology will contribute GE forms to this endeavor, at least in the short term.

The challenge of this chapter was to predict future developments. It is tempting to dream, but it is unlikely that there will be increased biological and environmental responsibility by proponents until such time when poor conformity has serious impact on the financial health of the industry. This is not at all unusual, nor is it necessarily bad, but GE crops raise new issues. It is in no one's interest to forbid access to new technologies, but it is also in no one's interest to leave a new and potentially devastating technology unregulated. It is very important to remember that while a faulty car or a drug with serious side effects can be recalled effectively, the overwhelming evidence is that environmental release of biological agents is irreversible.

Successful uses of GE technology in aquaculture, forestry, and environmental remediation, as well as in animal and human biology, require rigorous pursuit of long-term plans that include independent, probably precautionary, evaluation of possible benefits and pitfalls. The functions of government to ensure the welfare of the public interest will require political leadership that will be much easier to provide when the scientific community makes itself more available to undertake the rigorous research that society has a right to expect from the academic world.

NOTES

1. For example: The Arabidopsis Genome Initiative. 2000. Analysis of the genome sequence of the flowering plant *Arabidopsis thaliana. Nature* 408: 796-815; The Brazilian government announced the successful description of the coffee genome on August 10, 2004. Access and use of the information is regulated. The data will be kept by the Agronomic and Environmental Genomes Network of the São Paulo Research Assistance Foundation (Fapesp) and the Genetic Resources and Biotechnology Center (Cenargen) of the Brazilian Agricultural Research Company (Embrapa), in Brasília.

2. Rural Advancement Foundation Intl. 2000. *Global acreage of GE crops starts to fall.* http://www.etcgroup.org. Accessed October 11, 2005.

3. Reuters News Service. 2001. India to destroy illegal gene-altered cotton crops. See http://www.organicconsumers.org/Patent/IndiaCotton1101.cfm. Accessed October 11, 2005.

4. Until 2004, Brazil actually banned the sale of GE crops and faced a problem of finding legal ways to dispose of GE soya grown in the south of the country. See Science Development Network. 2003. http://www.scidev.net/News/index. Last accessed October 11, 2005. Recently, the government legalized production. See Nutti, M.R., Sampaio, M.J.A., and Watanabe, E. 2006. *GMO research and agribusiness in Brazil: Impact of the regulatory framework.* This volume. Chapter 14.

5. See, for example, Chaffey, N.J., ed. 2002. *Wood formation in trees: Cell and molecular biology techniques.* London: Taylor & Francis, 384 pp.

6. CSIRO Australia, 1997-2005. Acid soils—A ticking time bomb? Information sheet available at http://www.csiro.au/index.asp. Accessed October 7, 2005.

7. Ronen, G., Cohen, M., Zamir, D., and Hirschberg, J. 1999. Regulation of carotenoid biosynthesis during tomato fruit development: Expression of the gene for lycopene epsilon-cyclase is down-regulated during ripening and is elevated in the mutant Delta. *Plant Journal* 17: 341-351; Ye, X., Al-Babili, S., Klöti, A., Zhang, J., Lucca, P., Beyer, P., and Potrykus, I. 2000. Engineering the provitamin A (beta-carotene) biosynthetic pathway into (carotenoid-free) rice endosperm. *Science* 287: 303-305; Krishnan, H. 2005. Engineering soybean for enhanced sulfur amino acid content. *Crop Science* 45: 454-461.

8. Mor, T.S., Mason, H.S., Kirk, D., Arntzen, C.J., and Cardineau, G.A. 2004. Plants as a delivery vehicle for orally delivered subunit vaccines. In *New generation vaccines* (3rd ed.). Edited by Myron M. Levine, James B. Kaper, Rino Rappuoli, Margaret A. Liu, Michael, F. New York: Marcel Dekker. pp. 305-311; Moffat, A.S. 1998. Toting up the early harvest of transgenic plants. *Science* 282: 2176-2178; DaSilva, E.J., Baydoun, E., and Badran, A. 2002. Biotechnology in the developing world. *EJB Electronic Journal of Biotechnology.* http://www.ejbiotechnology.info/content/issues/01. [Last accessed August 22, 2006].

9. See several chapters in *Biology of molecular plant-microbe interactions, Vol 4.* Edited by I. Tikhonovich, B. Lugtenberg, and N. Provorov. St. Paul, MN: The American Phytopathological Society.

10. Mikkelsen, T.R., Andersen, B., and Jørgensen, R.B. 1996. The risks of crop transgene spread. *Nature* 380: 31.

11. Losey, J.E., Rayor, L.S., and Carter, M.E. 1999. Transgenic pollen harms monarch larvae. *Nature* 399: 214.

12. A scientific and editorial controversy rages concerning the paper by Quist, D., and Chapela, I. 2001. Transgenic DNA introgressed into traditional maize landraces in Oaxaca, Mexico. *Nature* 414: 541-543, which was later withdrawn by the editor of *Nature*. The editor wrote, "We published the paper 'Transgenic DNA introgressed into traditional maize landraces in Oaxaca, Mexico' by David Quist and Ignacio Chapela. Subsequently, we received several criticisms of the paper, to which we obtained responses from the authors and consulted referees over the exchanges. In the meantime, the authors agreed to obtain further data, on a timetable agreed with us, which might prove beyond reasonable doubt that transgenes have indeed become integrated into the maize genome. The authors have now obtained some additional data, but there is disagreement between them and a referee as to whether these results significantly bolster their argument. In light of these discussions and the diverse advice received, *Nature* has concluded that the evidence available is not sufficient to justify the publication of the original paper." See Campbell, P. 2002 Editorial note. *Nature* 416: April 4, 2002 (UK).

13. Ellis, B.E. 2006. Controversy around terminology and novelty: engineered, modified, and biotechnology transgenics. This volume, chapter 2.

14. For concise comment see Stephenson, J. 2004. Vioxx controversy. *JAMA* 292: 2827.

15. ICMJE 2005. *Uniform requirements for manuscripts submitted to biomedical journals: Writing and editing for biomedical publication.* http://www.icmje.org/index.html. Accessed October 11, 2005.

16. The Norfolk Genetic Information Network web site at http://ngin.tripod.com carried a report from the January 9th 2003 edition of the Sydney (Australia) Morning Herald that GE corn shipped to Zambia was refused by that country and then shipped to Australia.

17. See Chapters 10-16 in this volume.

18. Reuters. July 15, 2004. Canada sacks three whistle-blowing scientists. For text see http://www.organicconsumers.org/rbgh/whistleblower071904.cfm [Accessed October 11, 2005].

19. While documentation is often biased because of the specific interests of anti-GE lobby groups, there is a disturbingly large number of such allegations and the political responses from industry seem more concerned with deflecting criticism than with rebutting it with evidence.

20. Carson, R.L. 1962/2002. *Silent spring. 40th anniversary edition.* New York: Mariner Books, Houghton Miflin, 379 pp.

21. Barrett, K. and Brunk, C.G. 2006. A precautionary framework for biotechnology. See chapter 8, this volume; Raffensperger, C. 2006. The precautionary principle and biotechnology: Guiding a public interest research agenda. See chapter 9, this volume.

22. See Mor, T.S. et al. 2004. See note 8.

23. Devlin, R.H. et al. 1995. Production of germline transgenic Pacific salmonids with dramatically increased growth performance. *Canadian Journal of Fisheries and Aquatic Science* 52: 1376-1384.

24. Fletcher, G.L., Shears, M.A., Goddard, S.V., Alderson, R., Chin-Dixon, E.A., and Hew, C.L. 1999. Transgenic fish for sustainable aquaculture. In *Sustainable aquaculture: Food for the future?* Edited by N. Svennevig, H. Reinertsen, and M. New. Rotterdam: AA Balkema, pp. 193-201.

25. Hedrick, P.W. 2001. Invasion of transgenes from salmon or other genetically modified organisms into natural populations. *Canadian Journal of Fisheries and Aquatic Sciences* 58: 841-844.

26. Hu, W.-J., Harding, S.A., Lung, J., Popko, J.L., Ralph, J., Stokke, D.D., Tsai, C.J., and Chiang, V.L. 1999. Repression of lignin biosynthesis promotes cellulose accumulation and growth in transgenic trees. *Nature Biotechnology* 17: 808-812.

27. M. Isman, personal communication.

28. Kappers, I.F., Aharoni, A., van Herpen, T.W.M.J., Luckerhoff, L.L.P., Dicke, M., and Bouwmeester, H.J. 2005. Genetic engineering of terpenoid metabolism attracts bodyguards to *Arabidopsis. Science* 309: 2070-2072.

29. Bradshaw, A.D. 1965. Evolutionary significance of phenotypic plasticity in plants. *Advances in Genetics* 13: 115-155.

PART II:
ISSUES IN CURRENT GOVERNANCE

Chapter 8

A Precautionary Framework
for Biotechnology

Katherine Barrett
Conrad G. Brunk

INTRODUCTION

The precautionary principle, or the "precautionary approach," is commonly invoked by government, industry, and civil society organizations as a guiding value for technological development and risk regulation. This is increasingly true in the context of agricultural biotechnology, where issues of health and environmental risk have been in the forefront of public concern.

The precautionary principle advises that, in the face of scientific uncertainty or incomplete knowledge, it is better to err in favor of human and environmental safety than to err on the side of risks to these values. Put another way, if one's assessment of the risk and benefits turns out to be in error, it is better that the error results in the foregoing of unnecessary benefits than in significant harm to health or the environment.

The precautionary principle has emerged in recent years as a common component of national and international environmental law.[1] It

The conceptual framework developed in this chapter is based upon a position article, titled "The Precautionary Principle: Issues Involving Burden of Proof and Standards of Evidence," prepared by the authors at the request of Environment Canada in June 2001 as part of its public consultation on the precautionary principle (used with permission).

Genetically Engineered Crops
© 2007 by The Haworth Press, Inc. All rights reserved.
doi:10.1300/5880_08

is cited in more than twenty international laws, treaties, protocols, and declarations. It is also clearly invoked in international agreements affecting the regulation of plant and animal biotechnology in trade. For example, the Cartagena Protocol on Biosafety appears to allow countries to use the precautionary principle as a basis for denying import of genetically engineered (GE) ("living modified") organisms and food products. Articles 10.6 and 11.8 of the Protocol state:

> Lack of scientific certainty due to insufficient relevant scientific information and knowledge regarding the extent of the potential adverse effects of a living modified organism on the conservation and sustainable use of biological diversity in the Party of import, taking also into account risks to human health, shall not prevent that Party from taking a decision, as appropriate . . . in order to avoid or minimise such potential adverse effects.

However, because the treaty later states that a rejection must be based on "credible scientific evidence," the exact impact of precaution in the treaty remains unclear.[2] This proviso reflects a central unresolved issue in national and international invocations of the precautionary principle, specifically the appropriate role for scientific evidence in the application of precaution.[3]

The 1992 UN Conference on Environment and Development (The Rio Declaration) adopted language similar to the Cartagena Protocol. Principle 15 states: "Where there are threats of serious or irreversible damage, lack of full scientific certainty shall not be used as a reason for postponing cost-effective measures to prevent environmental degradation." The Rio and Cartagena formulations are widely cited as the governing, if not the definitive, statements of the precautionary principle as it can be legitimately invoked in the context of international trade.

In the Canadian context, the Canadian Environmental Protection Act explicitly invokes the principle, requiring ministers to "apply a weight of evidence approach and the precautionary principle" in conducting certain kinds of reviews.[4] The principle was invoked explicitly as an appropriate stance in the regulation of the risks associated with food biotechnology by the Royal Society of Canada Expert Panel on the Future of Food Biotechnology in its 2001 report.[5] More recently, the Government of Canada has published guiding principles for application of precaution in science-based regulatory decisions.[6]

THE DEBATE ABOUT
THE PRECAUTIONARY PRINCIPLE

The precautionary principle has been subjected to a wide variety of criticisms, perhaps nowhere more than in the context of agricultural biotechnology risk assessment and management.[7] Proponents insist that the principle is an essential tool in the protection of environmental and human health because it provides a corrective against biases inherent in risk-assessment science.

On the other hand, critics charge that precaution compromises the scientific integrity of risk assessment and management decisions by opening the door to inappropriate extra-scientific influences. For example, it has been suggested that the adoption of the precautionary principle in the Cartagena Protocol has the potential to "lead to arbitrary unscientific rejection of some products."[8]

Such critiques are premised, in part, on the notion that the precautionary principle is essentially flawed because it sanctions regulatory decisions with little regard for scientific information. That is, any amount of uncertainty—and there is always uncertainty in science—is sufficient basis to reject new technologies. By this argument, precaution becomes politics by other means, creating for example, non-tariff trade barriers to foreign imported agricultural products.[9,10]

In this chapter, we counter the criticism that the precautionary principle is simply an extra-scientific limitation imposed upon the otherwise scientific assessment of risk. Rather, we argue that more or less precautionary choices must *always* be made in regulatory decisions involving uncertainty and potential harm. We suggest, more specifically, that inherent elements of science, such as allocation of burden of proof, choice of standards of proof, and choice of safety standards are necessarily incorporated into all decisions. The question, therefore, is whether it is appropriate in a particular situation to handle these elements in a strictly precautionary, weakly precautionary, or more risk taking way. We show how different types of risk, with different levels and types of uncertainty, require various and highly nuanced ways of handling these issues. Our analysis thereby frames the precautionary principle as a set of guidelines for addressing issues endemic to all regulatory science.

In the sections that follow below we first introduce several fundamental, though not unequivocal, elements of the precautionary principle. We then relate several of these elements into a continuum of decision-making styles, ranging from strictly precautionary to explicitly risk taking.

The Fundamental Elements of the Precautionary Principle

It is possible to identify several key elements common to all, or most, interpretations of the precautionary principle:

Scientific uncertainty. The principle recognizes that science cannot accurately predict the full range of consequences within highly complex, interrelated, and "real-world" systems. Analysts have also noted that there are various kinds of uncertainty stemming from different limitations of science and knowledge.[11,12,13,14]

Nonnegligible risk. The precautionary principle further acknowledges that some technologies and human activities pose degrees (albeit uncertain) of nonnegligible risk.[15] The question of what levels of risk are acceptable is recognized as complex, the answer to which depends upon a number of variables, including the nature of the risk in question and the values held by those who benefit from the risk and those who are the bearers of the risk. Thus, appropriate thresholds of acceptability are almost always contentious.

Presumption in favor of health and environmental safety. When there are significant uncertainties in the estimation of risks and/or benefits, a choice must be exercised about which values are most important to protect if the risk estimation or the management strategy errs. Is it better to err on the side of safety or on the side of the benefits of a particular technology? The precautionary principle is generally understood to establish a refutable presumption in favor of protecting health and environmental values.[16]

Shifting the burden and standard of proof in favor of safety. A presumption in favor of health and environmental safety requires that the burden and standards of proof be managed in a way that places greater onus upon those who allege a product or activity to be safe than upon those who allege it to be unsafe. This recognizes the fact that, under conditions of uncertainty, the placement of the onus upon one party or the other may make the case difficult or impossible to substantiate, depending upon the standard of proof (i.e., the required level of scientific certainty) invoked. The implications of this element of precaution are discussed later.

Proportionality. Many interpretations state that the financial or opportunity costs of precaution should be factored into decision-making,[17] so that precautionary measures are in some way proportional to the potential risks.[18] Precaution, therefore, does not require the avoidance of risk regardless of these costs. Instead, levels of precaution should be proportionate to the potential severity of the risk as well as the costs. Many analysts have further noted that precautionary approaches can encourage innovation by supporting alternative technologies.[19]

Precautionary safety standards. Some standards of safety (i.e., standards of determining acceptable levels of risk) are more conservative (i.e., risk averse) than others, depending upon the character of the risk and its specific context.[20] As discussed later, the choice of a safety standard will also influence the issues related to the appropriate standards of evidence or proof.

A Precautionary Framework for Biotechnology

The Precautionary Principle As a Continuum

A further point of debate around the precautionary principle is whether precautionary measures should be applied only after available evidence indicates that a particular threshold of potential harm has been reached. For example, the Rio Declaration's invocation of precaution "where there are threats of serious or irreversible damage" can be interpreted in this way.

It has been argued that requiring such firm thresholds runs contrary to the goal of the precautionary principle, which is to sanction regulatory action *prior to* conclusive knowledge about the nature and severity of harm.[21] Critics have countered that the precautionary principle is untenable precisely because it appears to sanction regulatory measures regardless of the potential severity or probability of harm, and with little or no evidence of harm.[22]

A more robust and useful regulatory interpretation understands criteria such as "seriousness" and "irreversibility" not only as critical considerations for invocation of the precautionary principle, but also as involving matters of degree rather than a singular threshold. This position is consistent with many analysts who have argued that precaution should function along a continuum from weak to strong.[23]

In this interpretation the level of appropriate precautionary action should be a function of the following:

- The degree and type of suspected harm, incorporating the potential for reversibility, and the nature and distribution of harm (see articles by Dovers and Handmer[24] and Stirling[25] for examples of criteria to determine stakes); and,
- Levels of uncertainty or ignorance related to potential harm.[26]

Such a framework results in a two-axis matrix for determining the appropriate level of precaution (e.g., Wagner 2000,[27] and references therein). A third axis—availability of alternatives—has also been proposed and could be incorporated into the framework suggested here.[28]

We argue below that, as potential harm and uncertainty increase, the burden and standards of proof should be handled in increasingly precautionary ways.

BURDEN OF PROOF

Burden of proof is a legal term referring to the onus to provide evidence sufficient to shift initial presumptions. In the Canadian judicial system, like many others, the presumption is usually one of innocence, and, therefore, the burden rests on the prosecutor or plaintiff to demonstrate guilt. Similarly, in regulatory law, the burden is often, but not always, on those wishing to prevent or regulate an activity.[29] In the case of criminal and tort law in many jurisdictions, giving the onus of proof to the prosecutor/plaintiff establishes a refutable presumption of innocence on the part of the defendant, and thus acts in a *precautionary* manner with respect to the defendant's rights and interests. Where evidence is weak and uncertainties are high, this presumption protects the defendant's interests over those of the prosecutor/ plaintiff.

Insofar as the precautionary principle establishes a presumption in favor of human health and environmental values, analogously to the juridical context, it shifts the burden of proof (of safety) to the proponent of potentially hazardous technology. Again, where uncertainties are high this shifting of the burden of proof protects these values over those related to the benefits of the technology.[30]

Critics have argued that this approach stifles innovation because it places an impossible onus on developers to prove safety.[31] Yet there are precedents (at least in principle) for shifting the burden of proof in this way. For example, effective pre-market regulations and licensing systems require developers to demonstrate safety prior to approval.[32] In terms of environmental protection, this approach is perhaps most fully implemented in northern European legislation. For example, "German Nuclear Energy Act (1976) provides for 'conditionally suspended injunctions' rather than conditional permits thereby shifting the burden of proof to the operator."[33]

While this shift of the onus of proof is an important feature of the precautionary principle, we argue that it should be applied in different ways depending on the stakes involved and the standards of evidence used. If the stakes are high or moderately high the presumption to be rebutted should be that the technology is harmful. If the stakes are very low a presumption of safety may be more appropriate. Furthermore, the burden of proof may be more or less onerous, depending on the case. As described later (and in Table 8.1) the burden may require different standards of evidence to alter a presumption of harm, again depending on the stakes involved.

For products already approved by the regulatory process and in use, it may be appropriate for at least part of the burden of proof to fall back upon the regulator or the public,[34] *provided that the initial regulatory review invoked an appropriate degree of precaution.* The longer products have an established track record of use without realization of potentially serious adverse effects, the more the burden of proof may move from the risk producer/beneficiary to the potential risk bearer. This might include, for example, the burden of postmarket monitoring, which initially should fall on the developers/proponents, but may switch to the regulator and finally to the public (or its representatives) as a long-term safety record is established. This issue is independent of the question of continuing liability for adverse consequences, which remain with the risk producer and may provide strong incentives for maintaining strong postmarket monitoring.

Standards of Proof

"Standards of proof (or evidence)" refer to the amount and strength of evidence needed to establish a conclusion (e.g., of unacceptable risk or of safety). We suggest that standards of proof can be applied at two points in the decision-making process:

1. A particular standard of evidence is required to establish an initial presumption of harm, thereby placing the burden of proof on the proponents of the technology. The appropriate standard here will depend on the stakes involved (see Table 8.1).
2. A particular, perhaps different, standard of evidence will be required to *change* the presumption, that is, to establish safety. Again, the appropriate standard here will depend on the stakes, as indicated in Table 8.1.

The combination of stakes, presumptions, burden of proof, and standards of evidence determines the level of precaution applied, ranging from strict precaution to risk taking. Various standards of evidence are well documented in legal literature as well as in analyses of the precautionary principle. Evidentiary standards range from "no evidence" to "highly significant evidence."[35] A similar continuum is outlined later in Table 8.1.

Beyond Reasonable Doubt

The strictest standard of evidence is some form of high confidence, as represented in the typical court standard for establishment of criminal guilt "beyond all reasonable doubt." In contexts of science this is most closely represented by the so-called 95 percent confidence rule.[36]

Such high standards reduce the chance of false positives, known as the "Type I Error" (e.g., falsely convicting an innocent person in court or falsely affirming an hypothesis in science), and are consistent with norms for criminal trials and for peer-reviewed scientific

TABLE 8.1. Relationship among level of harm/uncertainty, standards of evidence, and burden of proof (see text for details).

Suspected "stakes"; level of harm and degree of uncertainty	Standard of evidence required to establish presumption of harm and place burden of proof on proponent	Burden of proof (assuming previous standard is fulfilled)	Standard of proof required to change presumption	Appropriate standards of safety	Level of precaution
Extremely high	*Prima facie* case Need little direct evidence	Developer establishes safety	Beyond reasonable doubt Proving negative is not possible but standard should include: high statistical power diverse forms of evidence outlier views	Zero increased risk or High margin of safety risk (threshold)	Strictly precautionary
High	Good evidence (but less than preponderance)	Developer establishes safety	Clear and convincing evidence	Reasonable margin of safety	Strongly precautionary
Moderate	Weight of evidence	Developer establishes safety	Weight of evidence	Actual acceptance ALARA	Moderately precautionary
Low	Clear and convincing evidence	Public establishes unacceptable risk	Weight of evidence	Comparative Balancing RCBA	Weakly-to-not precautionary
Very low	Beyond reasonable doubt	Public must demonstrate harm	Public must provide strong evidence of harm (~ 95% confidence level)	Comparative Balancing RCBA	Not precaution-ary/ risk taking

research, which advise caution in veering from the presumption of innocence or from the null hypothesis of "no effect."

However, applying these standards to the regulation of technologies—*without shifting the burden of proof away from the presumption of safety*—tends to favor strongly the approval of potentially hazardous products. This is because it is difficult to establish unacceptable risks of harm with such high confidence, especially in cases where adverse effects may be diffuse, delayed, of low frequency, or otherwise difficult to assess.[37] Precaution in these cases means reducing the chance of the "Type II Error" (e.g., failing to find a risk where there truly is one), while increasing the chance of the "Type I Error." Applying high standards of proof to the demonstration of harm may be justified only under explicitly "risk taking" regulations where the stakes are very low (see Table 8.1).

Equally, high standards of evidence can render the establishment of safety extremely difficult if the burden of proof is shifted to proponents of a technology. In cases where the potential risks are extremely serious, however, this strongly precautionary burden may be appropriate. While proving a negative is not possible, high evidentiary standards can still be employed by demanding, for example, that statistical power is measured[38] and that diverse forms of evidence including outlier views are considered. (See further discussion under the section "The Prima Facie Case.")

Clear and Convincing Evidence

The level of proof sometimes required of plaintiffs in civil cases is that of "clear and convincing evidence." In this instance, the judge or jury must be persuaded by the evidence that the truth of the claim is highly probable. The standard of proof invoked here is not as strong as the criminal law standard but neither is it simply a matter of tipping conflicting evidence in favor of the claim, as described in the "preponderance of evidence" standard. In scientific terms, the rule of confidence in this case is significantly higher than 50 percent, but not necessarily the 95 percent confidence of the "beyond reasonable doubt" standard.

Preponderance of Evidence

A less onerous standard of evidence, the one commonly used in civil trials, is the "weight" or "preponderance" of evidence. By this standard, a conclusion is sufficiently supported if the balance of evidence tips in its favor.[39] In regulatory contexts the analogous rule of "reasonable weight of evidence" is often invoked. This pushes the weight of evidence standard more strongly in the direction of the previously discussed "clear and convincing evidence standard."

As with stronger evidentiary standards, the weight of evidence standard can be used in more or less precautionary ways, depending on which party has the burden of proof. If the stakes are relatively high or moderate, a balance of evidence should be sufficient to shift the burden of proof to technology proponents. The latter would then have the burden of establishing safety, either by a weight of evidence standard if stakes are moderate or by a slightly higher standard akin to "clear and convincing evidence" if stakes are high. This is likely to be the most common regulatory implementation of precaution.

Weight of evidence may also be used under a less precautionary framework if the stakes are low. In this case, the presumption would be one of safety, and the public would be required to demonstrate harm. When a balance of evidence (rather than strict evidentiary standards) is sufficient to establish unacceptable risk, the framework could be characterized as weakly precautionary. When high confidence is required for this case, the framework is not precautionary.

The Prima Facie Case

In cases where the stakes are extremely high (great uncertainty; potentially large-scale serious and irreversible damage; very difficult to predict and test), it may be appropriate to make a presumption of risk based on only prima facie evidence of potential for the harm. Incomplete, contested, or preliminary scientific data or plausible but unsubstantiated scientific hypotheses or models, together with significant levels of uncertainty, may be sufficient to establish a reasonable prima facie case for the *possibility* of serious harm.[40]

In such a case, precaution requires proponents to demonstrate safety by high standards of evidence before a shift back from this presumption would be justified.

This is a very strict form of the precautionary principle, and the most controversial. However, as Cranor argues, "It is quite possible that there may be things so precious that we are prepared to take precautionary actions to protect them, even when the balance of evidence does not show that they are threatened or there is little or no rational basis in empirical evidence that they are threatened."[41]

The Report of the British BSE Inquiry clearly advocated the use of such a "prima facie" standard of proof in its critique of the manner in which the British Ministry of Agriculture, Fisheries and Food managed the early stages of the BSE crisis, when the possibility that BSE could cross the species barrier into humans as in the case of vCJD was considered "remote" on the basis of the existing scientific evidence. The BSE Inquiry concluded in retrospect, however, that, even when the available scientific evidence fails to establish a risk as anything other than "remote," where there is a prima facie case of serious risk, significant and, in this case, highly costly precautionary action is warranted.[42]

STANDARDS OF SAFETY

Precautionary standards of proof are integrally related to the standard of safety by which a given risk is judged acceptable. The chosen standard of safety reflects varying degrees of precaution, and it also dictates the kinds and levels of evidence required in the scientific assessment. "Zero-risk" is generally considered to be the most demanding, and hence the most precautionary, safety standard. At the other extreme, balancing approaches to safety such as risk-cost-benefit are generally considered to be the least demanding, and hence the least precautionary, insofar as they permit any health or environmental risk if there are compensating benefits, or if the costs of risk reduction outweigh the benefits. Various threshold standards fall in the middle of the precautionary scale, as illustrated in the following description.

"Zero-Risk" Standard

The "zero-risk" standard of safety stipulates that only the complete absence of a particular risk is acceptable. Under this standard, the risk producer/beneficiary has the burden to prove safety, and precaution would require a very high standard of proof. This is because a conclusion of "zero-risk" obviously requires a high level of confidence. "Weight of evidence" is not enough to provide confidence in zero-risk standard. Some might argue that even the "clear and convincing evidence" standard is not enough. Note, however, that if "zero-risk" is the standard of safety, the placement of the burden of proof on the risk bearer remains relatively precautionary, unless the standard of proof is high. This is because relatively weak evidence of risk would arguably still justify a conclusion that the risk is not confidently "zero."

Most commentators agree that in most contexts the complete absence of risk is practically unachievable.[43] However, it is important to point out that what technology opponents often appear to formulate as a demand for zero-risk is in fact a demand for "no additional risk" associated with a particular technology, beyond some assumed baseline or threshold, or a willingness to accept the opportunity costs of not proceeding. Such a demand is practically achievable simply by not proceeding with the technology. Understood either way, this safety standard is obviously the most precautionary.

Threshold Standards

Threshold standards of safety establish a nonzero level of risk as acceptable, using some rationale having to do either with the function of the technological or biological system at issue or with attitudes of acceptance within the at-risk population. They differ from balancing standards (see later) in that they set absolute limits on the acceptability of risk, absolute in the sense that these limits are not relative to considerations of cost or benefit. In this respect, threshold standards can be more or less precautionary than balancing standards, depending entirely upon how high or low the threshold bar is set by the particular standard, and on levels of certainty about threshold effects within the system at issue.

Common examples of threshold standards are as follows:

- *NOAEL (No Observable Adverse Effect Level).* This threshold can be interpreted as a form of "zero-risk" insofar as its objective is to maintain exposures to hazards at levels below which there is any probability of discernible adverse effect (e.g., risk).

- *ADI* (*Acceptable Daily Intake*). The rationale for this threshold can be assumptions about either biological thresholds for adverse effects or levels of risk acceptance in the risk population. In the former case it is really an application of the NOAEL standard.
- *Natural Background Risk.* Often the existence of "natural" background levels of risk, which are unavoidable or presumed to be accepted by the risk bearers, is used as the threshold of acceptable risk (e.g., background levels of radiation in the environment).
- *Actual Acceptance.* In many circumstances risks can be considered acceptable if the risk bearers *in fact* give their consent to the risks. In biomedical therapy contexts, informed consent is the "gold standard" of risk acceptability. In these circumstances, the risks are often high, but are judged acceptable by those exposed to them because they are voluntarily borne in the expectation of potential benefits or avoidance of even greater risks. This approach to safety is also put forward in many non-biomedical contexts. Demands for the labeling of genetically modified foods to allow consumers to choose whatever risks might be associated with the technology is an example of an invocation of the "actual acceptance" standard.
- *Presumed Acceptance.* By this standard, a risk is deemed acceptable if it is reasonable to presume the consent of risk bearers on the basis of their behavior or expressed preferences The use of the "GRAS" ("Generally Regarded as Safe") standard with respect to food additives in U.S. regulatory law, and the "substantial equivalence" standard in the Canadian approach to genetically modified foods are examples of a "presumed acceptance" approach to safety. In both cases it is presumed that the risk levels in long-accepted food additives or conventionally bred plants and animals, while not negligible, have been accepted by consumers, and, therefore, any products that do not exceed these risk levels are also acceptable.[44]

Balancing (Comparative) Standards

Balancing standards of safety define the acceptability of risks in terms of how they are balanced against other values, such as the costs of risk reduction, benefits of the risk taking activity, and the risks of alternatives. They are distinguished from other "absolute" standards in that they do not specify any level of risk as inherently unacceptable, regardless of other values at stake. Any level of risk is theoretically acceptable if the balancing factors are high enough.

A relatively precautionary application of this standard would demand a strong onus of proof upon the risk producer to establish with high levels of confidence that the risks do not in fact exceed the expected benefits. If the standard of proof, for example, weight of evidence were relaxed in this case, it would in most cases produce a strong non-precautionary bias in favor of benefits over risks. This is because the benefits, having been designed into a technology, are usually highly predictable, exhaustively understood, and highly certain. The risks, however, are not designed into the technology (i.e., they are unintended), and are usually highly diffuse, long-term, and uncertain. In other words, we often have asymmetric information about risks and benefits.[45] That is, the potential benefits are more certain or predictable than are unforeseen hazards. This "inadvertently favours keeping the substances in commerce or permitting them in commerce." The asymmetry is further exaggerated because the potential victims or risk bearers tend to be more geographically dispersed and less well organized politically than the beneficiaries (developers) of technologies. Therefore, risk-benefit safety standards, generally considered the least precautionary of the various safety standards, are rendered even less precautionary if the standard of proof on the risk side of the equation is relaxed.

The most common examples of balancing standards are the following:

- *Cost-Effectiveness.* This approach establishes the acceptable level of risk at any point where, according to some criteria (e.g., budget constraints), further costs of risk reduction are exceeded.
- *ALARA (As Low As Reasonably Achievable).* ALARA is one of the most common balancing standards in regulatory contexts. The criteria of "reasonableness" may vary from simple cost-effectiveness to "whatever is technically possible." Thus ALARA is more or less precautionary, depending upon the chosen criterion of "reasonableness."

- *Risk-Cost-Benefit (RCBA)*. This well-known economic model of safety defines any level of risk as acceptable if the total benefits of the risk activity are outweighed by benefits produced by the activity (minus the costs, including those of risk reduction). The RCBA standard makes several assumptions that are often controversial. Among them are the following:

  — The risks, costs, and benefits are distributed equitably among the population of risk beneficiaries and risk bearers.
  — There is a reliable common measure of all the risks and the benefits (e.g., monetary value).
  — The scientific understanding of the risk and costs is as reliable as the understanding of the benefits.

Some commentators argue that if RCBA analysis is done comprehensively, for example, meeting the aforementioned criteria and including Bayesian analysis, there is no value added by invocation of the precautionary principle.[46] However, insofar as RCBA weighs risks and benefits equally, even if it meets its ideal criteria, it runs contrary to one of the fundamental premises of the precautionary principle outlined earlier—the presumption in favor of environmental and human health values. For this reason, and because of serious doubts about the ability of RCBA to practically meet its own methodological criteria, it is considered by many to be among the least precautionary safety standards—or indeed not precautionary at all. Others have argued that RCBA can in fact be done in a precautionary way if the environmental and health risks and benefits are given appropriate extra weighting in the analysis.[47]

CONCLUSION

The foregoing discussion attempts to lay out the complex interrelation between issues involving burden of proof, standards of proof, and standards of safety, as these impinge upon the implementation of the precautionary principle in the regulatory context of environmental and health risks.

Our arguments suggest that applying the precautionary principle does not necessarily entail banning all risky and uncertain technologies. Nor does it mean that scientific evidence is ignored or distorted by inappropriate "unscientific" considerations. Rather, we suggest that the precautionary principle entails a complex continuum of values and choices about issues endemic to science: appropriate determination of burden of proof, standards of proof, and standards of safety. These choices must be made in all regulatory decisions involving uncertainty and potential risks to health and the environment. They are clearly value-laden, and as discussed in this chapter, will profoundly affect the outcome of the regulatory process. We suggest that such choices must, therefore, be made as explicit and open to debate as possible.

NOTES

1. Shipworth, D. and Kenley, R. 1999. Fitness landscapes and the precautionary principle: The geometry of environmental risk. *Environmental Management* 24: 121-131; Cameron, J. 1999. The precautionary principle: Core meaning, constitutional framework and procedures for implementation. In *Perspectives on the precautionary principle*. Edited by R. Harding and E. Fischer. Annandale, NSW: Federation Press., pp. 29-58; O'Riordan, T., Jordan, A., and Cameron, J. 2001. The evolution of the precautionary principle. In *Reinterpreting the precautionary principle*. Edited by T. O'Riordan, J. Cameron, and A. Jordan. London: Cameron May, pp. 9-34.

2. Helmuth, L. 2000. Both sides claim victory in trade pact. *Science* 287: 782-783.

3. Royal Society of Canada. 2001. *Elements of precaution: Recommendations for the regulation of food biotechnology in Canada*. Ottawa: The Royal Society of Canada.

4. Canadian Environmental Protection Act. 1999. Section 76.1.

5. Royal Society of Canada. 2001. See note 3.

6. Privy Council of Canada. 2003. *A framework for the application of precaution in science-based decision making about risk*. www.pco-bcp.gc.ca. Accessed May 5, 2003.

7. Royal Society of Canada. 2001. See note 3.

8. Mahoney, R.J. 2000. Opportunity for agricultural biotechnology. *Science* 288: 615.

9. Adler, J. 2000. More sorry than safe: Assessing the precautionary principle and the proposed International Biosafety Protocol. *Texas International Law Journal* 35: 173-205.

10. Foster, K.R., Vecchia, P., and Repacholi, M. 2000. Science and the precautionary principle. *Science* 288: 979-981.

11. Funtowicz, S.O. and Ravetz, J.R. 1992. Three types of risk assessment and the emergence of post-normal science. In *Social theories of risk*. Edited by S. Krimsky and D. Golding. Westport, CT: Praeger. pp. 251-273.

12. Wynne, B. 1992. Uncertainty and environmental learning. *Global Environmental Change* 2: 111-127.

13. Barrett, K. and Raffensperger, C. 2002. From principle to action. Applying the precautionary principle to agricultural biotechnology. *International Journal of Biotechnology* 4: 4-17.

14. Schettler, T., Barrett, K., and Raffensperger C. 2002. The precautionary principle: A guide for protecting public health and the environment. In *Life support: The environment and human health*. Edited by M. McCally. Cambridge, MA: MIT Press, pp. 239-256.

15. Cameron, D. 1999. See note 1.

16. Royal Society of Canada. 2001. See note 3.

17. Pearce, D. 1994. The precautionary principle and economic analysis. In *Interpreting the precautionary principle*. Edited by T. O'Riordan and J. Cameron. London: Earthscan., pp. 132-151; Graham, J.D., and Wiener, J.B. 1995. *Risk versus risk: Tradeoffs in protecting health and the environment*. Cambridge, MA: Harvard University Press; Cross, F.B. 1996. Paradoxical perils of the precautionary principle. *Washington and Lee Law Review* 53: 851-925; Miller, H., and Conko, G. 2001. Precaution without principle. *Nature Biotechnology* 19: 302-303.

18. Commission of the European Communities. 2000. *Communication from the commission on the precautionary principle*. Brussels.

19. Boehmer-Christiansen, S. 1994. The precautionary principle in Germany— enabling government. In *Interpreting the precautionary principle*. Edited by T. O'Riordan and J. Cameron. London: Earthscan, pp. 31-60; O'Brien, M. 2000. *Making better environmental decisions. An alternative to risk assessment*. Cambridge, MA: MIT Press; Kriebal, D. and Tickner, J. 2001. Reenergizing public health through precaution. *American Journal of Public Health* 91: 1351-1361.

20. Royal Society of Canada. 2001. See note 3.

21. Cameron, D. 1999, see note 1; Jordan, A. and O'Riordan, T. 1999. The precautionary principle in contemporary environmental policy and politics. In *Protecting public health and the environment: Implementing the precautionary principle*. Edited by C. Raffensperger and J. Tickner. Washington, DC: Island Press, pp. 15-35.

22. ExxonMobil. 2000. Unbalanced caution. *New York Times,* Sec. A, p. 2 (op-ed) (November 2); Miller, H. and Conko, G. 2001. See note 17.

23. Barrett, K. 1999. *Canadian agricultural biotechnology: Risk assessment and the precautionary principle*. PhD Thesis. Department of Botany. University of British Columbia. Vancouver, BC; Cranor, C.F. 1999. Asymmetric information, the precautionary principle, and burdens of proof. In *Protecting public health and the environment: Implementing the precautionary principle*. Edited by C. Raffensperger and J. Tickner. Washington DC: Island Press, pp. 74-99; Jordan, A., and O'Riordan, T. 1999, see note 21; Tickner, J. 2000. *Precaution in practice: A framework for implementing the precautionary principle*. PhD Thesis. University of Massachusetts. Lowell, MA; Royal Society of Canada, 2001. See note 3.

24. Dovers, S.R. and Handmer, J.W. 1995. Ignorance, the precautionary principle and sustainability. *Ambio* 24: 92-97.

25. Stirling, A. 2001. The precautionary principle in science and technology. In *Reinterpreting the precautionary principle*. Edited by T. O'Riordan, J. Cameron, and A. Jordan. London: Cameron May, pp. 61-94.

26. Funtowicz, J.O. and Ravetz, S.R. 1992 See note 11; Wynne, B. 1992. See note 12; Barrett, K. and Raffensperger, C. 2002. See note 13; Schettler, T. et al. 2002. See note 14.

27. Wagner, W.E. 2000. The precautionary principle and chemical regulation in the U.S. *Human and Ecological Risk Assessment* 6: 459-477.

28. Tickner, J. 2000. See note 23.

29. Cranor, C.F. 1999. See note 23; Jordan, A. and O'Riordan, T. 1999. See note 21; Applegate, J.S. 2000. The precautionary preference: An American perspective on the precautionary principle. *Human and Ecological Risk Assessment* 6: 413-443.

30. Royal Society of Canada. 2001. See note 3.

31. Holm, S. and Harris, J. 1999. Precautionary principle stifles discovery. *Nature* 400: 398; ExxonMobil. 2000. See note 22; Miller, H. and Conko, G. 2001. See note 17.

32. Cranor, C.F. 1999, see note 23; Applegate, J.S. 2000. See note 29.

33. Sand, P. 2000. The precautionary principle: A European perspective. *Human and Ecological Risk Assessment* 6: 445-458.

34. Cranor, C.F. 1999. See note 23.

35. MacGarvin, M. 2001. Science, precaution, facts and values. In *Reinterpreting the precautionary principle*. Edited by T. O'Riordan, J. Cameron, and A. Jordan. London: Cameron May, pp. 35-60.

36. Brown, D.A. and Zaepfel, P. 1996. The implications of scientific uncertainty for environmental law. In *Scientific uncertainty and environmental problem solving*. Edited by J. Lemons. Cambridge, MA: Blackwell, pp. 377-393; Cranor, C.F. 1999. See note 23.

37. Shrader-Frechette, K.S. 1991. *Risk and rationality*. Berkeley, CA: University of California Press; Cranor, C.F. 1999. See note 23; Barrett, K. 1999. See note 23; MacGarvin, M. 2001. See note 35.

38. Peterman, R.M. and M'Gonigle, M. 1992. Statistical power analysis and the precautionary principle. *Marine Pollution Bulletin* 24: 231-234; Conner, T. 1997. *Burdens of proof*. Columbia, SC: Energy Research Foundation; Royal Society of Canada, 2001. See note 3.

39. Tickner, J. 1999. A map towards precautionary decision-making. In *Protecting public health and the environment: Implementing the precautionary principle*. Edited by C. Raffensperger and J. Tickner. Washington, DC: Island Press, pp. 162-186; Tickner, J. 2000. See note 23.

40. Royal Society of Canada. 2001. See note 3.

41. Cranor, C.F. 1999. See note 23.

42. BSE Inquiry: The Report. 2000. *The inquiry into BSE and Variant CJD in the United Kingdom*. www.bseinquiry.gov.uk/report/index.htm. Accessed May 5, 2003. Royal Society of Canada. 2001. See note 3.

43. Graham, J.D. and Wiener, J.B. 1995. See note 17.

44. Royal Society of Canada. 2001. See note 3.

45. Cranor, C.F. 1999. See note 23.

46. Hammitt, J.K. 2000. Global climate change: Benefit-cost analysis versus the precautionary principle. *Human and Ecological Risk Assessment* 6: 387-398; Montgomery, W.D. and Smith, A.E. 2000. Global climate change and the precautionary principle. *Human and Ecological Risk Assessment* 6: 399-412.

47. Shrader-Frechette, K.S. 1991. See note 37.

Chapter 9

The Precautionary Principle and Biotechnology: Guiding a Public Interest Research Agenda

Carolyn Raffensperger

INTRODUCTION

Early work on the precautionary principle, including that of the conveners of the January 1998 Wingspread Conference on the Precautionary Principle, was based on the assumption that the principle was a regulatory backstop to bad decisions. The principle was seen as a way of making decisions to prevent harm after a product or activity was fully developed and ready for market—or even, in some cases, already in use. But much of the analysis since Wingspread has concluded that the regulatory process is too late a stage at which to exercise true precaution.[1] The emergence of problems that call for precautionary regulation indicates that the wrong questions have been asked, or that hard questions were omitted during product development or implementation, such as large-scale deployment of GE crops.

Few resources or other incentives exist to ask such questions. Money is not allocated either by proponents and product developers or by government to investigate systemic ecological or sociological questions. Few rules, procedures, and resources have been dedicated to scrutinizing the harmful potentials of products and William McDonough, the green architect, has argued that regulation reflects design failure.[2] Accordingly design success, as measured by the yardstick of

Genetically Engineered Crops
© 2007 by The Haworth Press, Inc. All rights reserved.
doi:10.1300/5880_09

safety, would indicate that a precautionary research policy reduces the necessity for regulation except as an extra layer of precaution.

The precautionary principle, used only as a regulatory device, becomes a risk-management tool, thereby diminishing the potential for proactive harm prevention. For instance, in the summary of the Communication from the European Communities Commission on the Precautionary Principle, the Commission devolved the principle into risk management when it said, "The precautionary principle should be considered within a structured approach to the analysis of risk which comprises three elements: risk assessment, risk management, risk communication. The precautionary principle is particularly relevant to the management of risk."[3]

If it is used as a risk-management tool, regulators are unlikely to undertake a full precautionary principle analysis and examine whether a product is really needed to meet a societal goal or whether less harmful alternatives are available. They are less likely to stop an activity that may cause harm if the product is examined only when the rush to market is well underway. It is fiscally irresponsible and inefficient, after all, to stop a product at the last hurdle when major resources have been invested in its development. The current system has no other mechanism for stopping or modulating harmful activities, and unfortunately regulation has been the only recourse for society to take some action to prevent harm.

The European Union (EU) and the United States seem to agree that the precautionary principle is a risk-management device, particularly for purposes of trade, but this subsumes the principle under risk assessment and substantially weakens its force.

I argue that the precautionary principle must be an overarching guide and is best incorporated as part of the research agenda. Early introduction of the principle provides a rubric that could help decision makers conserve resources and make better decisions during both private and publicly funded research and development. Furthermore, such precaution can be carried forward through the regulatory and judicial processes. If the principle guides the agenda as well as the work of individual scientists, institutions, and private entities, the result will likely be products that pose less harm and face fewer regulatory hurdles. The principle thus used would introduce increased reliability and clarity into administrative decisions.

Incorporating the precautionary principle into the research agenda would reflect a new social contract with scientists and more accurately reflect the public interest. A public interest precautionary research agenda will by definition aggressively defend the commons and, accordingly, have its largest impact in energy, medicine, and agriculture since these are areas where biotechnology is evolving most rapidly and has the largest potential for good and ill. As agriculture is such an enormous sector of the world of science, I use agriculture and biotechnology as the case study to explore this research agenda.

THE PRECAUTIONARY PRINCIPLE

The Wingspread Statement laid out several components required to implement the precautionary principle in any forum from research to regulation. These are (1) identifying alternatives to harmful activities, (2) shifting the burden of proof, and (3) democratic participation in decisions. They work best if they are anchored to goals that are in the spirit of the precautionary principle, specifically to protect human health and the environment. With a precautionary goal in sight, the products, technologies, processes to meet the goal are more likely to be in harmony with the underlying ethic of preventing harm to humans and environment and be consistent with its foundations in the precautionary principle's German precedent, Vorsorgeprinzip, which more literally means forecaring.

In the absence of such goals, regulatory agencies are too often faced with a fait accompli in the form of a potentially harmful product, such as a GE crop. Neither the agency nor any other public body has had the opportunity to ask whether and how this product meets a societal need and whether there are safer or more cost-effective ways to meet that need. Rather, the single, overarching implicit goal of short-term economic profit has dominated.

Once a goal has been established it is easier to identify alternative technologies or strategies for meeting that goal. One goal, which has been identified for agriculture and biotechnology, is to end hunger. However, the effectiveness of biotechnology to meet that goal has been fiercely contested.[4] Proponents have argued vociferously that opposition to biotechnology effectively denies food access to the poor.

Opponents of biotechnology have argued just as vociferously that biotechnology and patenting will deny the poor access to seed.

This debate came to a head with proposed aid to Zambia in 2002. Zambia faced a famine but rejected U.S. aid that was in the form of genetically altered grain. Part of the rationale for rejecting the grain was that Zambia had alternatives to transgenic whole seed corn, primarily corn that had already been milled. However, the United States refused to pay for the milling and accused the Zambians of starving their people.[5]

The second implementation step to prevent harm under the precautionary principle is to put the burden of proof on the proponents of an activity. This notion parallels the idea that the polluter must pay for damage, not the public. In the Zambian situation, the United States placed the burden on the Zambian government for rejecting the transgenic crops even though the U.S. risk assessment had not taken into account the Zambian context where unprocessed grain would probably form a large percentage of a starving child's diet. The precautionary approach would place the burden on the United States to demonstrate that there were no less risky alternatives.

Democratic participation may prove to be the most significant element of the precautionary principle for the research agenda because it could substantially change public support of research. The post–World War II social consensus on science[6] gave rise to a self-perpetuating structure for funding research (government, industry, and university partnerships) that has all but eliminated meaningful public participation. These partnerships proved particularly fruitful for generating advanced military technologies and transforming these technologies into items useful in domestic pursuits: the atomic bomb technology became nuclear power and nerve gas became pesticides. The public, however, had little say in whether these technologies were the ones they wanted.

ETHICS

Whether local or global, all research questions have an ethical dimension. The precautionary principle makes explicit the ethical considerations that are present in every stage of decision-making and directs a particular kind of ethical choice by requiring precautionary action to prevent harm and protect the commons. The kind of harm that we seek to prevent reflects the values of society. For instance, if the harm being considered is allergenicity in children arising from a biotechnology product, the ethical norm guiding precautionary action is that children come before profits.

Similarly, if there is scientific uncertainty about the nature and extent of that harm, experiments designed to resolve the uncertainty should err on the side of false positives (rather than false negatives that conclude something is safe when it is not) as a matter of ethics because children and the environment should get the benefit of the doubt, not the economy. False positives are to be avoided in normal science because they reduce scientific reliability.[7] However, a false negative generally stops an investigation since the research suggests that there is no cause-and-effect relationship between the agent and the subject, thereby reducing policy reliability. A false positive (finding a correlation when there is none) often generates more research and is easier to refute and test than a false negative. Normal science, by erring on the side of false negatives, misses problems that could have significant public health or environmental consequences. Precautionary science by erring on the side of false positives catches these potential harms in its broad sweep.

Precautionary actions, including rigorous research design, are an obvious place to employ ethical considerations. Both avoiding action and taking action to prevent harm reflect the values and ethics of those who have the power to act. If the entire chain is flawed, from the original research, which missed a harmful effect, through to the decisions based on that research, then the ethic of protecting the public and the environment from harm is thwarted.

Ethics guides the incorporation of precaution in the research agenda in other ways as well. It should prompt us to look carefully at forces such as financial conflicts of interest that militate against precaution. Unfortunately, as university research is increasingly funded by private industry, independent science is becoming an endangered enterprise. Recent privatization controversies, such as the contract between Novartis and the University of California, Berkeley, highlight the difficulty of maintaining any independence. Was the biotechnology research undertaken at Berkeley tainted by the Novartis contract?

Some professional societies have been explicit about the role of ethics in their disciplines. For example, the International Society for Ethnobiology's code of ethics contains a paragraph that says,

> Principle of Precaution: This principle acknowledges the complexity of interactions between cultural and biological communities, and thus the inherent uncertainty of effects due to ethnobiological and other research. The Precautionary Principle advocates taking proactive, anticipatory action to identify and to prevent biological or cultural harms resulting from research activities or outcomes, even if cause-and-effect relationships have not yet been scientifically proven. The prediction and assessment of such biological and cultural harms must include local criteria and indicators, thus must fully involve indigenous peoples, traditional societies, and local communities.[8]

This formulation of the principle is particularly interesting because it extends the notion of harm from environmental and public health threats to cultural harm such as research done by scientists bioprospecting among First Nations and native peoples.

PRECAUTIONARY RESEARCH: A NEW SOCIAL CONTRACT

In the largest sense, the assumption of publicly funded science is that it is serving the public interest. It is part of a social contract between scientists and the general public. However, there is a background assumption in publicly funded research that the major benefit to the public is innovation and increased economic growth. An overarching precautionary research agenda would extend its aims beyond commercial opportunities (and the few broader public goals: national security, territorial expansion and exploration, and treatments and cures for disease) that have dominated the research agenda from the founding of the United States until today.

Biotechnology science would be subsumed under this precautionary rubric and the focus of the research could be substantially different from the science designed to increase corporate wealth and consolidation. It might reflect something like the "new social contract" outlined by Jane Lubchenco, former president of the American Association for the Advancement of Science (AAAS). Lubchenco argued that the planet is now human-dominated and altered in unprecedented ways. The resulting environmental problems threaten the very life-support system of the earth. For these reasons she proposed that "the scientific community formulate a new Social Contract for science. The contract should be predicated upon the assumptions that scientists will:

1. address the most urgent needs of society, in proportion to their importance;
2. communicate their knowledge and understanding widely in order to inform decisions of individuals and institutions; and
3. exercise good judgment, wisdom, and humility."[9]

The Contract should "recognize the extent of human domination of the planet . . . and help society move toward a more sustainable biosphere."[10] Lubchenco thus frames the social contract for science both as a contract with at least two parties, science and the public, and as a goal for science and research. Such a contract could set the stage for full implementation of the precautionary principle particularly if society, fulfilling its side of the bargain, can redirect funding, advance a precautionary ethic, and develop appropriate public infrastructure. Similarly, science must develop institutions, training opportunities, epistemological tools, and other resources for this mandate as well as fully incorporating a precautionary ethic into research. While this is a tall order, it is feasible and would help accomplish the precautionary goal of both recognizing the extent of human domination and helping society move toward sustainability.

PUBLIC MONEY FOR PRIVATE INTEREST

If we are to adopt this new contract, the structure of research funding, particularly in the United States, will have to change from the trend toward direct public (government) support of private interests. A trilogy of interlocking laws has dominated research funding in the United States since the early 1980s: the Federal Technology Transfer Act (FTTA) was passed in 1986 to amend and build upon two laws passed in 1980, the Stevenson-Wydler Technology Innovation Act and the Bayh-Dole Act. These were geared to increase U.S. productivity in the face of foreign competition and essentially transfer the bulk of discoveries to private industry, even though the public has funded them. These laws also skew funding away from research that would provide a non-monetary public benefit.

Danielle Nierenberg[11] notes the following in an article on FTTA,

> According to the FTTA's authors, the Act was proposed in order to alleviate the impediments to technology transfers among Federal agencies, industry, public and private foundations, and non-profit organizations such as universities. The FTTA gives power to federal laboratories to enter into Cooperative Research and Development Agreements (CRADAs) with industrial organizations.

CRADAs work by encouraging the directors of federal agencies, such as the National Institutes of Health, the U.S. Department of Agriculture (USDA), or the Department of Defense, to collaborate with private institutions. These collaborations give companies of all sizes access to the resources of a federal laboratory. The highly sought-after resources are the brain trust of the federally employed researchers and scientists, as well as federally owned equipment. Federal labs may, and often do, waive ownership rights to any resulting inventions. In other cases the federal partner grants licenses to the inventions to the project collaborator.

The federal employees involved may also directly benefit from their work. Nierenberg points out that CRADAs allow federal employees or former employees to "commercialize the inventions they developed while on the government clock."[12] Moreover, the products of CRADAs, although these agreements are funded by public money, are not required to meet the broad public interest. The intent is to increase U.S. economic competitiveness by supporting private enterprise.

Perhaps the most egregious example of a technology transfer run amuck is the now infamous case of what is popularly called the "Terminator Technology." In August of 2001 the USDA announced that it had a licensing agreement (a legal instrument which sets the conditions under which a party can use something that has been patented) for the Terminator technology with the industry partner, Delta & Pine Land Co., which had helped develop it even though the public had paid for the research. As a result of joint research, the USDA and D&PL are co-owners of three patents on the transgenic technology that modifies plants to produce sterile seeds, preventing farmers from reusing harvested seed.

The current funding package for research in the United States sets up incentives for all parties to develop new products, like Terminator, with no one responsible for protecting the public interest. USDA essentially has a conflict of interest since it codeveloped the technology.

State legislatures, Congress, and other funders, such as foundations, must match the precautionary, public interest research agenda with funding. Funding should be coherent and adequate for the task at hand. It is important to support fully the necessary research so that scientists are not forced to seek private corporate money, which might create financial conflicts of interest.

At present, funding for biotechnology development far outstrips funding for monitoring, risk assessment, or developing alternative technologies. For example, in 2003 USDA solicited grant proposals to assist federal regulatory agencies in making science-based decisions about the environmental effects of introducing genetically modified organisms, including plants, microorganisms (including fungi, bacteria, and viruses), arthropods, fish, birds, mammals, and other animals excluding humans. Investigations of effects on both managed and natural environments are relevant. A paltry 3 million dollars was allocated for this research, which according to USDA, was about twice that in 2002. In contrast, Congress allocated to USDA $6.6 million for trade issues that specifically mentioned biotechnology.

In May 2003, the United States announced that it would challenge the EU's moratorium on transgenic crops. The United States makes the case that there is no science that indicates these crops could be harmful. However, the budget allocation indicates why there is no science. The funding for research into effects and consequences is so small that the science simply does not exist. Of course, USDA has more than double the amount available for risk assessment to carry out legal challenges against the science.

PUBLIC INTEREST RESEARCH

A first step in fulfilling the new social contract called for by Jane Lubchenco is to establish a public interest research agenda (see Exhibit 9.1). Scott Peters has defined public interest research as research that "aims at developing knowledge and/or technology that increases the commonwealth." He continues: "Such research requires complex problem-solving and will involve at least the economic, social, and environmental dimensions of people and natural resources. It will require that insights from these different ways of knowing be synthesized, and that an active citizenry be involved."[13] Public interest research will be identified by its beneficiaries, the public availability of its results, and public involvement in the research. The following key benchmarks identify public interest research:[14]

- The primary, direct beneficiaries are society as a whole or specific populations or entities unable to carry out research on their own behalf.
- Information and technologies resulting from public interest research are made freely available (not proprietary or patented); and
- Such information and technologies are developed with collaboration or advice from an active citizenry.

Most research done in the private interest is done for the financial gain of a limited, circumscribed group. Research done in the public interest will seldom involve such direct financial gain to the developers and will benefit a community or the commons.

A key word in this definition of public interest research is "commonwealth" and the underlying assumption that the goal of this kind of research is to invest in the public good. Public interest research is one process or method for understanding, protecting, and adding to the commonwealth. The commonwealth necessarily invokes the notion of the commons and the shared public good. The Tragedy of the Commons, according to Garrett Hardin,[15] the person who identified it as such, was the overuse of a resource because of lack of management. The favored solution to the tragedy has been privatization because it is the most efficient use of the resource and it promotes individual liberty and choice.

EXHIBIT 9.1. The key questions in defining public interest research

Identifying the beneficiary of research:

1. Whose problem is being addressed?
2. What new sources of economic and political power will emerge as a result?
3. Who benefits from any scientific uncertainty surrounding the solution?

Making results freely available:

1. How are the data and results of publicly funded research kept in the public domain? Are they made available through the Internet, public libraries, newsletters, and press releases for media stories?
2. Who decides how such results are used?

Involving citizens in research:

1. Has an active citizenry been involved in or signed off on the research?

Protecting the public from research that is not in the public interest:

1. Will new problems be created by solving an old one? Who may be harmed as a result?
2. Is science being used to delay or obfuscate action?
3. Will the citizenry and natural resources be protected by precautionary measures, if results are uncertain?

However other writers have proposed that, contrary to the Tragedy of the Commons, there is also an anti-commons, the tragedy of privatization. Heller and Eisenberg[16] describe the anti-commons as the "possibility of underuse when governments give too many people rights to exclude others. Privatization can solve one tragedy but cause another." In the anti-commons the resource is prone to underuse because of exclusion. The notion of the anti-commons is particularly relevant to seeds and the seed research that leads to patenting. Patenting is, of course, a legal tool for excluding others from use and access of the resource.

The related problem of privatization was described in the *New York Times* as a "new kind of Apartheid separating haves from have-nots"[17] because it permits the wealthy to exclude the poor from access to a life-giving resource. The story dealt with the privatization of water in South Africa and the impossibility of the poor paying for clean water. However, privatizing seed resources through technology transfers and patenting, essentially excludes the poor from access to patented foodstuffs.

Finally, privatization damages the commons because it forces problems out of the privatized arena and back into the commons as unmanaged harm (the mirror image of the Hardin Tragedy). It permits the private owner to externalize costs.[18] As such, what Michael Pollan and others have called genetic pollution (outcrossing into the commons' seed stock from transgenics) is a key example of an externalized cost and another case of the anti-commons.

The seed commons is one of the most important natural resources held in trust for all the people of the world. As a matter of international law, The Consultative Group on International Agricultural Research (CGIAR) through its Centers holds the world's largest international ex situ collection of plant genetic resources, more than 500,000 accessions that are vital for crop improvement worldwide. The seeds in question are plant genetic resources held "in trust for the benefit of the international community, in particular developing countries," and subject to conditions contained in agreements signed by the Centers and the UN Food and Agriculture Organization (FAO) in 1994 (see http://www-users.york.ac.uk/~drf1/866_sds2.htm—consulted June 11, 2003).

Significantly, the Centers do not have intellectual property rights over the material they hold in trust, and no one who uses this material can restrict access to the seeds. The agreements assure governments that the CGIAR Centers will maintain the designated collections in trust.

LOCAL TO GLOBAL

Perhaps the best place to start with a precautionary research agenda is at the local college and university level. Consider Wendell Berry's challenge to the universities to employ a specific ethical yardstick, the standard of the community's health, as the basis for its work.[19] Community health, understood as the local ecosystem (including humans), and health, understood as more than physical health, is a concrete way of recasting the new social contract into a research agenda with clear tasks and milestones. What is healthy now? What is damaged? What are the community's goals? Can the university undertake and participate in research toward these goals? Do the community's needs and goals point to a larger agenda? Do they serve the larger commons as well as the local commons? Does the research budget match the goals and priorities?

Universities are well situated to involve their communities in setting research priorities, which in turn could be the basis for budget allocations by state legislatures, and then are used to set national research priorities and funding.

Inviting engaged laypeople to help set the agenda, allocate funding, and participate in determining policies based on the results of the research might do the following two things: (1) produce better science in the public interest particularly when dealing with uncertain, complex risks and technologies; and (2) enhance public support for science. The reason for involving affected parties in the research is that this is likely to generate better science by using outside observations about emerging patterns to generate useful hypotheses or to reach a better sense of possible solutions to problems. For example, schoolchildren were the first to report the occurrence of deformed frogs in Minnesota, which has led to extensive research on frogs both in the United States and around the world.

The University of Wisconsin Center for Integrated Agricultural Systems (CIAS), which specializes in sustainable agriculture, attempts to engage the citizenry in all aspects of its research, from setting the research agenda through the research process to communicating and using the findings. Several new projects have emerged from this scientist–citizen dialogue, including the Grazing Based Dairy Systems and the Regional Food Systems studies.[20]

According to CIAS staff, their model for citizen involvement is predicated on a mutual respect of the citizens and faculty, early, frequent feedback, and academic freedom.[21] However, perhaps most interestingly, the Citizens Council reports independently to the governor and state legislature rather than to university officials. The Center's state funding depends on the report made by the Citizen's Advisory Council.

While CIAS has not done any direct research on biotechnology, the model might be useful for universities that are interested in following Berry's direction to consider the health of their communities when engaged in research.

Other institutions have developed methods for engaging the community in research. Modeled after the innovative Dutch "science shops," community-based research is a methodological tool to embed the precautionary principle in the research agenda. Community-based research is science done in collaboration between scientists and a community. It is research done for, and often with, the community.[22]

The Dutch Science Shops were created by universities as a response to community concern about social and technological issues. Professors and graduate students undertook research at the behest of labor unions, public interest organizations, and municipalities.[23] In some cases the Science Shops were storefront operations to encourage public access to research.

Community-based research satisfies the democratic participation requirement under the precautionary principle because the community helps to set the research agenda and, if appropriate, can participate fully in the research process, and benefit from the research results (Sclove 1998).[24] As Sclove and Scammell have said so eloquently, "In contrast with the prevailing undemocratic model of research, where expert concerns or market incentives drive research agendas, community-based research is rooted in the community, serves a community and encourages participation of community members at all levels."[25]

Another excellent example of community-based research is the farmer seed-breeding club in the Great Plains. This program stands in contrast to the technology transfer research method under which private companies develop patented and licensed products that are then sold by transnational corporations. It was set up to develop seeds that were adapted to the climate, soils, and agricultural practices in that region. Many companies are developing seeds that are available to farmers in any geographical region from Brazil to Iowa, but they are not bred for a specific region with its own diseases, insect communities, or soil types. The Northern Plains Sustainable Agriculture Society, seeking an alternative to hybrid or GE seeds, approached several scientists and asked for help in breeding seeds suited for organic practices on the Great Plains.

The farmers brought their own experiences and knowledge of seed breeding to the table. Many of these farmers had extensive experience breeding various varieties and species. The scientists from Canada and Minnesota had been working on some varieties of wheat, sunflowers, oats, and other crops for their entire careers and brought these seed banks and useful scientific protocols to the table. Deon Stuthman, an oat breeder, proved particularly open to working with laypeople. He went to meetings at inconvenient times, provided plant breeding information in clear, accessible language, and helped interpret the results. In many ways this was a "marriage made in heaven": farmers who were uncannily smart and scientists who were intellectually generous, all participating in public interest research. This project is still in its infancy but the collaboration between scientists and farmers promises to be an interesting alternative to biotechnology developed through CRADAs or the private domain. The collaboration had another interesting dimension; it expanded the sheer volume of research. For instance, the oat-breeding group expanded from one scientist to one scientist and fifteen farmers, all of whom could compare and contrast results of their work.

This is the heart of community-based research in a precautionary context: it tightly links goal, problem, and solution by bringing together the right people for each link in the process. It solves real problems and provides hope in an increasingly troubled world.

LEGAL CHALLENGES

A public interest precautionary research agenda will reduce many social costs of introducing a new technology. The costs of regulating and adjudicating technologies "gone bad" can be astronomical. The history of chemical pesticides suggests that these costs, particularly to the court system, can be reduced by doing the research up front.

At present, toxic agents such as tobacco, Agent Orange, or asbestos have been mired in massive class action lawsuits. The courts have not dealt with these problems effectively, partly because the science that was done was too late and too little to be useful. If we have not handled judicial disputes over chemical harm regulation how can we handle emerging technologies like biotechnology? Can we design a court system that will provide a robust justice in the twenty-first century? Can we design a court system that will be able to address emerging technologies like biotechnology and nanotechnology? I believe the answer can be "yes" if we do the research early and in a timely fashion. We cannot afford to wait to understand the potential harm of an emerging technology until after it has been in the public domain long enough for its alleged harms to become legal controversies.

The judicial community has displayed an unfortunate ignorance of the real and potential dangers associated with society's widespread use of chemicals, especially regarding the nature of scientific uncertainty. Scientific uncertainty is inherently problematic in situations that seem to call for fast and clear decisions. One result is that such decisions are easily manipulated. A judge with a particular political agenda can easily exploit the uncertainty of the science to dismiss an expert or evidence.

In the wake of a 1993 U.S. Supreme Court Case, *Daubert vs. Merrill Dow,* many federal judges are interpreting the rules of evidence to allow, or even require, them to apply strict standards to the kind of scientific evidence presented in courts. These standards do not account for real world uncertainty. By extension, the judges are taking it upon themselves to dismiss expert witnesses even before the trial begins, based on court-established and imposed standards for science.

Judges' interpretations of standards for scientific evidence suggested by *Daubert vs. Merrill Dow* and the rulings that followed are often arbitrary. In some cases they are not scientific, and in other cases they apply science inappropriately.

Legal scholar Margaret Berger argues that it is time to create a new toxic tort that would condition culpability "on the failure to develop and disseminate significant data." Berger says,

> In order to minimize risk in the face of uncertain knowledge, the law ought to concentrate on developing the required standard of care regarding a corporation's duty to keep itself reasonably informed about the risks of its products. If a corporation fails to exercise the appropriate level of due care, it should be held liable to those put at risk by its action.

It is essential to put this kind of legal mechanism in place before the consequences of new technologies cascade down through society and the environment.

According to Berger, in the cases of Agent Orange, asbestos, the Dalkon Shield, thalidomide, and tobacco, companies failed to test their products initially, failed to report problems as they emerged, and failed to do research to investigate these problems. As Berger notes, a system that encourages a "don't ask, don't tell" policy decouples liability from moral responsibility and thus threatens the basic underpinning of corrective justice.

The history of the U.S. courts and their misunderstanding of scientific processes and logic will translate into the international arena as trade challenges over biotechnology. On May 13, 2003, the United States and Cooperating Countries filed a legal case against the EU moratorium on GE foods and crops. The United States claimed that the EU moratorium was not based on science. In February of 2006 the WTO agreed with the United States.

In its legal challenge, the United States sought to use its judicial standards of scientific certainty rather than a weight of evidence approach. Since there is so little money spent on risk assessments and safety analyses are not readily incorporated into the research agenda, the science was missing in action, precisely when it was most needed.

Can we learn from the history of research and law on chemicals and develop a public interest precautionary research agenda for biotechnology that will help us understand, add to, and protect the commonwealth? We can, if we rethink our funding priorities, the role of ethics, and ground our decisions fully in a new social contract. We have not done well with our chemical policies. We cannot afford the same history of surprise and failure with biotechnology. We can avoid harm if we adopt the precautionary principle and revamp the research agenda accordingly. Our future depends on it.

NOTES

1. Raffensperger, C., Schettler, T., and Myers, N. 2000. Precaution: Belief, regulatory system, and overarching principle. *International Journal of Occupational and Environmental Health* 6: 266-269.

2. McDonough, W. 2001. Plenary Lecture, Proceedings of the Bioneers Conference, San Rafael, CA. Collective Heritage Institute. Unpublished.

3. Commission of the European Communities. 2000. *Communication from the Commission, on the precautionary principle.* Brussels, February 2, 2000. COM p. 3. http://europa.eu.int/comm/dgs/health_consumer/library/pub/pub07_en.pdf. Last accessed September 13, 2006.

4. Sharma, D. 2004. *Food for future: Trade, biotechnology and hunger.* Transcript of a talk delivered at an international conference on trade and hunger jointly organized by the National Farmers Union of Norway and the Development Fund. June 7-8, Oslo.

5. Thompson, C. 2002. Quoted on the Norfolk Genetic Information Network at http://ngin.tripod.com/270902a.htm. Last accessed September 13, 2006.

6. Lubchenco, J. 1998. Entering the century of the environment: A new social contract for science. *Science* 279: 491-497.

7. Barrett, K. and Raffensperger, C. 1999. Precautionary science. In *Protecting public health and the environment: Implementing the precautionary principle.* Edited by C. Raffensperger and J. Tickner. Washington, DC: Island Press, pp. 106-122.

8. Myers, N. 2000. Precaution and ethnobiology: Influencing the scientific agenda. *The Networker, the electronic newsletter of the Science and Environmental Health Network* 5 (4)., article III (September). http://www.sehn.org/Volume_5-4_3.html. Last accessed September 13, 2006.

9. Lubchenco, J. 1998. See note 6.

10. Ibid.

11. Nierenberg, D. 2000. Cooperative Research and Development Agreements (CRADAs) direct public science to private gain. *The Networker, the electronic newsletter of the Science and Environmental Health Network* 5 (2): (April). http://www.sehn.org/Volume_5-2_2.html. Last accessed September 13, 2006.

12. Ibid.

13. Raffensperger, C., Peters, S., Kirschenmann, F., Schettler, T., Barrett, K., Hendrickson, M., Jackson, D., Voland, R., Leval, K., and Butcher, D. 1999. *Defining public interest research: A white paper written for the science and environmental health network.* The Center for Rural Affairs and the Consortium for Sustainable Agriculture, Research and Education. http://www.sehn.org/defpirpaper.html. Last accessed September 13, 2006.

14. Ibid.

15. Hardin, G. 1968. The tragedy of the commons. *Science* 162: 1243.

16. Heller, M. and Eisenberg, R.S. 1998. Can patents deter innovation? The anticommons in biomedical research. *Science* 280: 698.

17. Ginger Thompson, May 29, 2003. Water tap often shut to South Africa poor. *New York Times.*

18. Cranor, C. 2003. What could precautionary science be? Research for early warnings and a better future. In *Precaution, environmental science, and preventive public policy.* Edited by J. Tickner. Washington, DC: Island Press, pp. 305-319.

19. Berry, W. 2000. *Life is a miracle: An essay against modern superstition.* Washington, DC: Counterpoint.

20. Stevenson, S. and Klemme, R. 1992. Updated summary of Advisory/oversight councils: An alternative approach to farmer/citizen participation in agenda setting at land-grant universities. Originally published in *American Journal of Alternative Agriculture* 7: 3. http://www.wisc.edu/cias. Last Accessed September 13, 2006.

21. Ibid.

22. Sclove, R.E. and Scammell, M.L. 1999. Practicing the principle. In *Protecting public health and the environment: Implementing the precautionary principle.* Edited by C. Raffensperger and J. Tickner. Covelo, CA: Island Press, pp. 252-265.

23. Sclove, R.E. 1995. *Democracy and technology.* New York: The Guilford Press.

24. Sclove, R.E. 1998. For U.S. science policy, it's time for a reality check. *The Chronicle of Higher Education, Issue* (October 23, 1998), pp. B1, B4-B5.

25. Sclove, R.E. and Scammell, M.L. 1999. See note 22 and p. 254.

26. Berger, M.A. 2001. Upsetting the balance between adverse interests: The impact of the Supreme Court's trilogy on expert testimony in toxic tort litigation. *Law and Contemporary Problems* 64(2 and 3), Spring/Summer. http://www.law. Duke. edu/journals/64LCPBerger. Last accessed September 13, 2006.

Chapter 10

Trade, Science, and Canada's Regulatory Framework for Determining the Environmental Safety of GE Crops

Elisabeth A. Abergel

Canada is a top exporter of agricultural goods and its economy depends largely on global market accessibility. This country's active participation in international and multilateral fora, relating to trade of agricultural biotechnology products, is mirrored in the relative importance of this sector within government. Canada is the third largest producer of genetically engineered (GE) crops and with the United States and Argentina generates 99 percent of global GE crop acreage. According to the International Service for the Acquisition of Agri-Biotech Applications (ISAAA), the global value of GE crops was estimated to be $5.25 billion. Numbers are expected to reach over $5.5 billion for 2006.[1] With such optimistic figures, the Canadian federal government is poised to capitalize on decades of public investments in this industrial sector. Key to Canada's global market expansion for GE crops is its regulatory system, namely its risk-based assessment for determining the environmental safety of transgenic crops.

This chapter examines the relationship between the demands of foreign trade and Canadian regulations regarding the environmental safety evaluation of agricultural biotechnology crops. It is in this con-

---

The author wishes to thank the Social Sciences and Humanities Research Council (SSHRC) for their support.

text that safety assessments and the regulatory principles that govern the commercialization of GE crops will be discussed in order to frame questions of decision making under conditions of scientific uncertainty. I have linked scientific uncertainties inherent in the Canadian regulatory system for evaluating the environmental release of GEOs to pressures to expand global markets for Canadian products. I conclude that since Canadian governance of agriculture and more specifically agricultural biotechnology are largely structured around the country's ability to open agricultural export markets, the principles used to regulate the commercial release of GE crops were derived for this purpose, placing emphasis on the design and implementation of scientific rules and standards leading to predictable biotechnology governance nationally and internationally.[2] Important factors to consider in this process are the levels of industry involvement and mobilization around issues of national and international regulatory regimes.[3] However, the present analysis focuses mostly on the tensions expressed in the Canadian federal government's open support for the commercial development of GEOs and its implementation of a regulatory regime that purports to be scientifically sound and precautionary.

Understood in this context, Canadian regulatory policy was designed to be operational, limiting comprehensive scientific investigations of potential environmental risks, and failing to anticipate different conceptions and perceptions of risk and ecological protection. By placing limits on the scientific understanding of ecological risks and discounting social and ethical considerations, the Canadian regulatory system has downplayed the role of scientific disputes about basic research needs for assessing the risks of GE crops as well as precautionary measures deliberated in other jurisdictions. In effect, this has meant that scientific debates about the ecological safety of transgenic crops, including scientific uncertainty, have remained caught between fundamental disagreements about the principles and norms of food security, environmental protection, and demands of trade liberalization.[4] One of the consequences of this approach has been that, instead of achieving the goal of market expansion, Canada's trade in GEOs has been compromised by the limited scope of its regulatory system.

REGULATORY AND ECONOMIC UNCERTAINTY

Canadian officials often claim that the success of their scientific risk-assessment framework lies in its capacity to expand export markets for GE products through "sound science" and a streamlined process of product approval unsurpassed internationally.[5] Yet the lack of public debate about important unresolved scientific and trans-scientific conflicts over the desirability of transgenic agriculture has had the effect of severely restricting Canada's trade of GE products. Exports of crops such as herbicide-tolerant canola have dramatically decreased while the demand for low-input agriculture has risen. Import bans, moratoria, strict labeling rules, public and political resistance to GE foods around the world are undermining the success of agricultural biotechnology. Meanwhile, countries heavily committed to this technology, such as Canada, are relying on the scientific basis of their regulatory systems, insisting on the equivalence status of GE crops and their decade of experience approving GE products, to press for the harmonization of international food and environmental safety standards. Since the creation of the regulatory framework governing GEOs, great emphasis has been placed on Canada's international commitments. Federal policymakers actively participate in worldwide negotiations concerning GE crops and have contributed in important ways to the establishment of regulatory standards developed by international organizations since the 1980s.[6] Canada is a "responsible member of the world community and a global leader in biotechnology" working to institute an international regulatory framework.[7] Since their inception in 1994, the basic scientific principles governing environmental releases of novel crops have been rooted in the language of economic growth and international competitiveness, leaving few opportunities for the Canadian public to debate the desirability of GE crops and the legislative framework governing their use. The links between Canada's regulation of GEOs and efforts to harmonize international safety rules provide an understanding of how regulatory space is conceived nationally and internationally. Exposing the types of uncertainty intrinsic to Canada's system means shedding light on global regulatory issues; especially, since the Canadian approach is promoted as a model for international biotechnology governance.

MARKET DISPUTES

Generally, the object of risk evaluation has been to allow large-scale environmental releases of GEOs for commercial purposes without specifically addressing issues associated with commercialization, such as ecological and socioeconomic consequences. Fundamental scientific questions and competing conceptions of environmental risks that GE crops might pose are not addressed by the current system, leading to greater levels of uncertainty. Decisions made by federal regulators regarding the ecological safety of transgenic crops, reflect high levels of scientific uncertainty that cannot be resolved through the current framework for assessing environmental safety. This has hindered the commercial exploitation of these crops and has led to economic insecurity for Canadian farmers about their ability to export GE products. GEOs have been commercially grown for more than seven years and cases of adverse environmental effects have been uncovered, documented, and scientifically investigated.[8] As the numbers and scale of transgene contaminations caused by commercial plantings of GE crops such as canola and maize increase, so do the levels of scientific uncertainty affecting trade in these products. Canadian farmers stand to lose the most when countries adopt protective measures that restrict shipments of GE and/or conventional crops containing GEOs. Europe has stopped importing canola seed from Canada, pending approval of GE varieties because purity cannot be guaranteed.[9]

In May 2003, under the World Trade Association (WTO) dispute settlement mechanism, Argentina, the United States, and Canada filed a formal complaint against the EU de facto moratorium on the import and cultivation of new GE crops in effect since 1998. In April 2004, the EU adopted the most stringent import rules for GEOs, requiring strict traceability and labeling provisions, allowing sale of transgenic crops. Rules demand that food and animal feed be labeled if they contain at least 0.9 percent of GE ingredients. The latest EU directives did not appease tensions between the EU and North America since they felt new rules formed a barrier against speedy GEO regulatory and market approvals. North American authorities have not required the establishment of effective identity preservation systems that would segregate conventional from GE crops, making product traceability costly and difficult. It appears that under U.S. and

WTO pressure, the Commission voted to approve a variety of GE sweetcorn (*Bt*-11) produced by Swiss giant Syngenta.[10] This ruling might signify a partial lifting of the EU ban on GEOs, allowing the sale for human consumption of clearly labeled GE corn. However, soon after this announcement, the company declared that the decision had no real financial impact and that it would "continue to sell conventional sweet corn to its customers."[11] The impact of this latest EU decision on Canadian farmers is difficult to assess; *Bt*-11 corn has been approved and sold in Canada since 1996 and has been marketed for both its insect resistance and herbicide-tolerance traits. In May of 2005, the European Food Safety Authority (EFSA) adopted a positive opinion regarding the safety of *Bt*-11 corn for use as food. Although Bt-11 sweet corn is now available for sale and purchase in the EU, it is still subject to consumer demand and cultivation of this crop remains unauthorized. What this implies is that sales of *Bt*-11 in the EU continue to be closely tied to public perceptions about GE crops in general.

The World Trade Organization panel's interim report released in February 2006 rules against the EU's restrictions on GEOs claiming that it violated WTO rules. Among some of its findings, the Panel determined that the precautionary principle, invoked in the European Commission's (EC) defense against accusations of "undue delay" in its commercial approvals of biotech products, did not constitute a strong enough principle in international public law to base panel rulings. According to the Institute for Agriculture and Trade Policy (IATP), a U.S.-based organization, this ruling is an indirect challenge to the authority of the United Nation's Cartagena Protocol on Biosafety, which strongly supports the precautionary principle for regulating GEOs in cases of scientific uncertainty. The WTO's dispute resolution panel's final report is expected in 2007, at which time, the EC may decide to launch an appeal.

To a certain extent, worldwide resistance to GE agriculture exposes the weaknesses of Canadian regulations, such as failure to address the socioeconomic consequences of mass contamination of crops like canola before approving them for commercial release. EU labeling and traceability requirements challenge Canada's deficiencies when it comes to the enforcement of a comprehensive system for labeling and segregating GE products. These rules have rendered Canadian agriculture more vulnerable to international pressure. Citizen resistance to GE foods and a lack of demand by key importers remain major hurdles to the expansion of global GE trade.

Transatlantic Battles over GE Products Highlight Market Uncertainties

Domestically, farmers and civil society organizations mounted an aggressive campaign against the introduction of herbicide-tolerant wheat, resulting in widespread public unease regarding GE foods. A broad coalition opposing the commercialization of Roundup Ready-(tm) wheat consisted of consumers, the Canadian Wheat Board, the National Farmers Union, the Saskatchewan Organic Directorate, food manufacturers, environmentalists, chefs, and others. In 2002, the Saskatchewan Organic Directorate launched a class action lawsuit against two biotech corporations, Monsanto and Bayer Crop Science, seeking compensation for damages caused by widespread contamination of organic canola crops. In addition, they filed an injunction to prevent the commercialization of Roundup Ready(tm) wheat, generating broad appeal for their cause. Outside the Canadian prairies, GE wheat also faced worldwide opposition. After an intense three-year campaign, activists claimed a partial victory when Monsanto announced in May 2004 that it would temporarily suspend plans to commercialize its product but would continue to seek regulatory approvals in Canada and elsewhere.

According to a Monsanto press release, a decision was made to defer introduction of Roundup Ready wheat for a period estimated to last four to eight years in order for the commercialization of new biotechnology traits in wheat to be achieved.[12] This decision can be interpreted in many ways: as a partial victory for activists given more time to organize resistance to GE wheat and/or as a deferral mechanism aimed at obtaining a kind of de facto global acceptance of this crop via introduction in other parts of the world. At this point, herbicide-tolerant wheat has obtained the necessary Canadian regulatory approvals. The Saskatchewan Organic Directorate is continuing its legal campaign against widespread GE contamination. Early on, the campaign against transgenic wheat got momentum from a review of the regulatory system prepared by the Royal Society of Canada

(RSC).[13] The RSC severely criticized the Framework for its lack of scientific rigor, transparency, and for its misuse of the precautionary principle. More recently, a government audit tested the capacity of the responsible agency to implement its own rules, revealing serious weaknesses.[14] These findings raise significant questions about the Canadian regulatory system's ability to protect public safety nationally and internationally. Since governments have been focusing their efforts on the establishment of risk-assessment regimes and their operationalization, they have been reluctant to strengthen their regulatory science and policy. Since the early 1980s, the federal government has maintained an explicitly pro-biotechnology stance, reflected in its policymaking. As subsequent generations of transgenic organisms of increasing complexity enter the environment and the food system, regulators will be confronted with new scientific and social challenges not addressed under current regulatory frameworks.[15] At the same time that issues of scientific complexity go unresolved and trade is further liberalized, potential environmental risks amplify. A brief review of how Canada's framework for assessing the environmental risks of Plants with Novel Traits (PNTs) facilitates commercialization follows.

CANADA'S RISK-BASED APPROACH

Early on during the development of a regulatory framework, it was agreed that existing legislation should serve as a basis for managing GEOs and that existing agencies with relevant expertise would be used. Amendments were made to existing legislation, the *Seeds Act,* to include provisions for environmental safety assessment. Originally, Agriculture and Agri-Food Canada (AAFC) was the lead department for agricultural biotechnology, but since 1997 the Canadian Food Inspection Agency (CFIA) has assumed all regulatory functions pertaining to the large-scale release of GE crops. Environment Canada plays a minor regulatory role in this process; its responsibility extends to providing a safety net for products that fall outside CFIA's mandate.

CFIA is responsible for the administration and enforcement of the Canada Agricultural Products Act, the Seeds Act, the Feeds Act, the Fertilizers Act, the Fish Inspection Act, the Health of Animals Act, the Plant Protection Act, the Plant Breeders' Right Act, the Meat Inspection Act, and the Agricultural and Agri-Food Administrative Monetary Penalties Act. The CFIA Plant Biosafety Office is responsible for regulating the environmental release of GE plants under the Plant Protection Act and the Seeds Act. In order to apply for unconfined release, crop developers must follow the guidelines contained in Dir94-08 "Assessment Criteria for Determining the Environmental Safety of Plants with Novel Traits" and the relevant species-specific companion document (see Dir94-09: Biology of *Brassica napus* L. (Canola Rapeseed); Dir94-10: The Biology of *Linum usitatissimum* L. (Flax); Dir94-11: The Biology of *Zea mays* L. (Corn/Maize); etc.).[16] Companion documents contain biological information about specific plant varieties regarding "major interactions of the unmodified or normal plant species with other higher life forms (such as predators, grazers, parasites, pathogens, competitors and symbionts and beneficial organisms and including humans where appropriate)" that exist within the production range of the novel crop and that can identify potential risks relative to a counterpart of the same species.[17] They provide technical information for use during the regulatory assessment of new varieties in terms of the biology of the plant or the novel trait.

The regulation process is product based and considers the following: the effect of the product on animal/human health; environmental impact of the product; and merit or efficacy of the product. It is designed as a step-wise process leading to the licensing, approval, and registration for commercial sale and use of all novel plant varieties. The developers of new crops must obtain authorizations for each developmental stage from laboratory, greenhouse, and confined field trials to unconfined field releases. CFIA oversees each step, providing guidelines for controlling and monitoring confined field trial sites, planting isolation, pollen containment, and postharvest land use, that can minimize environmental impact.[18] Confined field trials are conducted for a period of two years and are limited to one hectare per site and a maximum of five sites per province.[19]

Canadian regulations governing the unconfined release of GE crops are the same for all novel crop varieties whether they are derived from GE or traditional methods. Novel varieties are referred to as Plants with Novel Traits (PNTs), defined as "a plant variety/genotype possessing characteristics that demonstrate neither familiarity nor substantial equivalence to those present in a distinct, stable population of cultivated plant species in Canada and that has been intentionally selected, created or introduced into a population of that species through a specific genetic change."[20] The safety evaluation of PNTs relies on the principle that plant varieties obtained through molecular techniques do not differ substantially from their traditionally obtained counterparts. The Canadian system makes use of the concept of substantial equivalence, which considers plant species with similar novel traits (in terms of use and safety to the environment and human health) as types of the same species used in Canada and considered safe.[21] Determination of substantial equivalence is based on "valid scientific rationale" and a demonstration that the product will not differ in terms of potential adverse environmental and health effects from its traditional counterpart.[22] In cases for which species-specific companion documents exist, comparative data can be generated for a PNT and its conventional equivalent. Based on scientific information submitted by crop developers, novel plants that are substantially equivalent are considered sufficiently safe and do not undergo further safety assessments. The concept of substantial equivalence originated from food safety evaluations but was later adapted for environmental assessments by Canadian regulators.[23]

The Canadian framework draws upon regulatory principles established during the late 1980s and early 1990s. The concepts of familiarity and substantial equivalence are used to decide the level of risk posed by a new GE plant variety and whether a risk assessment will be conducted. Barrett and Abergel[24] provide a more complete history of these concepts. The term "plants with novel traits" is a regulatory category that describes new varieties developed through traditional plant breeding and/or through molecular techniques. This category was created in order to include all newly created varieties without particular reference to recombinant DNA technology and is, therefore, consistent with a product-based approach to regulations. The idea that GEOs are not inherently novel originates from a U.S. National Academy of Sciences report.[25] The framework also incorporates the concept of familiarity of GE crops with conventional crops, which is derived from a U.S. National Research Council report.[26] It was believed that familiarity could be inferred from the final characteristics of the plant and from small-scale releases, and that familiarity was generally predictable. The NAS-NRC reports made familiarity a central feature of regulations, allowing regulators to draw on past experience of introductions of plants and microorganisms into the environment.[27] Since the accumulated information from field tests provides increased knowledge about and the phenotypic expression and interactions of GEOs with the environment, it is believed that as experience of GEOs increases through approvals, entire classes of introductions will become familiar enough to avoid oversight.[28] This rationale is used to justify excluding the method of new gene introduction from regulatory scope, since it is not considered to be a source of risk. As such, risk evaluations could focus entirely on the phenotypic characteristics of new plants.

In 1993, the year of the Canadian regulatory framework, the OECD extended the notion of familiarity to large-scale releases when it issued a report on the scale-up of GE crops. Canadian regulations define familiarity as "the knowledge of the characteristics of a plant species and experience with the use of that plant species in Canada."[29] The second concept, substantial equivalence, originated from food safety concerns posed by GEOs during joint FAO and WHO consultations in the early 1990s but was elaborated further by the OECD in 1993.[30] In this case, foods resulting from molecular techniques were not considered "inherently less safe" than conventional foods, and safety comparisons between GE and conventional foods could be made.[31] The rationale for extending this principle for environmental safety assessments first came from Canadian regulators. Originally, both familiarity and substantial equivalence were not devised as replacements for comprehensive safety assessments. Instead, they function as regulatory thresholds during the approval process by defining levels of acceptable risk. Since familiarity alone provides no indication of the risks resulting from genetic modification, it deals exclusively with knowledge of a plant's characteristics and its agricultural use. In addition, since familiarity can only be gained by approving GE commercial releases, this concept directly links the industrial development of agricultural biotechnology to regulatory oversight.[32]

AAFC's definition of substantial equivalence indicates that it is a trigger point in the regulatory process for supporting the unconfined release of novel plant varieties. Plants that are neither familiar nor substantially equivalent must undergo a detailed safety assessment before being authorized for release. Conversely, plants that are determined to be familiar and substantially equivalent are considered safe. Regulators flag applications that are not familiar and/or substantially equivalent and subject them to oversight. Details about the actual process for conducting a safety assessment of plants that are neither familiar nor substantially equivalent are not provided in regulatory documents. Similarly, the scientific determination of familiarity and substantial equivalence remains vague and ill defined.

Developers of GE crops provide scientific data on the following criteria before being granted approval for unconfined release: increased weediness in agricultural settings; increased invasiveness into nonagricultural settings; potential for becoming a "plant pest" (based on an evaluation of toxicity and/or allergenicity); impacts on nontarget organisms; gene flow to related plants; impacts on biodiversity.[33] These categories provide guidelines for developers to consider during their investigation of the potential risks posed by their product in preparation for the submission of scientific data to regulators for unconfined release approval. CFIA does not specify scientific methodologies or experimental designs for collecting data about new varieties. In essence, this means that crop developers determine how products are assessed for environmental safety.[34] Since the end goal of regulation is to prove substantial equivalence, it is in the interest of proponents to focus their investigations on providing regulators with as much information regarding the similarities of their products to natural counterparts, likely failing to observe potential differences. To support their claim of substantial equivalence, applicants may submit confined field data, variety trial data, experimental data, and/ or data from the scientific literature to regulators. In some cases, if valid scientific rationale is provided, additional evidence of equivalence may not be necessary and an application can be approved without experimental data.

Regulators review the information supplied by proponents within a period of ninety days, establish the acceptability of the risks posed by the PNT, and issue a decision. Products can be approved with or without conditions or they can be refused approval with or without conditions.[35] Evaluators can call on scientific experts for guidance; they can also request additional information from proponents in case of a submission with insufficient or inconclusive data.

Each PNT submitted for regulatory review is evaluated on a case-by-case basis, each approval is considered independently; however, previously approved products that have been commercially released establish new safety baselines for newer crop varieties. Once a novel trait has been considered familiar and substantially equivalent, it is deemed to pose an acceptable risk. This allows the gradual elimination of whole categories of novelty, as they are no longer novel, but may also eliminate potential risks from regulatory scope as more GEOs are approved for commercial release because they are now seen not to pose risk. CFIA issues a decision document containing general conclusions about the assessment. Decision documents are made public, but the scientific data reviewed by regulators is considered confidential business information and remains undisclosed. This document does not provide specific details about the experimental data or the rationale used for arriving at a decision of safety, nor does it elucidate how risk acceptability is negotiated.

Variety registration is the last step before a novel crop can be marketed. It is not intended to provide information about environmental risks, and does not constitute an appropriate ecological assessment of complex ecological GEO interactions, but, in at least one documented case, data from varietal and field trials were used in the approval application for herbicide-tolerant GE canola. The variety trials used to test agronomic performance, assess disease, and merit quality provided field observations that led regulators to conclude that, in this particular instance, GE canola would not have any adverse effects on beneficial insects.[36] Although this may be an atypical case, seed from a PNT undergoing safety evaluation can enter plant variety registration trials before it has been approved for unconfined release and that registration data, which is required from the seed industry, can provide information about ecological impact.

The report by the Auditor General of Canada[37] suggests that no systematic mechanism for post-approval monitoring exists to verify compliance by growers and developers when conditions for unconfined GE crop release have been imposed by CFIA. Developers of commercially grown GE crops must report any adverse environmental effects to the agency, although it is not clear how contingency plans are provided to producers who remain liable for any environmental damage arising from use of GE crops. Under imposed conditions for unconfined release, as in the case of insect-resistant plants with novel traits, the Auditor-General's report found that the procedures for verifying compliance with insect-management plans were incomplete.[38] In addition, current gene-containment practices, such as refuge strategies for *Bt* corn, have not prevented the occurrence of transgene flow or insect resistance.[39]

HERBICIDE-TOLERANT CANOLA

A 1995 submission for the unconfined release of herbicide-tolerant (HT) canola was made to regulators as a test of the newly developed regulatory system. This particular variety of HT canola was produced using GE technology and was intended for use as animal feed, and subsequently in the production of oil for human consumption. Regulators evaluated the novelty status of this new PNT in terms of: an environmental safety assessment (Seeds Act); a livestock feed safety assessment (Feeds Act); and a food safety assessment by Health Canada (Food and Drug Act). The crop then had to go through a variety registration step in order to be commercialized. In spite of the fact that this was the first variety of HT canola to go through the Canadian regulatory system, it was not considered novel. Herbicide tolerance was considered familiar even though the genes responsible for its expression were originally derived from bacteria. In addition, this variety was considered substantially equivalent in spite of the fact that it contained bacterial gene sequences that differed from those found in non-GE HT canola plants that were known to be safe. As such, this GE variety was approved for unconfined release and later for commercialization. This approval provided the basis for the approval of additional lines of HT canola with different genetic sequences expressing the same trait.[40] Since then, Canada has approved sixty GE crop varieties for unconfined release.

SCIENTIFIC UNCERTAINTY
AND THE CANADIAN REGULATORY FRAMEWORK

The ecological concerns raised by the use of GEOs in agricultural production can be summarized in two broad categories: the potential effects of transgenes and their phenotypes on the environments where they are released, and the possible consequences of transgene escape.[41] Disagreements about the most appropriate risk-assessment methods for accurately predicting long-term ecological impacts persist in the scientific literature.[42] According to Wolfenbarger,[43] "At issue are the quantity and quality of data collected prior to commercial release, and . . . the extent to which these data will predict negative consequences." Questions about experimental design, data needs, the impact of noticeable environmental changes, and the degree of uncertainty for predicting the risks of commercial releases remain controversial. No consensus exists about the types of risk that should be evaluated by assessment procedures. Experimental designs for measuring both high- and small-magnitude effects contain high degrees of scientific uncertainty.[44]

The uses of familiarity and substantial equivalence for the environmental risk assessment of GEOs in the Canadian system raise important scientific questions. The design of a product-based approach, the decision to use existing legislation, and the regulation of all new plant varieties under a single category of novelty, PNTs, combine to establish GE as an extension of traditional breeding practices. As a result, the risks posed by GE technologies evade regulatory oversight. Since the late 1980s when both NAS and NRC reports providing the rationale for the Canadian approach were published, knowledge about the intrinsic risks of genetic modification has expanded.[45] Since GE produces a greater range of phenotypes and genotypes previously unseen in nature, these can lead to unpredictable environmental consequences. Since 1996, large-scale commercial GEO releases have provided valuable information, documented in the scientific literature, about ecological effects.[46] Although their significance and magnitude remain contested, these findings point to the need for research in this area.

In an unprecedented way, Canada extended the use of substantial equivalence to environmental releases. This concept establishes baselines of hazard acceptability by comparing GEOs with those already used and considered to be safe in Canada. It is claimed that familiarity and substantial equivalence provide a scientifically sound basis to evaluate the safety of biotechnology products in an efficient step-wise process. Knowledge gained about the crop plant, the trait, the environment, previous modifications of the crop or trait, and interactions among the trait, the crop, and the release environment can all be used to justify increasing the scale of GE crop releases through a step-by-step process from small-scale field trials to large-scale unconfined release. This approach is necessarily limited in scope since it fails to address the unintended consequences of GEO releases, focusing instead on prior knowledge; besides, as previously mentioned, the implementation of substantial equivalence and familiarity remains scientifically vague.

The Royal Society of Canada's Expert Panel on the Future of Food Biotechnology, Canada's premier expert scientific body, was extremely critical in its 2001 report about Canada's regulation of food biotechnology. In particular, it criticized the government's use of substantial equivalence on the grounds that it is ambiguous and ill defined. Serious doubts about the science of establishing substantial equivalence were raised, since no clear methodologies define it. Moreover, the lack of independent evaluation of either the quality of the data or the statistical validity of the experimental design used to collect data makes it impossible to determine whether information requirements are actually met during approval process.[47] The Panel also found it impossible to ascertain the degree to which the regulatory process was founded on sound science. On questions relating to the transparency of regulatory procedures, the Auditor General of Canada's report, another critical assessment of the Canadian system, states: "The Canadian Food Inspection Agency should consider clarifying the regulatory framework for plants with novel traits to strengthen its ability to effectively deliver its regulatory program."[48]

The focus on substantial equivalence and establishing equivalence scientifically means that many assumptions made about the safety of GEOs and their unanticipated characteristics are left unverified.[49] Novel traits assessed independently from the plant variety may not pose safety threats in themselves. Most plant varieties tend to be familiar. It is the combination between new traits and the plant that need to be evaluated for determining safety. The Panel adds that there is logical confusion between familiarity and substantial equivalence; substantial equivalence is seen as the "most critical element of the regulatory process" yet its scientific determination as a proof of environmental safety remains unclear.[50] This is disconcerting since risk evaluation hinges on this principle.

The Royal Society claims that because expectations about GE plants derive from experience with traditional breeding, comparison of GE phenotypes with check varieties says little about how genetic modifications may be expressed phenotypically. This is because successful breeding programs essentially yield similar plants with expected characteristics where "barley is barley is barley." In addition, no simple method exists to assess the specific contribution of each genetic difference to the new phenotype. Comparisons between new genotypes and existing varieties (referred to as "test" or "check" varieties) have become established practice in order to determine the quality and performance of new plant varieties against established agronomic standards. Testing normally includes laboratory analyses of harvested plant parts, as well as comparative field performance data from test plots grown at multiple sites over several seasons. The general assumption is that "while the new variety will not be identical to existing germplasm (otherwise no improvement would have occurred), it does meet the expectations for the crop in question, and offers some enhancement of one or more trait."[51]

The scientific content of risk-assessment regimes adds to the uncertainty of evaluating unconfined releases, particularly because the validity and comprehensiveness of the scientific data submitted by developers for establishing biosafety is unclear. Details of these scientific studies were discussed in the Royal Society report and in a study of information obtained through an access to information request, where safety claims used for the unconfined release of HT canola were examined.[52] This data generated by industrial developers lacked consistency and some of the methodologies used to answer safety claims were questionable. Scientific data suffered generally from intrinsic weaknesses due to small numbers of replicate studies and limited spatial and temporal scales used to detect subtle environmental effects. Unreliable statistical analyses regarding sample size, variability, and sampling techniques exacerbated these problems, and there were uncertainties about data collection methods used to determine risk factors, such as effects on nontarget organisms, impacts on biodiversity, and how increased invasiveness into natural settings could lead to conclusions of safety. In other instances, required scientific data for specified criteria were not included in the submission. Generally, risk-assessment results tended to be very broad and conclusions of acceptable risk drawn from extremely limited and inconsistent data sets. What is more, complex ecological interactions and cumulative effects could not have been addressed by the scientific studies presented to regulators. Known risks, such as canola's potential for invasiveness, outcrossing, and weediness, were not considered extensively in the assessment.[53] It seems that the Canadian case is not isolated, since studies performed by Marvier[54] and Gurian-Sherman[55] in the United States corroborate some of these findings. The Canadian system did not use a sufficiently precautionary approach in its science-based risk-assessment regime.[56] This means that risk science was defined very narrowly, that the decision-making process was constrained by a lack of openness, transparency, and inclusiveness and, more importantly, by economic priorities. These procedural weaknesses did not prevent the approval of this GE variety for unconfined release. The Royal Society of Canada report recommended that, at the very least, the scientific data reviewed by Canadian regulators should be subject to independent peer review to assess and maybe improve the data, experimental design, and interpretation of results used to argue that GE crops are as safe as the non-engineered varieties. Evidently, ontological positions asserting that GEOs are the same in kind as conventional crops and that GE does not pose new risks do constitute barriers to more comprehensive and precautionary science. While this case may not be representative, it provides important insights regarding scientific uncertainty, in particular, about the interpretation of scientific evidence in the decision-making process. Most importantly, it reveals how the regulatory system restricts scientific knowledge at the expense of expediency and economic commitments.

Another important factor contributing to uncertainty in the regulatory process deals with the costs of preparing regulatory submissions. Since developers bear the responsibility of establishing the substantial equivalence of their crops, they may favor short-term evaluations of possible environmental impacts.[57] This means that the scientific studies submitted for regulatory review pit the protection of a developer's investments against the need for comprehensive scientific analysis.

As a result of the Royal Society report, the federal government promised to address some of the scientific issues raised by the Panel. In 2002, the Canadian Biotechnology Advisory Committee (CBAC) expressed concerns, echoed by the Royal Society, regarding the lack of transparency of regulatory decisions. However, CBAC endorsed the use of substantial equivalence stating that it should be used as a guide for identifying differences between GE and conventional crops that can then be assessed for human and environmental safety considerations.[58] Various civil society groups have also challenged the Canadian government's regulatory process on the grounds that no meaningful public debate and involvement about the commercial use of GE crops has taken place. They note the speed at which GE food has spread; the lack of comprehensive mandatory labeling of GE foods and products; segregation of GE crops; coexistence with conventional crops; and concern for the survival of low-input agriculture. Like the Royal Society, this broad coalition of environmental, social justice, consumer groups, scientists, and organic farmers criticize the close relationships between regulators and industrial developers. Lack of public confidence in the regulatory process and critical reports published by the Royal Society, CBAC, and the Auditor General of Canada have prompted responses by CFIA, which is still addressing some of their findings. As previously mentioned, Monsanto's decision to pursue regulatory approval of HT wheat, while abandoning plans to suspend commercialization for four years, has also been a catalyst for mobilizing Canadian civil society organizations as well as groups of commodity producers concerned about losing export markets. This constitutes an important case for the regulatory system since both Canadian wheat producers and major international buyers have declared their refusal to purchase this product.

INTERNATIONAL TRADE
AND REGULATORY HARMONIZATION

Canada's regulatory framework for assessing the relative safety of GE crops was created in 1993. During the early phases of the development of the Canadian regulatory framework, Canadian government officials met regularly with their U.S. counterparts. Terry Medley, then a USDA official involved in designing U.S. regulations, provided Canadian regulators with important advice.[59] Medley advocated the development of strategic regulations, based upon logical reasoning rather than empirical methodologies that would balance regulatory oversight with the needs of the biotechnology industry.[60] He also supported a system whereby performance standards rather than rigid controls or specific designs for compliance should be used to govern biotechnology innovations. This forms the basis of the risk-based approach adopted by Canadian regulators, permitting regulation and state promotion of agricultural biotechnology for trade purposes while allowing public and environmental safety to be reconciled.

Clearly, Canadian policymaking in the area of biotechnology and the speed of introduction of GE crops encouraged their rapid commercialization long before social and ethical issues could be publicly debated. In fact, the government makes it clear that the purpose and scope of regulations is part of the "continuum of research, development, evaluation and commercialisation of plants with novel traits."[61] This strategy failed to recognize and anticipate public unrest[62]; it also underestimated the impact of "scientific-ethical discontent"[63] in Canada and abroad.

In addition, by delaying public discussions about the potential risks of GE agriculture and eliminating the possibility of postponing large-scale environmental releases, policymakers, in their rush to approve GE products, downplayed the potential environmental and human health risks posed by the technology and the lack of scientific studies about them. The main focus of policy efforts from the early 1980s to the late 1990s was to reconcile "scientifically defined risks with competitiveness in international markets" at the expense of identifying key issues associated with the introduction of GE technology.[64] The effect of this policy approach has been to marginalize any discussion relating to questions of scientific as well as nonscientific uncertainty. It is for this reason that many questioned the Canadian government's use of science to regulate the safety of GEOs. The Royal Society Panel points to the high degree of scientific uncertainty of the system in place, particularly as it pertains to the concept of substantial equivalence. It made several recommendations for improving Canada's regulatory system for GE crops and public confidence, including a rigorous scientific assessment of potential environmental and human health risks, an open consultation process with scientific experts and a more precautionary approach that does not view GE products as safe until proven otherwise.[65] The CBAC, the government appointed advisory body made up of stakeholders, disagrees with the Panel's findings concerning substantial equivalence; however, it does ask for more transparency during decisionmaking.[66] The CFIA has begun public consultations on ways to improve regulations of GE crops.[67] In 2002, CFIA held a public consulation aimed at reviewing and updating its guidelines and directives for approving novel foods, plants with novel traits, and livestock feeds derived from plants with novel traits. Since then, it has held consultations on plant molecular farming and animal biotechnology. The outcome of these expert and multistakeholder talks, while not explicitly clear, is the periodic amendment of existing legislation and guidelines for regulating GE products.

CFIA is committed to periodic revisions of its regulatory framework as new developments arise. CBAC is the main channel for public input into biotechnology policy.

In 2000, a Government of Canada task-force report[68] on the consolidation of departmental regulatory and inspection functions stated:

> At the international level, harmonisation is critical to facilitate trade in global markets. Harmonisation initiatives vary from setting common standards to agreeing on common technical documents for pre-market submissions to mutual recognition agreements, where the parties will accept the results of a review and approval process in another country. Progress is very slow in these areas but nevertheless the contribution from Canadian regulators is significant.

The report also urges regulators not to focus solely on scientific data when assessing risks; they are told that other factors such as environmental and trade impacts must be considered in the evaluation of GE foods. Although left unclear, this signals an expanded definition of GE risk assessment, which is no longer simply science based.

The Department of Foreign Affairs and International Trade (DFAIT) plays a strong leadership role in the promotion of Canadian biotechnology "in the global marketplace."[69] In particular, the Technical Barriers and Regulations Division works to identify "international barriers to trade as they arise and works to reduce those barriers through negotiation, making representations in various international fora and organisations, and by pursuing outreach efforts that help build fair, transparent regulatory capacity in other countries."[70]

Industry Canada is another key departmental agency that oversees the development and marketability of agricultural products derived from recombinant DNA technology. It is responsible, with DFAIT, for planning the Canadian Biotechnology Strategy (CBS), which guided government policy as biotechnology products evolved from the laboratory to confined and unconfined field trials in order to reach commercialization. Canada currently chairs the committee on development of international standards for foods derived through biotechnology and their labeling through *Codex Alimentarius*. Its participation in Codex is managed by an interdepartmental committee made up of senior officials from Health Canada, CFIA, DFAIT, AAFC, and Industry Canada. It is also the host country for the Codex Committee on Food Labelling (CCFL). The Canadian Codex Committee is involved in a project to create procedures for the alignment of Canadian requirements for food safety and public health to Codex standards. Currently, nine international bodies coordinate efforts related to biotechnology governance.[71] Canada has long been a supporter of free trade regionally and globally; it is a member of several trading blocs such as the North American Free Trade Agenda (NAFTA), the Asia Pacific Economic Council (APEC), and the World Trade Organization (WTO). Canada has sought to export its regulatory system along with its GE products, claiming that the environmental and health safety framework it developed is a model for the international community. Fittingly, its policies have been orientated toward world trade, and its economy has had to adjust to the demands of the global market. Like its NAFTA partner the United States, Canada has been a key player, promoting and reinforcing the position of its biotechnology-based industries, disputing the acceptance and desirability of GE crops at the global level.

As expected, Canada has played an important role, defending the worldwide development and commercialization of agricultural biotechnology in different international regimes such as the WTO, the Convention on Biodiversity (Cartagena Protocol on Biosafety), the Organization for Economic Cooperation and Development (OECD), the United Nations Food and Agricultural Organization (FAO), as well as the UN World Health Organization (WHO), including their joint safety standard body *Codex Alimentarius.*

In addition to its membership in major international organizations, Canada is signatory to several bilateral agreements on issues related to trade in biotechnology such as the Canada-Chile Free Trade Agreement (CCFTA). Fittingly, its policies have been orientated toward the market liberalization and market accessibility of biotechnology products. Canada has argued for the worldwide standardization of private intellectual property rights (IPRs) for biotechnology and its products.[72] The IPR rules under the WTO stipulate that members enact patent legislation for the products and processes of biotechnology. Microorganisms cannot be excluded from patentability but plants and animals may be exempt. Unlike the United States, Canada has not yet permitted the patenting of Higher Life Forms despite repeated challenges to its patent legislation. In May 2004, the Supreme Court of Canada ruled in favor of Monsanto Co. in a case involving a Prairie farmer accused of infringing patent rights when canola plants grown on his farm were found to contain Roundup Ready genes. Critics feel that this judgment carefully avoided the issue of transgene contamination, focusing mostly on the rights of patent holders and their opposition to Canada's current IPR rules.

Canada's involvement in various international fora facilitates access to international markets, "which in turn fosters the development of Canadian biotech industries."[73] Clearly, the ability to open markets has been a key component of biotechnology policy and regulatory development. As Newell suggests, it is the relationship between state regulators and market actors that structures the governance of biotechnology products and their development.[74] This directly impacts the kind of science used to construct regulations. The scientific authority of regulations is used to justify seemingly conflicting mandates: maintaining a balance between increased trade of GEOs and precaution. Science for trade purposes (or "trade science") satisfies the demands of industry, while appearing to be responding to public concerns. The adoption of a risk-assessment framework for biotechnology by focusing attention on *acceptable* risks routinely allows the commercialization of GEOs in the face of uncertainty and insufficient scientific evidence.[75] GE crop approval processes are streamlined and developers of new crops essentially define the scientific processes for establishing the safety of their products, restricting their experimental design to the determination of substantial equivalence.

Since the largest market actors tend to have a transnational focus, biotechnology regulations have been globally orientated. According to Newell, patterns of globalization have influenced the governance of biotechnology and have uncovered how ideas such as the environmental risks posed by GE products might be dealt with at a policy level. Agricultural biotechnology regulations generally involve "a complex mix of advisory bodies, committees, professional bodies and industry associations operating at international, national and subnational levels."[76] Despite this institutional confusion, he claims that regulatory functions such as risk management, under pressure from industry, are designed specifically to keep markets open to GEOs and provide regulatory predictability for GE products.[77] This in turn has meant that important scientific assessments have been rendered unnecessary and perceived as barriers to commercialization.

In terms of international commitments, Canadian regulatory officials routinely participate in international standard setting bodies, which, according to CBAC, should not be confused with Canada's trade functions. The CBAC report[78] suggests that

> CFIA also plays a role in international negotiations. However, it is important to distinguish, on the one hand, between the involvement of Canadian regulators in CODEX, OECD, FAO/ WHO and other international bodies in the development of scientific and technical standards related to environmental or food safety assessment and, on the other hand, international activities such as trade missions or negotiations that deal with matters bearing on the nature and direction of Canada's regulatory policies.

While the report acknowledges that participation in international trade negotiations might potentially be construed as being in conflict with the primary role of regulators, it is not clear how those two activities differ. According to Newell,[79] "Global trade rules appeal to industry because they effectively narrow the menu of regulatory choices open to governments." Indeed, pressure to commercialize becomes apparent in the scientific principles used in regulations.

SMART REGULATIONS

The government of Canada has been promoting the idea of "smart government" and "smart regulations" as a means of developing a consolidated approach to regulatory decisions in the area of food, drugs, and environmental protection. According to the Canadian Treasury Board: "The Government of Canada has challenged regulatory departments and agencies to seek ways to use their regulatory frameworks to contribute to health, sustainability, innovation and economic growth, while at the same time reducing the regulatory burden on businesses. This is known as the Smart Regulation Strategy."[80] New legislation, known as the "CFIA Enforcement Act" introduced in Parliament in November 2004, seeks to streamline regulatory approvals of new plant varieties. In an effort to harmonize food, drug, and environmental standards with the United States, these so-called smart regulations would provide a single review and approval process for new biotechnology products for the North American territory. Bill C-27 would consolidate the inspection and enforcement powers of CFIA by integrating Canadian and U.S. regulatory processes for food and agriculture. Reviews of current Canadian practices are planned for the next three to five years in order to attain consensus standards with U.S. authorities. This initiative was prompted by food crises such as Bovine Spongiform Encephalopathy (BSE) and avian influenza and the "continuing possibility that animals or food may be targets for bioterrorism."[81] Although the proposed Bill was abandoned when a new Conservative government was elected in 2006, Smart Regulations remain as a broad-based, cross-sectoral policy objective now known as the "Smart Regulation Initiative." In principle, Smart Regulations would allow the CFIA to accept testing and certification results from other countries. With regard to GE crops, this initiative would not redress weaknesses present in the regulatory system. In fact, these are likely to be compounded because of CFIA's entrepreneurial mandate.[82] Under this strategy, a variety of acts, regulations, standards, and other forms of governance in the area of food safety would be streamlined and consolidated to be more in line with

Canada's largest trading partner, the United States. It is not clear how the current system of GEO regulation would be affected; however, it is clear that Smart Regulations, when implemented, will influence the design of new regulatory regimes for second- and third-generation biotechnology products as well as GE animals. Smart Regulations will also strengthen Canada's efforts to push for international regulatory harmonization in areas such as biotechnology by eliminating interjurisdictional inconsistencies and gaps. This regulatory streamlining will set up a system whereby potential security threats from outside could be identified and anticipated. This would be based on linkages and information sharing between decision makers in foreign governments and international organizations. Presumably, regulatory decisions could be based on foreign decisions and approval processes, avoiding lengthy delays for businesses. While domestic GEO regulations tended to focus on managing risks from within, with an emphasis on public health protection, smart regulations, by removing national standards, may erode Canadian sovereignty in the area of food safety and regulations. This may also eliminate the safety checks that come when multiple assessments are carried out by different jurisdictions, which dictate different context-specific conceptions of GE risk.

Clearly, international trade and the complex integration of the global food system have become a major source of unintended and potentially devastating hazards. Trade protectionism of agricultural goods, based on food safety concerns, has been a major source of disagreement between countries.

Framing food safety concerns under a security agenda does nothing to hinder trade; on the contrary, it is believed to eliminate trade barriers by weakening national protection standards and establishing more efficient product movement and tracking. Accordingly, it will be increasingly difficult to invoke scientific uncertainty as a precautionary mechanism against potentially harmful practices, especially when trade-inspired regulatory principles become the cornerstone of multilateral and bilateral agreements. The federal Government Directive on Regulating, the body charged with implementing Smart Regulations, stresses the importance of trade within the design of technical regulations. It explicitly guards against adopting technical regulatory requirements that would restrict trade "any more than necessary to fulfil the intended policy objectives" and more specifically, it promotes regulations that focus on performance rather than "design or descriptive characteristics."

PRECAUTION AND SCIENCE-BASED REGULATIONS

The Canadian regulatory approach, because it relies on the OECD-inspired concept of substantial equivalence, is consistent with a WTO version of precautionary action that places trade foremost in safety considerations. This also explains why many of the principles and standards used in development of international biotechnology governance originated from market-orientated organizations.[83] As a signatory to the Cartagena Biosafety Protocol (CBP) and an active biotechnology player, Canada's regulatory framework for assessing the risks of GEOs may not fully reconcile some of the disagreements existing between the CBP and WTO rules. The following section highlights the main areas of controversy.

The precautionary principle generally provides guidance on decision making under circumstances of uncertainty. As such it contains "an admonition to prevent damage to public health and the environment under conditions where (i) potential harm has been recognised, and (ii) there remains scientific uncertainty regarding the possible extent and nature of harm that may occur."[84] The use of the precautionary principle has been central to international biotechnology governance. In particular, its relationship to risk assessment for GE crops has been included in WTO rules (Article 5.7 of the Sanitary Phytosanitary Agreement, SPS)[85] and the Convention on Biological Diversity (CBD) also known as the Cartagena Biosafety Protocol (CBP).

Under WTO agreements, both the Sanitary Phytosanitary Measures (SPS) and Technical Barrier to Trade (TBT) apply to GEO import regulations. Also relevant is the General Agreement on Tariffs and Trade (GATT), which covers trade provisions such as nondiscrimination of imported products that relate to GEOs. The SPS Agreement contains rules about how governments can apply food safety and animal and plant health measures as they relate to contaminants, but also product labeling and GE foods. Moreover, SPS builds on previous GATT rules to restrict the use of unjustified sanitary and phytosanitary measures for the purpose of trade protection.[86] Article 5.7, while implicitly recognizing the need for precaution, does not adequately recognize it as a rational and scientific response to unacceptable risk and scientific uncertainty.[87] The SPS Agreement is consistent with international standards set by *Codex Alimentarius* and the International Organization for Standardization (ISO). The TBT Agreement deals with mandatory product regulations, voluntary standards, certification procedures, and so on and, developing procedures for assessing their conformity if they are not covered by the SPS Agreement. Both SPS and TBT are designed to limit the adoption of national trade-restrictive measures. For instance, the 2003 WTO challenge against the European GEO moratorium charged that the EU's latest labeling and traceability rules constitute trade-restrictive measures akin to protectionism The CBP, on the other hand, governs the transboundary movement of living GEOs as it relates to biodiversity, and is based on the precautionary principle.[88] Provisions for Advanced Informed Agreement (AIA), risk assessment, and decision-making procedures allow an importing country the right to assess the safety of GEOs. It can restrict entry of potentially harmful GEOs even when scientific evidence of environmental damage is lacking. Canada is a signatory to the CBP and ratified it in 1992.

Clearly, the relationship between the WTO and the CBP remains ambiguous, and this tension is reflected in the Canadian system. The SPS Agreement and CBP illustrate different interpretations of precaution. Coleman and Gabler[89] identify three areas of disagreement and incompatibility between the norms and principles included in the WTO and the CBP that compete in terms of risk-assessment regimes for declaring GEOs safe. Disputes include the use of precautionary science or strict science-based risk analysis for establishing food safety and "the relationship between the trading rules and trade measures adopted in pursuit of biodiversity objectives" in various international agreements.[90] The CBP questions the use of science-based risk assessment as the only legitimate approach for determining safety, it advocates strong precautionary measures for maintaining biodiversity, while the WTO, on the other hand, relies solely on a restricted definition of sound science based on a different version of the precautionary principle.[91] While the CBP can use scientific uncertainty, lack of scientific evidence, to postpone or restrict the use of GEOs, the WTO SPS Agreement, requires "sufficient scientific evidence" in order to prohibit trade.[92] Moreover, under the CBP, biodiversity over-rides trade in terms of its guiding norms and principles while the converse is true under WTO rules.

According to Thomas, "the Biosafety Protocol is an improvement on the SPS Agreement because it does *not* assume that precautionary decisions are provisional and therefore does *not* require those taking precautionary action to seek further information and review the decision within a 'reasonable period of time.'" In addition, the Protocol places the burden of seeking additional information not on regulators but on those proposing an activity, unlike the SPS Agreement, thus augmenting the status of the precautionary principle in international law.[93] Coleman and Gabler[94] suggest that the European system of risk assessment is closer to the biodiversity view "that sees scientific determinations of risk thresholds and related assumptions about what constitutes harm as highly subjective" while the Canadian framework is closer to the WTO definition. The CBP places the burden of proof on the exporters of biotechnology products who tend to be the strongest supporters of liberalized trade and the largest of GE crop producers. Nevertheless, even the CBP has been criticized for its market-orientated approach, in particular concerning benefit-sharing provisions resulting from the commercial exploitation of genetic resources.[95]

Trade disputes involving European restrictions of North American GE products and their regulatory approval highlight the different conceptions of risk assessment and regulatory review for determining the environmental safety of GE crops.

One of the challenges of globalized GEO trade has been to reconcile the need for international agreements and the need for flexible regimes that would acknowledge diverse social, cultural, environmental, and economic conditions and values.[96] The different positions adopted by countries concerning the safe use of GEOs has sparked global battles against biotechnology and the U.S. hegemonic governance of biotechnology has led to conflicts.[97] Developing countries have been caught in these controversies; the recent WTO court challenge pitting the United States, Canada, and Argentina against the EU is believed to act as a disincentive for developing countries to adopt the European regulatory approach.[98]

CONCLUSION

This chapter establishes links between regulation and trade, arguing that regulations were designed principally to assist trade liberalization of GE agricultural products. The CFIA is the lead agency responsible for determining the environmental safety of GE crops. Its mandate is to ensure both public health and safety and to facilitate trade by encouraging food exports through the advancement of free-trade agreements. In doing so, the CFIA performs GE safety evaluations applying a "science-based" risk-assessment approach broadly designed not to impede commercialization while seemingly addressing public concerns. However, guiding principles such as familiarity and substantial equivalence, while providing the technical rationale for Canadian regulations, have been severely criticized for their lack of scientific rigor and consistency. The free-trade focus of Canadian regulations explains the limited scientific scope of the approval process and the high degree of uncertainty contained in the principles used to reach environmental safety decisions. Since the early implementation of biotechnology, market-orientated institutions, such as the OECD, have played a major role in defining the terms of biotechnology governance,[99] and Canada has actively participated in these deliberations.

The market reliability of the Canadian system enables CFIA to perform its dual mandate and makes it an attractive model for international harmonization. Given that much of the controversies surrounding the global spread of transgenic crops involve trade-related issues, it is apparent that the success of GE crops relies on persuading trading partners of the expertise of Canadian regulators as a guarantee of the safety of exported agricultural products. By promoting sound science-based decisions and policies at the international level, the federal government hopes to "combat discriminatory regulatory action in foreign markets against Canadian exporters and strengthens Canada's reputation as a supplier of safe, high-quality products throughout the world."[100]

However, the federal government's unwillingness to address fundamental issues associated with the use of GEOs, through control of public biotechnology discourse, reliance on ambiguous and opaque scientific principles and guidelines, exacerbates questions of scientific and social uncertainty, damaging Canada's reputation. Allowing commercial trade to prescribe the scientific boundaries of GEO rules undermines the advancement of knowledge in the areas of environmental protection and food safety. Furthermore, it contributes to the gradual erosion of governmental capacity in the area of GEO risk mitigation and prevents the constructive use of scientific uncertainty as an entry point for developing comprehensive and democratically responsible technology policy. By implementing a "risk-management" framework rather than a "risk-prevention" approach that uses the precautionary principle to determine acceptable levels of uncertainty (scientific, social, etc.), Canada may be trading away its ability to protect both its public and environmental health.

NOTES

1. James, C. 2005. Executive Summary of Global Status of Commercialized Biotech/GM Crops: 2005. ISAA Briefs No. 34. ISAAA: Ithaca, NY.

2. Coleman, W.D. and Gabler, M. 2002. Agricultural biotechnology and regime formation: A constructivist assessment of the prospects. *International Studies Quarterly* 46: 481-506.

3. Newell, P. 2003. Globalization and the governance of biotechnology. *Global Environmental Politics* 3: 56-69.

4. Coleman, W.D. and Gabler, M. 2002. See note 2.

5. Canadian Biotechnology Advisory Committee. 2002. *Improving the regulation of genetically modified foods and other novel foods in Canada.* http://cbaccccb.ca/epic/internet/incbac-cccb.nsf/vwGeneratedInterE/ah00186e.html. Accessed November 2003.

6. Barrett, K. and Abergel, E. 2000. Breeding familiarity: Environmental risk assessment for genetically engineered crops in Canada. *Science and Public Policy* 27: 2-12.

7. Canadian Food Inspection Agency. 2004. http://biotech.ic.gc.ca/epic/internet/incbs-scb.nsf/en/by00208e.html. Accessed May 2004.

8. Rissler, J. and Mellon, M. 1996. *The ecological risks of engineered crops.* Cambridge, MA: The MIT Press; Sears, M.K., Hellmich, R.L., Stanley-Horn, D.E., Oberhauser, K.S., Pleasants, J.M., Mattila, H.R., Siegfried, B.D., Dively, G.P. 2001. Impact of *Bt* corn pollen on monarch butterfly populations: A risk assessment. *Proceedings of the National Academy of Sciences USA* 98: 11937-11942; Snow, A.A., Pilson, D., Rieseberg, L.H., Paulsen, M., Pleskac, N., Reagon, M.R., Wolf, D.E., and Selbo, S.M. 2003. A *Bt* transgene reduces herbivory and enhances fecundity in wild sunflowers. *Ecological Applications* 13: 279-286; Dale, P.J., Clarke, B., and Fontes, E.M.G. 2002. Potential for the environmental impact of transgenic crops. *Nature Biotechnology* 20: 567-574.

9. Ag-West Biotech. 2003. *Genetically modified canola information for producers.* www.agwest.sk.ca. Accessed November 2003.

10. BBC News Online. 2004. European Union lifts GM food ban. www.news.bbc. co.uk/nolpda/ukfs_news/hi/newsid_2727000/3727827.stm. Accessed May 2004.

11. Syngenta. 2004. European Commission approves food use of imported B-11 sweet corn. Media Release. http://www.syngenta.com/en/media/article.aspx?article_ id=404, accessed May 2004.

12. Monsanto Company. 2004. Monsanto to realign research portfolio, development of Roundup Ready wheat. Deferred decision follows portfolio review, Consultation with Growers. http://www.monsanto.com/monsanto/layout/investor/news& events/2004/05-10-04.asp. Accessed May 2004.

13. Royal Society of Canada. 2001. Elements of precaution: Recommendations for the regulation of food biotechnology in Canada. Ottawa, ON: Government of Canada.

14. Auditor General of Canada. 2004. *2004 Report of the Auditor General of Canada: Regulation of plants with novel traits* (Chapter 4). www.oag-bvg.gc.ca/ domino/reports.html/20040304ce.html. Accessed May 2004.

15. Kapusciniski, A.R.. Goodman, R.M.. Hann, S.D., Jacobs, L.R., Pullins, E.E., Johnson, C.S., Kinsey, J.D., Krall, R.L., La Vina, A.G.M., Mellon, M.G., et al. 2003. Making "safety first" a reality for biotechnology products. *Nature Biotechnology* 21(6): 599-601.

16. Canadian Food Inspection Agency. 2003. www.inspection.gc.ca/english/ pbo/dir/dir9408e/shtml. Accessed May 2004.

17. Ibid.

18. Ibid.

19. Royal Society of Canada. 2001. See note 13.

20. Canadian Food Inspection Agency. 2003. See note 16.

21. Agriculture and Agri-Food Canada. 1994. *Assessment criteria for determining environmental safety of plants with novel traits.* Regulatory directive 94-08, Ottawa, ON: Government of Canada.

22. Ibid.

23. Barrett, K. and Abergel, E. 2000. See note 6.

24. Ibid.

25. National Academy of Science. 1987. *Introduction of recombinant DNA-engineered organisms into the environment: Key issues.* Washington, DC: National Academy Press.

26. National Research Council. 1989. *Field testing genetically modified organisms: Framework for decisions.* Washington, DC: National Academy Press.

27. National Academy of Science. 1987. See note 25.

28. National Research Council. 1989. See note 26.

29. Agriculture and Agri-Food Canada, 1994. See note 21.

30. Barrett. K., and Abergel, E. 2000. See note 6.

31. Ibid.

32. Ibid.

33. Barrett, K. and Abergel, E. 2002. Defining a safe genetically modified organism: boundaries of scientific risk assessment. *Science and Public Policy* 29: 47-58.

34. Barrett, K. and Abergel, E. 2000. See note 6.

35. Barrett, K. and Abergel, E. 2002. See note 33.

36. Ibid.

37. Auditor General of Canada. 2004. See note 14.

38. Ibid.

39. Chilcutt, C.F. and Tabashnik, B.E. 2004. Contamination of refuges by *Bacillus thurigiensis* toxin genes from transgenic maize. *Proceedings of the National Academy of Sciences USA* 101: 7516-7529.

40. Barrett, K. and Abergel, E. 2000. See note 6.

41. Wolfenbarger, L.L. 2004. GE crops: Balancing predictions of promise and peril. *Frontiers in Ecology and the Environment* 3: 154-155.

42. Wolfenbarger, L.L. and Phifer, P.R. 2000. The ecological risks and benefits of genetically engineered plants. *Science* 290: 2088-2093; Snow, A.A. et al. 2003, see note 8.

43. Wolfenbarger, L.L. 2004. See note 41.

44. Ibid.

45. Kohli, L.A., Leech, M., Vain, P., Laurie, D.A., and Christou, P. 1998. Transgene organization in rice engineered through direct DNA transfer supports a two-phase integration mechanism mediated by the establishment of integration hot spots. *Proceedings of the National Academy of Sciences USA* 95: 7203-7208; Pawlowski, W.P. and Somers, D.A. 1998. Transgenic DNA integrated into the oat genome is frequently interspersed by host DNA. *Proceedings of the National Academy of Sciences USA* 95: 12106-12110; Tinland, B. 1996. The integration of T-DNA into plant genomes. *Trends in Plant Sciences* 1: 178-184.

46. Letourneau, D. and Burrows, B. 2001. *Genetically engineered organisms: assessing environmental and human health effects.* Boca Raton, FL: CRC Press; Dale, P.J., et al. 2002. See note 8; Chilcutt, C.F and Tabashnik, B.E. (2004). See note 39; Wolfenbarger, L.L. 2004. See note 41; Mikkelsen, T.R., Andersen, R.B., and Bagger-Jorgensen, R. 1996. The risks of transgene spread. *Nature* 380: 31.

47. Royal Society of Canada. 2001. See note 13.

48. Auditor General of Canada. 2004. See note 14.

49. Royal Society of Canada. 2001. See note 13.

50. Ibid.

51. Ibid.

52. Barrett, K. and Abergel E. 2002. See note 33.

53. Ibid.

54. Marvier, M. 2002. Improving risk assessment for nontarget safety of transgenic crops. *Ecological Applications* 12: 1119-1124.

55. Gurian-Sherman, D. 2003. *Holes in the biotech safety net: FDA policy does not assure the safety of genetically engineered foods.* Washington, DC: Center for Science in the Public Interest. www.cspinet.org. Accessed November 2003.

56. Barrett, K. and Abergel, E. 2000. See note 6; Barrett, K. and Abergel, E. 2002. See note 33; Royal Society of Canada. 2001. See note 13.

57. Wolfenbarger, L.L. 2004. See note 41.

58. Canadian Biotechnology Advisory Committee. 2002. See note 5.

59. Hollebone, J. (1998). Interview. Ottawa, Canada.

60. Medley, T.L. and McCammon, S. 1995. Strategic regulations for safe development of transgenic plants. *Biotechnology: Legal, Economic and Ethical Dimensions.* Vol. 12, pp. 197-211. Edited by D. Brauer. New York: Wiley-VCH.

61. Agriculture and Agri-Food Canada. See note 21.

62. Abergel, E. and Barrett, K. 2002. Putting the cart before the horse: A review of biotechnology policy in Canada. *Journal of Canadian Studies,* 37: 135-161.

63. Wynne, B. 2001. Expert discourses of risk and ethics on genetically manipulated organisms: The weaving of public alienation. *Politeia* 17(62): 51-76.

64. Abergel, E. and Barrett, K. 2002. See note 62.

65. Royal Society of Canada. 2001. See note 13.

66. Canadian Biotechnology Advisory Committee. 2002. See note 5.

67. Canadian Food Inspection Agency. 2003. See note 16.

68. Government of Canada. 2000. Building a regulatory and inspection community. http://ricommunity.gc.ca/documents/task_force_reports/building_community_e.asp?template=no&. Accessed April 2005.

69. Department of Foreign Affairs and International Trade of Canada. 2004. www.biotech.ci.gc.ca/epic/internet/incbs-scb.nsf/en/by/00211e.html. Accessed May 2004.

70. Ibid.

71. Canadian Biotechnology Advisory Committee. 2002. See note 5.

72. McAfee, K. 2003. Neoliberalism on the molecular scale. Economic and genetic reductionism in biotechnology battles. *Geoforum* 34: 203-219.

73. Government of Canada. 2000. See note 68.

74. Newell, P. 2003. See note 3.

75. Tickner, J., Raffensperger, C., and Myers, N. 1998. The precautionary principle in action: a handbook. http://www.sehn.org/rtfdocs/handbook-rtf.rtf. Accessed April 2005.

76. Newell, P. 2003. See note 3.

77. Ibid.

78. Canadian Biotechnology Advisory Committee. 2002. See note 5.

79. Newell, P. 2003. See note 3.

80. Treasury Board Secretariat. 2005. Report on plans and priorities 2005-2006: Canadian Food Inspection Agency. http://www.tbs-sct.gc.ca/est-pre/20052006/CFIA-ACIAAr5601_e.asp. Accessed April 2005.

81. Ibid.

82. Prince, M.J. 2000. Banishing bureaucracy or hatching a hybrid? The Canadian Food Inspection Agency and the politics of reinventing government. *Governance* 13: 215-232.

83. Barrett, K. and Abergel, E. 2000. See note 6; Newell, P. 2003. See note 3.

84. Barrett, K. and Abergel, E. 2002. See note 33.

85. World Trade Organization. 2004. SPS Article 5.7 states: In cases where relevant scientific evidence is insufficient, a Member may provisionally adopt sanitary or phytosanitary measures on the basis of available pertinent information, including

that from the relevant international organizations as well as from sanitary or phytosanitary measures applied by other Members. In such circumstances, Members shall seek to obtain the additional information necessary for a more objective assessment of risk and review the sanitary or phytosanitary measure accordingly within a reasonable period of time. http://www.wto.org/english/tratop_e/sps_e/spsagr_e.htm. Accessed May 2004.

86. Ibid.

87. Hardstaff, P. 2000. A discussion paper for the European Commission Consultation on Trade and Sustainable Development. http://www.rspb.org.uk/Images/precautionaryprinciple_tcm5-31097.pdf. Accessed May 2004.

88. See Principle 15 of the Rio Declaration.

89. Coleman, W.D. and Gabler, M. 2002. See note 2.

90. Ibid.

91. Ibid.

92. Thomas, U. 2003. *WTO Public Symposium 2003: Session XIV- GM Food/ Biosafety Protocol.* Organized by the University of Geneva. http://www.cid.harvard.edu/cidtrade/geneva/sessions/session14.html. Accessed May 2004.

93. Ibid.

94. Coleman, W.D. and Gabler, M. 2002. See note 2.

95. McAfee, K. 2003. See note 72.

96. Thomas, U. 2003. See note 92.

97. McAfee, K. 2003. See note 72.

98. Winickoff, D., Jasanoff, S., Busch, L., Grove-White, R., and Wynne, B. 2005. Adjudicating the GM food wars: Science, risk, and democracy in world trade law. *The Yale Journal of International Law* 30: 81-123.

99. Newell, P. 2003. See note 3.

100. Treasury Board Secretariat. 2005. See note 80.

Chapter 11

Principles Driving U.S. Governance of Agbiotech

Kathleen A. Merrigan

In 1987, I was hired as a congressional staff aide for the U.S. Senate Agriculture and Judiciary Committees to manage emerging issues surrounding agricultural biotechnology (agbiotech). The antifreeze bacterium, *Pseudomonas syringae* "Ice Minus" had just been released, bovine somatrotropin (bST) was approaching commercialization, and the Senate was convening hearings to discuss scientific and legal impediments to agbiotech. During my first week on the job, I was flown to the headquarters of a major biotechnology company and briefed on the remarkable science that the company was developing. It was explained that the technology would help to feed the world while simultaneously reducing environmental degradation associated with conventional production methods. The industry lobbyists worked diligently to ensure that I understood that biotech is the great elixir that would not only end starvation, but also cure cancer and infuse great wealth into the American economy. However, they did raise a caution and urged my cooperation. Biotech would only succeed if the government did not overreact to unscientific alarmists who were using rhetorical powers and subversive tactics against the industry. From the onset of my Senate career, industry and government leaders alike advised me that if biotech were to fulfill its promise, the appropriate roles for the government were to (1) recognize that biotech is not remarkably different from conventional breeding and to regulate it accordingly; (2) invest in biotech research and education and remove

Genetically Engineered Crops
© 2007 by The Haworth Press, Inc. All rights reserved.
doi:10.1300/5880_11

disincentives to commercialization; and (3) provide adequate property rights to reward and encourage invention. Then and now, these three principles undergird American biotech policy.

In this chapter, I elaborate on each of these principles, offer conjectures on potential shifts in U.S. biotech policy, and suggest strategies to improve American governance of biotech consistent with these principles. I question whether there is a sufficient difference between the "sell" of biotechnology by the industry and the softer, but equally devoted "sell" by the U.S. government. One of the many occasions when the fusion of industry and government interests in biotech was abundantly clear occurred in 2003 when Secretary of Agriculture Ann Veneman welcomed participants to the first meeting of her new agbiotech advisory council, assuring them that the U.S. Department of Agriculture (USDA) would do anything in its power to facilitate adoption of this technology.[1]

PRINCIPLE 1:
NO "SPECIAL TREATMENT"

Most Americans were first introduced to agbiotech in 1987 when newspapers across the nation published photographs of scientists wearing what looked to be space suits as they sprayed the genetically engineered (GE) "Ice Minus" on California strawberries to inhibit frost formation. This was the first "deliberate release" into the environment of a GE product—a GE pesticide approved by the Environmental Protection Agency (EPA) for a field trial.

Times have changed, so much so that young people only know "Ice Minus," as a British rock band! No longer seen as alien, GE is largely ignored by the American public. Periodically, industry and NGO groups poll Americans for their sentiments on biotech, but at best the results are confusing.[2] Overall, when attention is directed to GE, Americans are leery of the technology but left unprompted, they hardly give it a second thought. Is it merely the passage of time that has led to this societal adjustment? No, I argue that a series of strategic actions taken by the government facilitated public acceptance. The first such action occurred when government scientists, widely perceived as "neutral" experts, participated in the second deliberate GE crop release in 1988. Researchers at the Agricultural Research

Service of the USDA, in partnership with Crop Genetics International, injected 2,000 corn plants in the state of Maryland with a GE microbe to combat the corn borer. This was reported in the media, but the news was almost all good and government regulators sighed with relief. The next, and to this day, the most significant action occurred when government defined the technology as safe by analogy, which effectively supported the concept of substantial equivalence and the Coordinated Framework.

Substantial Equivalence

Prior to government regulation, scientists had established voluntary guidelines that governed biotechnology for almost fourteen years, from Paul Berg's insertion of the SV40 monkey virus to *Escherichia coli* at Stanford University in 1972[3] until declaration of the Coordinated Framework in 1986. Initially, many in the scientific community viewed the technology as novel and urged caution. Letters by prominent scientists calling for a moratorium on certain types of experiments and stringent containment procedures were published in the journal, *Science,* in the early 1970s, and similar recommendations were echoed by the National Academy of Sciences in its first biotechnology report published in 1974.[4] The U.S. National Institutes of Health (NIH) formally became involved, when it established the Recombinant Advisory Committee (NIH RAC) in 1974 to assist the NIH in evaluating recombinant DNA (r-DNA) for its promise as well as "hypothetical hazards to public health and the environment and significant ethical, legal, and societal issues."[5] At a landmark meeting in Asilomar, California, in 1975 the scientific community came together and drafted a set of r-DNA research guidelines that scientists nationwide pledged to follow.[6] Many of the Asilomar guidelines, including a prohibition on any release of r-DNA organisms, were adopted by the NIH in 1976 and mandated for all scientists receiving federal funding.

Scientific understanding of r-DNA technology evolved over time, particularly knowledge about gene transfer between species, which occurred in nature, but also between lower and higher life forms in ways not previously thought possible. These discoveries led scientists to reassess the novelty of this technology and, ironically, as the scientific community became ready to relax r-DNA research requirements, the public was beginning to worry about the technology.[7] During the early 1980s, media stories, a series of U.S. congressional hearings, and most importantly, local disputes over construction of laboratories for r-DNA research, generated public alarm. While none of the disasters predicted by biotech critics occurred, supporters of the technology began to devise a strategy to ensure its acceptance.

Proponents began to argue that the technology was not really new and hence that there was no cause for alarm. A book published during this period, not atypical, placed the start of the agbiotech timeline in 1750 BC, with the oldest known recipe for beer that was written on Sumerian tablets.[8] The Augustinian monk Gregor Mendel's study of pea genetics in the mid-1800s, familiar to all students of high school science was said to be a form of GE through the use of selection in plant breeding. Even the term "biotechnology," it was emphasized, dated back to 1919, when a Hungarian scientist coined it to describe the interaction of technology and biology.[9]

Proponents argued, as well, that releasing foreign biological material into the environment is nothing new. Between 1862 and 1923 alone, 50,000 foreign plants were introduced into the United States by USDA scientists.[10] The example of the 1939 USDA eradication program to combat the Japanese beetles[11] by spraying *Bacillus popilliae* over the course of several years on 110,000 acres in five states was used as an illustration that the government had been deliberately releasing organisms into the environment successfully for decades. While there are examples of biological control introductions that went awry, biotech proponents calculated that it is far better to associate with the downsides of a known practice than to declare agbiotech releases as something new.

By the late 1980s, the position that biotech is just a more sophisticated and accelerated form of Mendel's selection work had become entrenched. In 1987, the NAS proclaimed, No evidence that unique hazards exist either in the use of r-DNA techniques or in the transfer of genes between unrelated organisms. The risks associated with the introduction of rDNA-engineered organisms are the same in kind as those associated with the introduction into the environment of unmodified organisms and organisms modified by other genetic techniques.[12]

As the technology advanced, state and federal government regulators began contemplating regulations. If the risks of biotechnology were considered "the same in kind" it followed that regulations should be concerned with the specific application, rather than the technology of GE. In 1989, the NAS declared that the "[a]ssessment of the risks of introducing r-DNA engineered organisms into the environment should be based on the nature of the organism and the environment into which it is introduced, not on the method by which it was produced."[13] In the same year, the Ecological Society of America issued its report on risks of release, stating: "Genetically engineered organisms should be evaluated and regulated according to their biological properties (phenotypes) rather than the genetic techniques used to produce them." Thus, the U.S. concept that biotech should be regulated on "product, not process" was established and it remains the mantra that shapes biotech regulation.

Adherence to the idea that biotech is simply more of the same has so permeated the U.S. regulatory system that the U.S. Food and Drug Administration (FDA) does not require a review of food derived from GE before it is sold for human consumption. In 1992, the FDA published guidelines that set out its policies for reviewing biotech food. The FDA declared that new proteins introduced by GE are likely to be "substantially equivalent" to proteins already found in food and classified them presumptively as "generally recognized as safe" (GRAS), a designation given to food that does not require regulatory analysis.[14] In response to a consumer petition, in 1992, the FDA affirmed its position that the safety of whole foods derived from biotech crops would be judged in the same manner as other whole foods that are assumed to be safe unless adulterated.[15] Alarmed biotech critics took the FDA to court over this policy statement, arguing that biotech foods should be considered food additives under the Federal Food Drug and Cosmetic Act (FDCA); however, the courts affirmed FDA's power to issue this discretionary policy statement.[16]

Few people understand that there is no mandatory premarket FDA review of biotech foods. This is largely because inventors of biotech foods have voluntarily sought FDA consultation prior to introducing their products to market. They do so because FDA encourages these voluntary consultations as a means of facilitating the agency's understanding of the products entering the marketplace.[17] Inventors sometimes seek these consultations because they misunderstand the regulatory structure and believe they are mandatory, but most use consultations to support their claim that the FDA has declared a particular food to be "safe." In reality, FDA never declares food of any kind "safe," and the most a company can hope for is a letter from the agency stating that it has no further questions and reminding the inventor that safety is their responsibility. Nevertheless, these letters seem to go a long way in assuring investors and the public that biotech products are assessed as suitable for public consumption.

The current, very rapid, technological advances suggest that FDA soon will be faced with products that are radically different from conventional foods and that it will be forced to evaluate them as food additives. Adherence to the doctrine of substantial equivalence, however, will likely prevail. It is very important to recognize that, while it forms the basis of U.S. regulation, substantial equivalence is a policy and not a scientific principle. Law professor Thomas McGarity[18] illustrates this point using a hypothetical example. Suppose, he says, that Starlink corn was found to contain a human allergen and is removed from the market. If substantial equivalence were a scientific principle, then all other crops engineered to express the *Bacillus thuringensis* protein (*Bt*) should be removed from the market, something that would be highly unlikely since regulators have the discretion to do otherwise. Other scientific voices have argued that the concept of substantial equivalence should be abandoned in favor of biological, toxicological, and immunological testing to assess risk, declaring "the degree of difference between a natural food and its GE alternative before its 'substance' ceases to be acceptably 'equivalent' is not defined anywhere, nor has an exact definition been agreed by legislators."[19] Still others have extended the concept and introduced the notion of "ethical equivalence" to push society to evaluate the moral value contained in food products.[20]

Coordinated Framework

The NIH RAC oversight of biotech experiments was curtailed in the late 1970s when the courts ruled that the agency had no jurisdiction over the research activities of private biotech companies. This decision prompted several states and localities to propose regulation.

Biotech supporters became alarmed at the prospect of a patchwork of regulations across the country. Reluctantly they began to discuss the need for federal intervention to preempt state and local action. Initially there seemed to be little consensus on how to develop a federal regulatory structure except that no one wanted an open debate in the Congress because it may provide a highly visible venue for the various anti-biotech groups to alarm the public and lead to nonscientific, even draconian biotechnology legislation.

The alternative to new legislation was to search through statutes and regulations to see whether agbiotech might somehow fit into existing structures. Such an approach was consistent with the principle that biotech is not significantly different from conventional breeding. But this was not easily done. None of the major federal departments or agencies had sufficient jurisdiction or expertise to govern all aspects of biotechnology nor did any existing statute provide adequate authority for the various kinds of oversight envisioned. For example, it did not make scientific or regulatory sense for a specialist in pesticides at the U.S. EPA to be responsible for determinations of food safety. Equally, it was not sensible to make a food safety specialist at the FDA responsible for evaluating environmental consequences of releasing a GE crop.

The U.S. president's Office of Science and Technology Policy (OSTP) decided that there was no need to establish new agencies or statutes. Rather, an effort should be made to sort out what agency and statute would govern particular agbiotech applications. This matrix of sorts was named the "Coordinated Framework" and it was officially announced as a policy statement in the Federal Register in 1984 and fully implemented in 1986.[21] The Coordinated Framework was based on three principles: existing statutes are sufficient; safety assessments and other regulatory questions should be based on the products themselves, not the process that produced them; and a lead federal agency would be designated for regulation of products when it seemed that more than one federal agency need be involved.

It is apparent that very few Americans know that the Coordinated Framework exists and even those most directly affected in the biotech industry admit that its detailed workings continue to elude them. This is understandable, given that authority for biotech regulation is derived from twelve different statutes, all written prior to the modern biotech era. Under each statute, the standard for regulatory approval differs and, to make things even more complex, within each statutory scheme, the way a biotech product is defined affects the way in which it is evaluated. Hence, the FDA, under authority granted to it by the Federal Food Drug and Cosmetic Act (FDCA), may require a novel biotech product to meet a "reasonable certainty of no harm" test, a whole food to be "as safe as" conventional food, or GE substances added to food to be "generally recognized as safe." A few streets away, regulators at the EPA, under the Federal Insecticide Fungicide and Rodenticide Act may require biotech products to present "no unreasonable adverse effect on the environment," while regulators at USDA, under the Plant Pest Act, may ponder whether the proposed biotech product will "injure, damage, or cause disease in any plant or plant product."[22] Use of these pre-biotech statutes, has required agencies to stretch their original meanings to cover current innovations; none of these sometimes-precarious interpretations has been legally tested.

Three agencies have now emerged with primary responsibilities for biotech regulation. The Animal Plant Health Inspection Service (APHIS) of USDA has responsibility for the safety of meat, poultry, and egg products; for regulating potential agricultural plant pests and noxious weeds; and for the safety and efficacy of animal biologics. APHIS governs biotech under authority granted to it by the Federal Plant Pest Act. APHIS has chosen to define GE plants as "plant pests" thus triggering the requirement for APHIS review and approval for interstate commerce.[23] The FDA of the Department of Health and Human Services has responsibility for safety of food and animal feed and for safety and efficacy of human drugs and biologics, and animal drugs. FDA primarily governs biotech under authority granted to it by the FDCA. The FDCA requires review of substances deliberately added to foods that are not generally recognized as safe (GRAS). As described earlier, while FDA does treat GE products as GRAS, the agency, nevertheless, encourages voluntary premarket consultation for all GE food products.[24] Finally, the EPA has responsibility for the use of pesticides and tolerances of pesticide residues in food, and for the regulation of non-pesticidal toxic substances, including microorganisms. EPA governs biotech primarily under the Federal Insecticide, Fungicide, and Rodenticide Act (pesticide registration) and the

FDCA (tolerance setting), although other statutes, such as the Toxic Substances Control Act, could be used and EPA has asserted that everything can fall under the broad TSCA authority, which many refer to as the "gap-filling statute."[25] Since most first-generation biotech products have been plants genetically engineered for pest control, EPA has pioneered much of the current regulation.

What does this mean? APHIS Science Advisor, Sally McCammon, explains that a new GE plant could be reviewed by one or all three regulatory agencies depending on its composition: "A Bt gene in a food crop would be reviewed by APHIS, EPA, and FDA; a plant with modified oil content for food would be reviewed by FDA and APHIS; and the modified flower color in a horticultural crop would be reviewed by APHIS alone."[26] This maze of regulatory agencies and statutes is cumbersome and opaque,[27] yet proposed legislation to place biotech regulation under a single statute and/or federal agency never seems to receive serious consideration.

Labeling

The labeling of GE food has been a hotly debated topic in the United States as well as around the world. Labeling advocates claim that people have a "right to know" about the source of food and, therefore, labeling should be mandatory. Others claim that people want to know and, therefore, the government should ensure that "absence claims" (e.g., "No GEOs") are accurate.[28] Still others claim that labeling should be market driven, since the benefits of providing this information to the few who care are outweighed by the additional food costs all consumers will pay to cover the cost of setting up a segregated, identity-preserved food system.

The no-special treatment principle carries into the labeling debate. Since the United States regulates biotech based on product, not process, the corollary is that government should focus on safety and not the right to know. The FDA does not require labels because biotech foods are substantially equivalent and the courts have upheld the agency's position that labeling for biotech foods is not necessary.[29] In 2001, FDA declined to institute mandatory labeling and instead issued a draft Guidance on how companies can voluntarily label whether or not their products contain biotech ingredients. The voluntary guidance focused on what is referred to as labeling deception (e.g., labeling a wheat cereal as GEO-free when biotech wheat is not on the market), and it is not expected to result in any labels that declare GEO content. It seems likely that when biotech foods with significant differences in composition, nutrition, or food allergens are introduced, the aversion to mandatory labeling may ease. In 2004, Secretary of Health and Human Services, Tommy Thompson, summarized the government position in a speech before the Grocery Manufacturers of America, saying biotech labeling "would only frighten consumers and play into the hands of those who exploit fear rather than deal in fact."[30]

PRINCIPLE 2:
INVEST IN RESEARCH AND EDUCATION

A Monsanto executive predicted in the mid-1980s that "a thousand years from now, when many of today's other technologies—microprocessors, robots, lasers—are old stuff, biotech will still be at the center of much that we do."[31] Indeed, the federal government has invested heavily in biotech, envisioning it as the foundation for the country's future economy. By 1993, biotech was designated as one of six major research and development initiatives of the U.S. government. At that time, the President's OSTP declared, "The Biotechnology Initiative aims to sustain and extend US leadership in biotechnology research, enhance the quality of life for all Americans, and spur the growth of the US economy."[32]

The USDA sponsors most of the government-financed agbiotech research, although EPA, NIH, and the National Science Foundation also contribute small sums to such undertakings. Overall, USDA's annual research budget is approximately $2.3 billion. Of this amount, approximately $250 million is spent on projects classified by USDA as solely biotechnology. Most of USDA's research budget is consumed for federal salaries and facilities, so this is a large share of the discretionary project allotment and it excludes project grants for many other relevant projects that are not exclusively biotech. Almost all of this dedicated funding is spent on a combination of basic science and new product development. The USDA website promotes the department's biotech achievements, providing examples such as development of transgenic trees that have more cellulose and less lignin, designed to reduce the environmental impact of paper manufacturing.[33]

For many years, federal funding was allocated for basic research as well as applied research that had clear public benefit but no foreseeable commercial application. Such research was funded by taxpayers and carried out by government and university scientists. Development of products was seen as the domain and responsibility of the private sector, but this all changed as the enormous potential of biotech became apparent. The Bayh-Dole Act (1980) and the Technology Transfer Act (1986) changed the U.S. framework fundamentally by eliminating barriers and providing incentives for research collaborations between industry, university, and government scientists. Today, private firms frequently enter into agreements with public and private universities and government laboratories, and provide financial and other resources in return for research and product testing. Some policymakers claim that this new framework encourages innovation and early application of new technologies. Indeed, Calestous Juma advises developing countries that want to benefit from biotech to reorganize their research systems based on the U.S. model so as to encourage close interactions between industry, government, and universities and use of federal funding for commercial purposes.[34]

Critics of this new research model argue that it places greater emphasis on commercial applications and diverts government and academic researchers from research that may have significant public benefits but little immediate commercial potential. A 2004 agbiotech study conducted by scientists at Portland State University found conflicting evidence and a shortage of data on this issue, but the authors concluded that "Society is largely 'flying blind' on one of the most important revolutions on our agricultural research system."[35] My own experience during the often tumultuous review of bovine somatotropin (bST) by the FDA in the late 1980s left me pondering the wisdom of our current research structure. When consumer and environmental advocacy groups and a FDA scientist "whistle blower" involved in the regulatory action questioned the adequacy of FDA's review, I was given the task of guiding Congress, and the Congressional Government Accountability Office[36] (GAO), through a secondary review to determine whether, indeed, use of bST posed any risk to humans. Together with GAO staff, we searched but could not find a single university expert in North America who had not, at one time or another, been on the payroll of one of the four companies developing versions of bST. I worried then and continue to worry that the future of biotech will be in jeopardy if we fail to restructure government research spending so to maintain a cadre of independent experts who compel public trust.[37]

A very small portion of the USDA biotech funding supports inquiries related to human and environmental safety and related risk assessments. While an exact number is not clear from the department's budget documents, it is in the vicinity of $8 million. The most obvious source of such investments is found in the Biotechnology Risk Assessment Research Program (BRARP) established by Congress in 1990 to "help researchers make sound judgments about the overall impacts of genetically engineered organisms in order to minimize risks and to help regulators make scientifically sound and timely decisions."[38] Concerned that all USDA biotech research dollars were being spent on product development and none on risk evaluation, the 1990 law stipulated that not less than 1 percent of the total amount of USDA research dollars spent on biotechnology research be spent on research related to the potential environmental risks posed by biotech, with particular attention to issues concerning unintended gene transfer. In 2002, Congress amended the law, raising the set-aside allocation to 3 percent of all biotech research spending. At first glance, this seems like increased commitment for this kind of research, but a corresponding shift in the BRARP mission has meant that environmental safety inquiries are now just one of several kinds of inquiries that compete for BRARP funding.[39] Biotech supporters succeeded in broadening the mission of BRARP to require risk-assessment research to compare the risks of agbiotech products to traditionally bred plants and animals, including comparison of biotech systems with other production systems such as organic and low-input farming. In an ideal world, every assessment would include such comparative analysis, a concept endorsed by the National Academy of Sciences, but the reality is that even a tripling of the BRARP budget is insufficient to encompass comparative inquiries. Indeed, the true purpose of this statutory change was to demonstrate that biotech is the same as other production techniques (substantially equivalent) and hence in no need of special regulation. BRARP was also broadened to encompass research on cost-benefit analysis of developing identity preservation systems for genetically modified agricultural products; establishment of international research partnerships on biosafety; and multidisciplinary inquiries into nutritional enhancement opportunities.

A scientific literature review published in *Science* in 2000[40] asserted that there had been insufficient investigation into the risks and benefits of releasing biotech organisms into the environment. That same year, an EU–U.S. biotech consultative forum released a report that called for greater investment in basic research to address safety concerns.[41] A 2004 Pew Initiative on Food and Biotechnology report found many gaps in the understanding of food allergies, including the causes, mechanisms, and trends of disease, leaving "food safety regulators without some of the critical tools they need to assess fully the potential allergenicity of novel food products." At a 2002 conference on biotechnology, I argued for additional research funding to aid risk assessments. To my surprise, the audience took my call for this kind of research as an inherent attack on biotech, which was not my intent. People like me, said one participant during the question session, are responsible for the thousands of children dying each day from malnutrition because they are blocking progress in agbiotech. In the drive to increase U.S. biotech research and development funding, consensus on the appropriate balance between new product development and research to inform safety assessments seems impossible.

Education goes hand in hand with USDA's research mission. The Extension Service is administered through thousands of county and regional offices and has, since 1914, been responsible for providing farmers and the public at large with expert information and assistance with agricultural related projects. The Extension Service has a great deal of information on biotech, much of it promotional in tone. Beginning in kindergarten, children are taught that biotechnology is a part of their lives in the production of basic foods such as bread and cheese. The agency's Agriculture in the Classroom program, established to provide school teachers with agricultural curriculum, promotes a three-minute classroom video for students of all ages, which according to the program's website, "shows wonderful uses of biotechnology and the potential it has to reduce the use of pesticides and chemicals and to make the world a better place. It is positive in its approach and will evoke excellent responses from its audience."[42]

PRINCIPLE 3:
STRONG PROPERTY RIGHTS

Property rights are so fundamental in the United States that our founding Constitution grants Congress the power to enact laws relating to patents. Patents are a property right extended and enforced by the government. Patents give their holders the right to exclude others from making, using, or selling their invention in the United States. This is intended to provide an inventor with the incentive to invest in and develop an idea to its fullest commercial potential. In exchange for what is a currently a twenty-year monopoly right, the patent holder is required to disclose his invention fully so that other would-be inventors are inspired to pursue greater research and discovery. BIO, the U.S. biotechnology trade organization, underscores the importance of property rights to their industry, proclaiming that "we believe that the biotechnology industry is on its way to becoming a tremendously successful industry because of the patent system in the United States."[43]

While patenting has been engrained in our culture from the beginning, it was first intended to extend only to inanimate objects, like machines. Later on, as advocates such as former President and agrarian, Thomas Jefferson, argued for the extension of property rights to living matter, people did pause, but only momentarily. The Congress passed the Plant Patent Act in 1930 after little debate and by voice vote. Plant patents are available to those who invent or discover, and asexually reproduce, any distinct and new variety of plants, excluding tuber-propagated plants. In 1970 Congress passed the Plant Variety Protection Act (PVPA) following only one hour of debate. The PVPA provides patent-like protection to sexually reproduced and tuber-propagated plants. In 1980, the Supreme Court ruled in *Diamond v. Chakrabarty* that patents are to be allowed on all living matter, in this particular case, a utility patent was approved for bacterium genetically engineered to consume crude oil.[44] At last, the Supreme Court said, "Everything under the sun could be patented."

Agbiotech inventors may seek protection for their intellectual property under the three different schemes: plant patents, plant variety protection certificates, and utility patents. The choice depends upon the invention and the level of protection sought.[45] One example of important ways in which these schemes differ is that, under the PVPA, an exemption is granted to researchers to allow them free use of plant materials that have been issued a plant variety certificate. However, no such exemption exists under the Plant Patent Act, or under utility patent law under which most biotech inventions are protected.

It is important to understand that U.S. patent protection does not extend beyond the U.S. borders. Due to this, the U.S. government has fiercely fought for international agreements that recognize U.S. patents and a harmonized global property rights regime. It makes no difference what political party is in power, as defense of the principle of property rights pervades U.S. action in international forums. The first Bush administration fought for intellectual property rights at the Earth Summit; the Clinton administration fought for interpretative statements to accompany the Convention on Biological Diversity; and the second Bush administration promoted terminator technology before the International Union for the Protection of New Varieties of Plants (IUPOV).

This is not to say that everyone in the United States is happy about our property rights system, although the objections critics offer have not generated much public concern. A patent applicant must show that an invention is novel, nonobvious, and useful. There is some debate over whether the U.S. Patent and Trademark Office should issue utility patents only on the final product rather than on individual components or processes (e.g., genes or transformation methods). There is debate over the appropriate breadth of a patent, with the concern being that some are too broad (e.g., species-wide claims) or inappropriate (e.g., altering only one amino acid in a protein). Some critics say that many biotech patents have failed the test of utility (e.g., gene sequences). In all these cases, the criticisms do not challenge the basic premise of property rights but rather the specific application of those rights by bureaucrats. Perhaps the hardest-hitting criticism of patents is that they have skewed university and government laboratory research. Public sector researchers are under pressure to supplement dwindling budgets with industry-sponsored projects and to pursue collaborative efforts with industry. Nowhere is this pressure more evident than in the field of weed science. Weed science departments receive a minimal share of university budgets since scientists are expected to raise their own funds from industry sources eager to "buy" university time to develop herbicide-tolerant crops. As a result, weed scientists have few resources to dedicate to nonchemical weed control research. The pressure to find profit-sharing arrangements with industry is so strong, that some universities now evaluate patent holding as a factor in promotion and tenure decisions.

In the future, property rights may be largely negotiated within contractual agreements. For example, four of the world's largest agriculture companies have agreed upon free sharing of some specific biotech technology with African scientists. An institute was established in Nairobi to identify crop problems in Africa and to negotiate with companies for assistance and patent licenses.[46] Technology may also be used to protect property rights, as was the case with the controversial Terminator gene, developed jointly by USDA and Delta and Land Pine Company (and subsequently purchased by Monsanto) in collaboration with government scientists, which inhibited crop seed germination if part of the harvested seed was replanted the next season, forcing farmers to buy seeds annually.

Several years ago, the U.S. National Academy of Sciences released a report, Global Dimension of Intellectual Property Rights in Science and Technology, claiming that other countries have moved to adopt a Western cultural view of the concepts of ownership and rights. Yet other cultures and legal traditions, including those in Asia and throughout the Islamic world, have different approaches to encourage creative participation in society. Following the *Diamond v. Chakrabarty* decision, the USDA commissioned a study to evaluate the implication of the Supreme Court's ruling for agriculture. The study uncovered little relevant data and recommended that the USDA undertake additional analysis; work that has never been done. The first patent was issued by the U.S. government in 1790; today, this country has approximately 3.5 million active utility patents alone.[47] U.S. companies spend an estimated $4 billion a year on patents and patent litigation, a sum, one critic points out, that equals one-third of the combined public and private agricultural research budgets in developing countries.[48] While there may be other cultural traditions and conceptions of property, it can be said with great confidence that the U.S. scheme is here to stay, although its international impact may remain controversial.

FUTURE GOVERNANCE

The many novel biotech products on the horizon (e.g., GE insects, pharmaceutical plants), the resolution of current disputes such as what constitutes an acceptable threshold of GEO contamination (also known as "adventitious presence"), and pending lawsuits may all create shifts in agbiotech policy. Given the adherence to the three principles described in this chapter, no major shifts in policy are likely unless prompted by state action or by an unanticipated and large-scale crisis.

Biotech crops of one variety or another are now grown in every U.S. state. That is not surprising, considering that in the United States in 2006, 83 percent of upland cotton; 89 percent of soybeans; 61 percent of corn were genetically engineered.[49] States see their economies intricately tied to the biotech industry, both in the agriculture and health science sectors. Policy statements from the National Association of State Departments of Agriculture reinforce the principles outlined in this chapter. The department Web site says:

> Humankind has used the traditional techniques of "biotechnology"—breeding and selection—for thousands of years to domesticate wild species. . . . The potential benefits to the world from future discoveries in biotechnology are almost too vast to comprehend. . . . Government has a vital role in the commercialization of biotechnology products for the future.[50]

While the U.S. Congress has been relatively inactive in the biotech arena, this has not been true for the states. In 2005, 117 bills were introduced in 33 states.[51] Bills related to agbiotech continue to increase in number. At this writing, most state and local government legislation is aimed at stimulating economic growth, such as providing tax incentives for biotech companies and more initiatives in biotech research and education, but that may not always be the case. Some states have pursued aggressive regulatory and labeling legislation.[52] Most worrisome for biotech supporters is passage of local and county regulations aimed at placing limitations or prohibitions on agbiotech crops (e.g., a local ballot initiative in Mendocino County, California in 2004 prohibits propagation, cultivation, or growth of any biotech crop in the county). By 2004 as well, over seventy-nine different towns in the state of Vermont supported nonbinding town hall resolutions calling for moratoria on GEOs. Michael Rodemeyer, Executive Director of the Pew Initiative on Food and Biotechnology, wonders whether state and local actions will lead to a "checkerboard of inconsistent legislation where rules for agbiotech vary from state to state."[53]

Public policy scholars have shown that shifts in policy are often preceded by crises of one kind or another. We have had a few in the United States. University of Illinois scientists producing GE pigs sent some of the pigs' offspring to market without the FDA's permission. ProdiGene, a now defunct Texas biotech company, contaminated soybeans intended for the food supply with an experimental corn that was engineered to produce pharmaceuticals. Pioneer and DowAgroSciences were fined by EPA for growing experimental corn in Hawaii in ways that could have pollinated nearby conventional crops. However, so far these incidents have been handled quickly and quietly by policymakers and industry leaders, resulting in little public alarm. This could, however, all change very quickly, if human health is seriously threatened by an agbiotech product.

CONCLUSION

The three principles of U.S. agbiotech governance described in this chapter are fundamental to how American policymakers have and will continue to approach biotech issues. Rather than try to change these principles, such as arguing against the issuance of patents for biotech innovations, advocates for system change should work within them. Rather than arguing that biotech is inherently different and requires a new regulatory regime, advocates should consider focusing on gaps in regulatory oversight (e.g., genetically engineered fish) or weaknesses in the system (e.g., confidential business information protections that prevent the public from gaining knowledge about releases in their own backyard). Rather than arguing that biotech research investments would be better spent on alternative systems such as organic agriculture, advocates should argue for dedicated public sector funding of environmentally friendly product development with no industry involvement.

NOTES

1. Veneman, A. 2003. Presentation to U.S. Advisory Committee on Agricultural Biotechnology. June 16; Caplin, R. (US PIRG representative to advisory council) 2004. Personal communication.

2. National Opinion Poll on Labeling of Genetically Modified Foods. 2002. *Environmental savior or saboteur? Debating the impacts of genetic engineering.* February 4. www.cspinet.org/new/poll_gefoods. Accessed March 12, 2003; Gallop Organization. August 12, 2003. *Genetically altered foods: hazard or harmless?*; Gallup Organization. August 13, 2002. *Bioengineered food? Sure, if it's not fattening.*

3. Berg, P. 1980. Dissections and reconstructions of genes and chromosomes. Nobel Lecture. Stockholm: Nobel Foundation. http://nobelprize.org/chemistry/laureates/1980/berg-lecture.pdf. Accessed June 10, 2005.

4. Singer, M. and Soll, D. 1973. Guidelines for DNA hybrid molecules. *Science* 181: 1114; Berg, P., Baltimore, D., Boyer, H.W., Cohen, S.N., Davis, R.W., Hogness, D.S., Nathans, D., Roblin, R., Watson, J.D., Weissman, S., et al. 1974. Potential biohazards of recombinant DNA molecules. *Science* 185: 303; Alberts, B. 1999. *Testimony before the U.S. Senate Committee on Agriculture*, Nutrition and Forestry. October 7. Washington, DC.

5. Recombinant DNA Advisory Committee Charter, http://www.4.od.nih.gov/oba/Archives/RACCharter.htm. Accessed September 9, 2006.

6. While sixteen countries were represented at Asilomar, the vast majority were American.

7. Asilomar restrictions eased by NIH in 1979.

8. Witt, S.C. 1990. *Biotechnology, microbes, and the environment.* San Francisco, CA: Center for Science Information.

9. Ibid.

10. Ibid.

11. Ibid.

12. National Research Council. 1987. *Introductions of recombinant DNA-organisms into the environment: key issues.* Washington, DC: National Academy Press.

13. White paper as cited in *Crossroads of science and regulation for genetically engineering organisms used in agriculture: The Role of the national academies.* Paper presented for the Pew Stakeholder Initiative, February 25, 2002. Washington, DC; this was reaffirmed in: National Research Council, National Academy of Sciences. 2000. *Genetically modified pest-protected plants: Science and regulation.* Washington, DC: National Academy Press.

14. Food and Drug Administration. 1992. Statement of Policy: Foods derived from new plant varieties. *Federal Register* 57: 22984. May 29.

15. FDCA section 402(a)(1); 21 USC section 342(a)(1).

16. Alliance for BioIntegrity v. Shalala, No. 98-1300 (CKK) (DC. Super. Ct. 2000) (dismissing case).

17. Food and Drug Administration. 1997. *Guidance on Consultation Procedures: Foods derived from new plant varieties*. Rockville, MD: FDA. http://vm.cfsan. fda.gov/~/rd/consulpr.html . Accessed June 10, 2005.

18. McGarity, T. 2001. Presentation to Pew stakeholders. August 28. Sun Valley, UT.

19. Millstone, E., Brunner, E., and Mayer, S. 1999. Beyond "Substantial Equivalence." *Nature* 401: 525-526.

20. Pouteau, S. 2000. Beyond substantial equivalence: ethical equivalence. *Journal of Agricultural and Environmental Ethics* 13: 273-291.

21. Office of Science and Technology Policy, The White House. 1984. Proposed coordinated framework for the regulation of biotechnology. *Federal Register* 49: 50858. December 31.

22. Pew Initiative on Food and Biotechnology. 2004. *Issues in regulation of genetically engineered plants and animals*. Table 1.4, p. 14.

23. At this point, APHIS is defining most as plant pests but this inventive move has not been challenged in the courts.

24. The FDA has proposed mandatory premarket review, but the proposal issued in 2001 has yet to be made final and many question whether it ever will be. See FDA. 2001. Proposed rule: Premarket notice concerning bioengineered foods. *Federal Register* 66: 4706. January 18.

25. Aidala, J. 2001. Presentation to Pew stakeholders. November 29. Warrenton, VA. Aidala is a former Assistant Administrator, Office of Prevention, Pesticides, and Toxic Substances.

26. McCammon, S.L. 1999. *Regulating the products of biotechnology*. http://www.usia.gov/journals/ites/1099/ijee/bio-mccammon-z.htm. Accessed December 14, 1999.

27. It would take more than this chapter to explain the US regulatory scheme with any precision. The most comprehensive description is: *Issues in the regulation of genetically engineered plants and animals*. Pew Initiative on Food and Biotechnology. April 2004.

28. Caswell, J.A. 2002. *Information policy for GM and Non-GM foods*. Presentation at Pew Stakeholder Forum Meeting. June 14. Chantilly, VA.

29. Alliance for Bio-Integrity et al. v. Shalala, 116 F. Supp. 2d 166 (D.D.C. 2000).

30. Thompson, T. 2003. Presentation before the Grocery Manufacturers of America. January.

31. Schneiderman, H. 1987. *Biotechnology: A key to America's economic health in health care and agriculture*. Presentation to Second Annual Society for Microbiology Conference on Biotechnology, June 25-28. San Diego, CA.

32. Gibbons, J.H., Assistant to the President for Science and Technology. 1997. Presentation to Council on Competitiveness, February 24. www.ostp.gov. Accessed July 30, 2003.

33. USDA's Biotechnology Website, www.usda.gov. Accessed June 10 2005.

34. Merrigan, Kathleen and Irv Rosenberg. Agricultural biotechnology: The road to improved nutrition and increased production? 2003. *Nutrition Reviews* 61: S99 June.

35. *University-Industry relationships: Framing the issues for academic research in agricultural biotechnology.* Portland State University and the Pew Initiative on Food and Biotechnology. November 2003.

36. Formally known as the Governmental Accounting Office.

37. I was the lead person for the US Congress at the time spearheading the review and responsible for finding independent experts to help Congress in its investigation.

38. Food, Agriculture, Conservation, and Trade Act. 1990. Report of the Committee on Agriculture, Nutrition, and Forestry, United States Senate, 101st Congress, 2nd Session, Report 101-357.

39. Sec. 1668. Biotechnology Risk Assessment Research of the 2002 farm bill set the following research priorities for funding in subsection C: "4) Environmental assessment research designed to provide analysis which compares the relative impacts of animals, plants, and microorganisms modified through genetic engineering to other types of production systems, and 5) Other areas of research designed to further the purposes of this section." Additionally, Congress stated that "funds made available under this subsection shall be applied, to the maximum extent practicable, to risk assessment research on all categories identified in subsection (c)."

40. Domingo, J.L. 2000. Health risks of GM foods: many opinions but few data. *Science* 288: 1748-1749; Wolfenbarger, L. and Phifer, P.R. The ecological risks and benefits of genetically engineered plants. *Science* 290: 2088-2093.

41. The EU-U.S. Biotechnology Consultative Forum, Final Report, December 2000. http://Europa.eu.int/comm/external_relations/us/biotech/report.pdf. Accessed June 5, 2005.

42. Agriculture in the Classroom, USDA, www.agclassroom.org. Accessed August 3, 2004.

43. *BIO applauds House measure to correct U.S. tax code.* Press release. July 28, 2003. www.bio.org. Accessed July 28, 2003.

44. Diamond v. Chakrabarty 447 U.S. 303 (1980).

45. For more discussion see my article in *Genes for the future: Discovery, ownership, access, National Agricultural Biotechnology Council,* Report 7, 1995, CNABC, Ithaca, NY.

46. This project is a collaboration between USAID, its counterpart in Britain, the Rockefeller Foundation and Monsanto, DuPont, Syngenta, Dow AgroSciences. Cowpeas, chickpeas, cassava, sweet potatoes, bananas, and corn are under development.

47. US Patent and Trademark Office. 2002. *TAF Special Report.* All patent numbers taken from Economic Research Service website. Agricultural Biotechnology Intellectual Property. www.ers.usda.gov/DATA/AgBiotechIP. Accessed September 9, 2004.

48. Shand, H. 2003. Is there a solution to the current controversies surrounding IP in AgBiotech? *Agbiotech Buzz* 3: March 3. Pew Initiative on Food and Biotechnology.

49. *Genetically modified crops in the United States*. Fact sheet. August 2004. Pew Initiative on Food and Biotechnology.

50. NASDA Policy Statement. *Biotechnology: A key to agriculture's future*. June 30, 2003.

51. *State legislative and local activities related to agricultural biotechnology continue to grow in 2003-2004*. Fact sheet. May 2005. Pew Initiative on Food and Biotechnology.

52. Ibid.

53. Rodemeyer, M. 2005. *Pew initiative finds state legislators focused on "next generation products."* News Release, May 26, Pew Initiative on Food and Biotechnology.

Chapter 12

Biotechnology Policy
in the European Union

Armin Spök

INTRODUCTION

When asked to predict the next wave of technological revolution for key and core technologies, policymakers on both sides of the Atlantic would probably include biotechnology as one of the most promising sectors. Since the 1980s both Europe and the United States have consistently acknowledged the enormous economic potential of biotechnology. Two decades later their status of commercialization of agricultural biotechnology is very different. In the European Union (EU) eighteen genetically engineered plants (GEPs) have been authorized so far for commercial release (including use in animal feed) and a total of sixteen GEPs for use in human food.[1] In contrast, the United States has approved fifty-eight GEPs for environmental release and for food/feed purposes.[2] These differences are more striking when looking at the total area used for commercial cultivation of GEPs. In the EU so far only Spain is growing GE plants. Only about 32,000 hectares comprising only one genetically modified maize variety are involved whereas total area of GEP in the United States amounts to 42.8 million hectares, comprising mainly various cultivars of maize, soybean, canola, cotton.[3]

The notion of the United States, Europe's main economic competitor, being constantly ahead in biotechnology has triggered many attempts in Europe to catch up. Despite launching large research pro-

Genetically Engineered Crops
© 2007 by The Haworth Press, Inc. All rights reserved.
doi:10.1300/5880_12

grams, promoting university spin-off companies and patenting, creating favorable conditions for biotechnology start-up companies, and establishing biotech clusters and whole regions dedicated to biotechnology, the gap between the United States and Europe has widened.

Policy analysts frequently point to fundamental differences in the regulatory regime and in consumer acceptance as causes of this divergence. While in the early 1980s the policy and commercial environment was largely similar on both sides of the Atlantic, since 1990 not only commercialization but also the regulatory regime of agricultural biotechnology has pursued different courses. In contrast to the United States, the political agendas in European countries have increasingly been dominated by issues of health and environmental risks and their appropriate management. The public debate and many policymakers have been preoccupied with these issues ever since, without being able to establish a feasible and stable regulatory regime that would allow for commercialization of agricultural biotechnology (as frequently noted by certain U.S. commentators). U.S. commentators have also criticized the European preoccupation with "hypothetical" risks as a reason for delaying market entrance of GE crops and food. EU politicians have even been accused of being hostile toward innovation and of hampering free trade. The EU regulatory regime has been alleged not to be based on sound science and even as irrational, while the European Commission (EC) has sometimes been depicted as captured by consumer and environmental lobbyists. The recently updated EU legislative framework is unfavorably perceived as very strict or even as overregulation.[4] The need for importers to comply with the comprehensive set of EU rules on risk assessment, labeling, traceability, and product segregation has led to a decline in agricultural exports from the United States and Canada to the EU and has indirectly impeded U.S. exports to so-called third world countries because some of them have adopted the EU position (for economic reasons). Recently the United States has formed and led an alliance of large exporters of agricultural products to file a case against the EU at the WTO Panel.

This chapter is inspired by these divergent and increasingly diverging biotechnology regimes on both side of the Atlantic, in spite of basically similar economic interests, and by commentators who swiftly identify Brussels' apparent obedience to the consumer lobby as a main cause of conflict. I go beyond consumer acceptance to explore some of the issues underlying the formation and the characteristics of the present biotechnology regime in the EU. EU biotechnology governance is one policy arena where the bigger issue of European divergence and integration is being negotiated. This chapter argues that European biotechnology governance is complicated and even constrained by fragmentation arising from divergent European political and cultural traditions, economic interests, and public support or resistance toward biotechnology. The emergence and further development of a European biotechnology regime can, therefore, be viewed as another case of integrating European diversity. Given the different grades of EU harmonization in promotion of R&D on one hand and of setting environmental and health standards on the other, biotechnology governance in the EU is more likely to emerge from the risk side. I argue that even within an agreed GEO regulatory regime, EU-specific institutional and legal constraints of market authorization procedure foster decision-making diversity more than they forge consensus. A second structural precondition is the scientific uncertainty inherent in GEO risk assessment. The regulatory regime of the 1990s eventually collapsed in the wake of a series of health scandals and scares in Europe. These were not directly related to the GEO issue but dramatically affected public trust in policymakers, legislative frameworks, and regulatory institutions on both national and EU levels, and highlighted persisting problems in dealing with scientific evidence and uncertainty in risk assessment. The current regulatory regime puts even more emphasis on precaution, transparency, labeling, and stakeholder participation as well as on a clear-cut separation of science and policy in risk-assessment procedures. Centralized risk assessment for streamlining the decision-making procedure is—to a certain extent—already embedded in the new regime.

EUROPE VERSUS THE EUROPEAN UNION

Biotechnology has been on the agenda of national and European supranational institutions since the early 1980s. The EU has become the major arena for European public policy, although it is not the only one. Biotechnology governance has been complicated by activities of the Council of Europe (COE),[5] the European Patent Office (EPO), and the European Science Foundations (ESF), which are international bodies independent of the EU each with different subsets of European membership. While the COE and the ESF played roles in the early phase of biotechnology in Europe, they became less important in the legislative phase from the mid-1980s onward. In recent years, the Council has focused more on biomedical aspects and bioethics. The EPO works with intellectual property rights and has worked with the EC since 1986 to develop the Directive dedicated to biotechnological interventions. While the EU and the European Commission are the most important supranational arena and actor in terms of biotechnology, national policies pursued by EU member states have been crucial in establishing the present regime.

THE LIMITED IMPORTANCE
OF EE R&D POLICY

The European Council at Lisbon set the goal for the EU to become "the most competitive knowledge-based economy in the world" and the life sciences and biotechnology are considered key in order to meet that goal.[6] Developing a competitive position in biotechnology has been a strategic priority in the EU for more then two decades. Starting from the fact that the EU was significantly behind its primary competitors, the United States and Japan, major efforts have been made to close this gap. The General Directorate (GD) XII (Science, Research, and Development) was the first initiative and was followed in the 1980s by others, including the Biotechnology Framework Programme (BFP), the Biotechnology Engineering Programme (BEP), the Biotechnology Action Programme (BAP), the Biotechnology Research and Innovation for Development and Growth (BRIDGE), the Euro Collaborative Linkage of Agriculture and Industry through Research (ECLAIR), and Food-linked Agro-Industrial Research (FLAIR).[7] Biotechnology has played a major role within the EU Framework Programme for Research and Technological Development ever since. A number of policy documents have been issued since the 1980s, which sought to develop for the EU a comprehensive biotechnology strategy that would improve its competitiveness. Progress was modest and the strategy was recently reiterated in a communication that set out an EU strategy for biotechnology and the life sciences.[9] Recent evidence suggesting a decline of R&D activities[10] led to the strategy being given even higher priority.

Even with these important long-term attempts to generate a European research agenda, it is important to consider the role of EC initiatives in the overall promotion of R&D. For instance, the Sixth Framework Programme, the EU's principal research policy instrument, which will allocate most of the EC's funding over the period from 2003 to 2006, comprises only about 5 percent of total public expenditures on R&D in the EU.[11] Obviously, national R&D policies are more important. Not surprisingly, there have been numerous initiatives at member-state level to boost biotechnology research.[12] A comparison of public funding, however, revealed very varied commitment to biotechnology R&D, from 13.8 percent of overall R&D (Belgium) to as little as 0.4 percent (Italy).[13] Comparisons of another widely applied indicator of innovative activity, the number of dedicated biotechnology firms (DBF) per capita, showed a similar divergence. The number in Sweden exceeds those in Spain and Italy by a factor of 25-30. The DBFs per GDP in Germany differed from Austria by a factor of 158. Four member states, Sweden, the United Kingdom, France, Germany, are home to about 75 percent of the more than 2,000 DBFs registered in the EU in 2000.[14] While these numbers do not provide a comprehensive picture of biotechnology R&D, they illustrate the divergence and clustering that relate to different starting times and different policies in member states. In fact, "the absence of a shared vision of what is at stake [. . .] [and of] common objectives and effective coordination" was also diagnosed in an EC communication proposing a community strategy for biotechnology and the life sciences.[15] The communication criticized the lack of coherence in community legislation and policy and acknowledged that the proposed strategy can "only be successful if it is accompanied by additional activities by the individual Member States."[16]

In essence both an EU R&D policy and particularly the promotion of biotechnology are rather fragmented. A substantive "European Research Area," a goal that is at heart of the Sixth Framework Programme of GD Enterprise, is, therefore, unlikely to emerge soon and is unlikely to have a major influence on EU biotechnology governance.

OVERALL REGULATORY APPROACH

In contrast to the United States and Canada, the EU has established a body of separate legislation covering the application of genetic engineering (GE) techniques in research, development, and commercial contexts. GEO legislation has been dynamically evolving since 1990. EC Directives and Regulations covering containment as well as deliberate release (import, cultivation, and processing) of GEOs and their application for food and feed were introduced; initially Directive 90/220/EC for deliberate release in 1990 and the Novel Food Regulation in 1997. The rationale of the EU regulatory approach is what is called "horizontal" or "process-based" legislation. This means that the process or technology itself is in the focus of regulation rather than a specific product. Under process-based legislation, all innovative activity is subjected to regulation from the R&D stage to the commercialization stage as permission is sought to sell, grow, and/or market the crop and the final product. This type of legislation was introduced in 1990 despite contradicting advice from OECD[17] and industry, and in spite of an earlier consensus within the EC and the CoE that there was no need for a specialized legislation. The reasons given were to harmonize EU legislation, reduce risk, and deal with uncertainty. First, there was a perceived need to protect the common market from idiosyncratic national legislation and to allow for a harmonized regulatory framework for biotechnology.[17] Second, it allowed the EU to deal somehow with emerging criticism, especially from countries that had already conducted or were in the midst of public debate and had already devised national legislation (especially Germany and Denmark). Third, to provide for the protection of human health and the environment. At the time there was still uncertainty about the criteria for GEO risk assessment in case of deliberate release into the environment.[19] Pressure from industry and the United States prompted the EC to launch the Directive 90/220/EEC (on deliberate release and placing on the market of GEOs) to facilitate and even to move the regulatory oversight to future sectoral legislation.[20] In fact the EC immediately started to work on what later became the Novel Food Regulation.[21] A key feature of the EC regulatory approach outlined in Directive 90/220/EEC was an implicit precautionary approach.[22] Releases for R&D purposes required consent from a national competent authority (CA). For market approvals, national CAs were to decide by majority on EU-wide authorizations. The Directive provided for a step-by-step approach of progressively decreasing physical containment (from the laboratory to the greenhouse and then to the field) and increasing the dimensions of the release from small scale to large scale. Furthermore, each case had to be dealt with separately (case by case).

RISK ASSESSMENT AND DECISION MAKING

Until recently GE crop or food risk assessments included in an application were evaluated by the national CA where the application was initially filed. If the national CA recommended an approval, a consultation process followed where all member states were provided with the complete or an abridged dossier[23] together with an initial assessment report of the national CA (Novel Food Regulation). On the basis of the information circulated, national CAs reviewed the risk assessment of the applicant and that from the national CA.[24] At this stage any national CA could raise questions and ask the applicant for further information. Unresolved problems are addressed following consultation with the Commission's Scientific Committee on Plants (SCP) or Scientific Committee on Food (SCF)[25] through a "comitology" procedure that presents a proposed decision to the respective Regulatory Committee.[26] This Committee is composed of representatives of national CAs who should eventually decide on the application by a qualified (two-third) majority vote.[27] Apart from minor differences[28] this procedure applied for applications according to Directive 90/220/EEC and Novel Food Regulation. Furthermore, it still applies to applications under the present Directive 2001/18/EC for deliberate release and placing on the market of GEOs.[29]

In the absence of a qualified majority in the Regulatory Committee, the proposed measure is referred back to the European Council (Council of Ministers), which also takes a decision by qualified majority. However, if the Council does not take a decision, the Commission eventually adopts the proposed decision.[30] Member states can thus exert considerable influence on the decision-making process either in the Regulatory Committee or in the Council. In case of politically sensitive decisions, this procedure is far from being a technical exercise. Only in three of thirteen EU market approvals under Directive

90/220/EEC—for ornamental plants—no objections were raised by member states. Eight cases were resolved by a qualified majority in favor, and two cases had to be decided by the Commission because a qualified majority could not be achieved in either the Regulatory Committee or the Council, in spite of favorable opinions of the Commission's own Scientific Committees.[31] In cases of applications for GE food market authorization under the Novel Food Regulation (after 1997) only those products were authorized that had initially been considered as substantially equivalent to conventional counterparts and that were thereby eligible to a shortcut procedure. No decision could be reached on full-fledged applications for not fully equivalent products before the de facto moratorium was invoked in 1998. Authorizations are in principle valid throughout the EU. The only way for member states to temporarily restrict or even prohibit the marketing of an authorized product is to invoke the so-called safeguard clause, which is another general feature in EU directives and regulations. Directive 90/220/EEC, for instance, provided that in case a product does "constitute a risk to human health or the environment, it may provisionally restrict or prohibit the use and/or sale of that product on its territory" (Article 16). Similar provisions are included in the Novel Food Regulation (Article 12). So far, safeguard clauses have been invoked for five authorizations granted under Directive 90/220/EEC and one under the Novel Food Regulation. The scientific evidence provided as justification given by the national authorities was evaluated by the EC's Scientific Committee and was dismissed in all cases.[32] According to statutory requirements included in both pieces of legislation, national safeguard measures have to be decided following the comitology procedure. Nevertheless, no formal decision was taken by the EC over the past years in relation to these national safeguard measures.

Discussions on market authorization of GEO products between member states, the EC, its Scientific Committees, and the applicants have been very contentious. In most cases the whole comitology procedure was exhausted leaving the EC to decide on the market authorization. This apparent inability of member states to agree on market authorizations based on scientific risk assessments and the frequent invoking of national safeguard measures on scientific grounds may surprise an outside observer because these decisions are supposed to be on technical and scientific matters. Furthermore, these conflicts raise a number of questions about the efficiency of EU harmonization and on the practicality of the decision-making procedure. These questions are dealt with next.

DIRECTIVE 90/220/EEC—THE SHORTCOMINGS OF A HARMONIZATION TOOL

Since the inception of EU institutions, their characteristics and remits as well as the fine-tuning of the balance of powers charactering their interactions have changed. Public policy outcomes at the European level have been strongly influenced by the very nature of these institutional arrangements and their evolution.

It should be recalled that the EU was initially established to facilitate intra-European trade. It brought together countries with sometimes quite different political, legislative, and cultural backgrounds and traditions. While integration of the EU was initially framed in terms of trade liberalization, many efforts are being made to facilitate a political integration. Thus the EU has to cope with a multiple integration process: to continue developing a European economic area while pursuing the aim of a stronger political union. At the same time new member states have been integrated during several rounds of enlargement.

The EU seeks to organize transport, industry, agriculture, fisheries, energy, and tourism in such a way that they can be developed without destroying our natural resources. Harmonization of environmental policy, therefore, started quite early. Meanwhile more than 200 environmental protection directives have been adopted, most of which are designed to prevent air and water pollution and to encourage environmentally sound waste disposal. Other major issues include nature conservation and the supervision of dangerous industrial processes. Harmonization of general food policy, in contrast, has been lagging behind, and only with the recent implementation of Regulation 178/2002 have the cornerstones of a general EU food law have been established. The most recent move was to establish the European Food Safety Authority (EFSA), which is commissioned to conduct scientific risk assessment of food and feed products and ingredients, especially in the context of market authorization procedures.[33]

Despite these efforts toward legislative and institutional harmonization within the EU, a considerable number of EU laws are still directives, which means that they are binding on the member states as an objective to be achieved but the national governments must decide on how the Community objective is to be incorporated into their domestic legal systems, thereby providing considerable leeway to the member states.

This was also the case with Directive 90/220/EEC, which did not permit socioeconomic criteria to be considered in the assessments of deliberate releases and market introductions. Only risk issues were to be dealt with in order to restrict the range of permissible arguments and to keep the regulatory process firmly within the scope of scientific risk assessment. Already at the stage of incorporating the Directive into national law, this turned out to be a matter of interpretation. Norway,[34] and, to a certain extent, Sweden, and Austria introduced such criteria in their national laws.[35] The competent CA is different in each member state and may change over time. For instance the lead CA in France is the Ministry of Agriculture, Food, Fisheries and Rural Affairs, in the Netherlands it was the Ministry for Environment, in Germany it was formerly the Ministry for Health, now the Federal Ministry of Consumer Protection, Food and Agriculture.[36]

The Directive also included reference terms such as "evidence for safety," "environmental harm," and "adverse effects on human health and the environments" without providing definitions for these terms. Thus the actual meaning of these terms and what constitutes appropriate precaution has been left open to interpretation[37] as illustrated in the following section.

DIVERGING FRAMINGS
IN GEO RISK ASSESSMENTS

According to the procedure for decision making laid out in previous sections, if anyone challenges either the applicant's risk assessment or the evaluation carried out by a national CA, or invokes a national safeguard measure, he or she must present scientific arguments to support the objections. However, in case of GEO risk, the science behind the risk assessment is not straightforward and is scattered—filled with uncertainties and even ignorance. This is determined by the very nature and the perceived irreversibility of potential negative environmental effects of GEPs and their associated risk factors (invading ecosystems, affecting ecological balances, harming nontarget species, spreading genes to other/related species, non-intended effects of single gene modifications etc.). On the other hand, the explicit preventive (or precautionary) character of the legislation determines that risk must be assessed before there is any direct evidence of harm. Hence, risk assessments have to face the complexity of ecosystems and the molecular complexity of organisms at the same time. Scientific knowledge on effects on both the molecular and ecosystem levels is scarce, and causal models and risk indicators are contested or lacking. Even the relevance of particular effects, risk factors, and the questions of defining environmental harm remains unclear and contentious.

As a result new problems with risk-assessment procedures and the science behind the risk assessment on various levels become apparent during authorization procedures. These include the overall scope of the risk assessment, the appropriate baseline of acceptable risk, and the particular requirements for risk-relevant information or studies to be carried out. It is not surprising that different interpretations on the scope soon become apparent. A ten-country comparative study conducted by Levidow and co-workers[38] revealed such differences in framing assumptions. Some of the results are provided in Table 12.1.

Under Directive 90/220/EEC, present agricultural practice was accepted as a baseline for evaluating environmental effects of GEPs. Hence, it was deemed acceptable that a herbicide (e.g., due to extensive cultivation of GE herbicide tolerant rape) would become ineffective to control weedy herbicide-tolerant rape or that *Bt*-toxin would become ineffective for controlling insect pests in conventional agriculture due to occurrence of resistance triggered by extensive cultivation of GE *Bt* crops. This was a view initially held by the U.K. CAs. Other member states took a different approach.[39] Denmark, for instance, argued that the risk assessment should encompass the implications for the overall herbicide usage and future weed-control options, especially given that oilseed rape can hybridize with weedy relatives. Sweden argued that broad-spectrum herbicides would damage wildlife habitats by eliminating all vegetation and thus demanded that such effects should be included in the evaluation of herbicide-tolerant GEPs. France and Italy pointed to the possible occurrence of multiple-

TABLE 12.1. Diverging framings in the context of Directive 90/220/EEC risk assessment.

Framing type	Austria	Germany	United Kingdom
Scope Directive 90/220/EEC	Safety/biodiversity, agronomic effects	Safety	Safety
Risk Assessment	Preventing adverse effects	Prevent risks according to the state of S&T	Comparison with conventional agriculture
Baseline	Organic agriculture	Conventional agriculture	
Herbicide impacts on agricultural practice and environment	Considered	Considered	Initially not considered
Bt crops	Considered	n.m.	n.m.
Antibiotic-resistance marker genes	Ampicillin resistance opposed	RKI advises to avoid antibiotic-resistance marker genes	Ampicillin resistance initially not considered.
Gene flow to weeds	Considered	Considered	Initially not considered

Source: Adapted from Levidow, L. and Carr, S. 2000. UK: Precautionary commercialization? *Journal of Risk Research* 3: 261-270; Torgersen, H. and Seifert, F. 2000. Austria: Precautionary blockage of agricultural biotechnology. *Journal of Risk Research* 3: 209-217. Dreyer, M. and Gill, B. 2000. Germany: "Élite precaution" alongside continued public opposition. *Journal of Risk Research* 3: 219-226.

Note: Herbicide use: Regulatory gap assumed as Directive 91/414/EEC authorizes the EC only to approve or to ban specific pesticides. The overall herbicide usage cannot be addressed within the scope of this Directive. n.m., not mentioned; RKI, Robert Koch Institute.

resistance weeds that might result from large-scale commercial use of various herbicide-tolerant GEPs.

In the case of *Bt* crops, the EC considered that the generation of *Bt*-resistant pest insects was not an adverse environmental effect under Directive 90/220/EEC because conventional means for managing resistant pest insects are still available.[40] By contrast, an increasing number of member states see the occurrence of *Bt*-resistant insects as adverse effects. Therefore, Denmark demanded that resistance-management measures be included in the *Bt* maize dossier. Sweden requested additional studies on whether a reduction of the target-organism population would affect insects that feed on them or would affect plants that they pollinate. France, while initially adopting the view that insect resistance would not cause an adverse effect under Directive 90/220/EEC, did not sign the authorization before resistance-management measures had been negotiated between the notifier and its CA.

In the course of these and other authorization procedures of GEPs[41] the normative baseline was challenged. The United Kingdom and the EU Scientific Committee on Plants, for instance, considered conventional chemical intensive farming as the normative baseline for assessing the risks. Austria, in contrast, considered organic farming as the baseline and included agronomic effects in the scope of its assessment; meaning that any increase of herbicide usage is unacceptable. The Netherlands evaluates whether persisting effects on the natural vegetation may arise from GEO release; Belgium will not accept a release that could aggravate existing environmental problems. The Italian authorities attached a monitoring requirement to a permission for transgenic maize, whereas Denmark included agronomic effects by using the criterion of "sustainable social development," which was even included in the national law.[42]

Diverging interpretations of what constitutes harm and what endpoints and approaches are considered appropriate in the course of risk assessment are not restricted to the general scope, normative baselines, and environmental impacts. Similar problems became evident in the course of assessing human health effects of both GEPs under Directive 90/220/EEC and GE food under the Novel Food Regulation as revealed by a review of risk-assessment practices in the EU that scrutinized Directive 90/220/EEC and Novel Food Dossiers.[43] In the course of the authorization procedure for sweet corn *Bt*11 under the Novel Food Regulation, for instance, the applicant did not provide any toxicity test in order to substantiate the claims that the newly introduced protein is safe for human consumption. An acute toxicity study in mice was only carried out after a request from the Dutch CA (who acted as rapporteur at the EC). Several member states considered an acute toxicity study alone as insufficient. France demanded a ninety-day sub-chronic toxicity study of the newly introduced protein as well as whole-food feeding studies. Greece and Austria demanded a whole-food feeding study, and Austria emphasized that it should be a long-term investigation. Austria demanded subacute twenty-eight to sub-chronic ninety-day toxicity studies for evaluation of maize GA21 under the Novel Food Regulation, and Greece requested whole-food toxicity studies instead of being content with only studies of the protein. Austria's requirement was refused by the applicant, but an additional ninety-day toxicity study of maize kernel in rats was submitted as a response to the Greek authorities.[44]

These studies revealed that the scientific database on possible toxic actions of proteins is very narrow or even nonexistent. Applicants frequently refer to anecdotal evidence and tend to interpret the absence of evidence of harm as evidence of the absence of harm.

Shortcomings in Risk-Assessment Practice

The same study also revealed a number of shortcomings and inconsistencies in GEP and GE food risk-assessment practice and raised questions on reliability and verifiability of certain approaches and methods, for instance:[45]

Substantial equivalence. This plays a key role in risk assessments, but in contrast to its conceptual role as a starting point in risk assessment, it is frequently a terminal stage. Claims of substantial equivalence are frequently based on field trials and compositional analysis that are not properly designed and are often not backed up by thorough and consistently applied statistical analysis.

Verifiability. Risk assessments and safety conclusions cannot be verified or may not be verifiable at all on the basis of information presented in the dossiers, because of lack of details in the description of tests, approaches, data display, and the tendency not to include full reports.

Consistency. Although overall approaches to risk assessment are similar in most applications, there were many detailed differences, especially between dossiers, pertaining to the same plant species and/or proposing similar applications. These point to differences in the assessment procedures and to a lack of details in guidance documents.

Safety conclusions are

- often based on indirect evidence and/or assumption-based reasoning while direct testing of toxic or allergenic properties is limited if conducted at all;
- partly based on methods, approaches, and assumptions that are questionable (e.g., homology and in vitro digestibility studies in toxicity assessment; studies and assumptions on which the decision tree approach in allergenicity assessment is based);
- not backed up by thorough compliance with safety and quality assurance protocols;
- largely focused only on the novel proteins introduced;
- not concerned enough with protection against unintended effects of genetic modification and may even be dismissed. Such effects are only considered if expressed as conspicuous alterations of morphological or agronomical properties or in key plant compounds. In some cases, significant differences in compositional analysis have been disregarded without attempts to investigate further the likelihood of unintended secondary effects. Whole-plant feeding studies included in the dossiers are usually feed conversion studies and cannot be considered as valid assessments of toxicity.

These examples from the two studies by Spök et al.[46] lead to the conclusions that (1) in the absence of shared norms for study scope and the effects that are considered as negative, divergent views from national CAs are inevitable; (2) due to the fact that the details of risk assessments must be put into concrete terms, each individual application only presented divergent views of national CAs on selected methods and endpoints in risk assessment; (3) some details in the risk-assessment process are interspersed with open questions and conflicting scientific evidence. In short, scientific risk assessment of GEOs and GE food is neither a straightforward nor a purely "technical" exercise.

Some preliminary conclusions on the characteristics and problems of GEO risk policy can be drawn. In the absence of a comprehensive scientific basis on potential environmental and health impacts of GEOs and GE food, CAs must deal with scientific uncertainty and ignorance. This explains why particular safety requirements are frequently contested. The present approach for risk assessment and decision making, especially the comitology procedure, provides opportunity to articulate and deliberate on conflicting views. On the other hand, the lengthy procedures seem inefficient, time consuming, and difficult to predict both in terms of effort and outcomes.

Improved authorization procedures require reduction of the inherent uncertainty in risk assessment. Alternatively, the decision-making procedure can be reengineered to ensure that divergent interpretations and views cannot delay or even block decision making. This might "improve" the efficiency of the process but not necessarily the quality of the outcomes. Currently the EC is trying to follow both paths.

EUROPEAN PUBLICS: RELUCTANCE AND DIVERGING ATTITUDES TOWARD BIOTECHNOLOGY

The regulatory regime established in the 1990s must be viewed against the backdrop of both the diversity of attitudes among the general public of member states and the overall reluctance and even outright rejection of GE crops and food, as shown by the Eurobarometer Surveys on Biotechnology (providing longitudinal and cross-country comparisons) in several influential member states. This backdrop is important in understanding certain aspects of past regulatory practice and the new regulatory regime.

According to the most recent Eurobarometer survey,[47] support for GE crops can be as high as 91 percent in Spain and as low as 54 percent in Greece and 57 percent in Austria. Similarly, Spain shows the highest support for GE food in the EU (74 percent), in contrast to Austria (54 percent), France (30 percent), and Greece (24 percent). It seems that support is declining moderately over time (data from 2002). Public perception was similar or even more diverse in 1996, which led the authors of the survey to conclude as follows:

> In terms of public policy, we have found different European countries dealing with modern biotechnology over very different timescales and in very different ways. In terms of media coverage, we have found that a commonly held discourse of progress and benefits is paralleled by rather different patterns of

media reportage in the European countries. [. . .] the different European countries tend to have widely differing levels of engagement with, knowledge about and attitudes towards biotechnology.[48]

In an attempt to explain this diversity Bauer et al. argued that particular biotechnologies occupy very different places in the different national economies and that there are considerable cultural differences that go far beyond the biotechnology issue. Such diversity also applies to the level of particular product authorization and its associated public debate, for example, the approval of transgenic soybean in the EU and arrival of the first shipments served to reopen seemingly dormant debates on biotechnology in Denmark and the United Kingdom and introduced new and critical discourses in Italy and Sweden.[49]

A Crisis of Trust and Legitimacy

The public debate on GE crops and food in Europe was paralleled by and interlinked to the advent of a crisis in European governance of food policy and to attempts at crisis management. In the last decade an unprecedented series of food incidents and scandals have shaken European consumers, national and EU institutions, and politicians. Some of these scandals were largely restricted to individual countries, such as the contaminated blood scandal in France, the Belgian dioxin scandal, and issues in the United Kingdom that included *Escherichia coli,* antibiotics and hormones in meat, salmonella in eggs, *Listeria* in cheese, and pesticide residues in food and phthalates (benzene-related compounds) in cosmetics, toys etc. Other incidents affected several countries such as the contamination of synthetic medroxyprogesterone acetate in animal feed and soft drinks. By far the biggest concerns that affected all of Europe were the linking of Bovine Spongiform Encephalopathy (BSE) in cattle to Creutzfeldt-Jakob Disease in man. These scandals and the BSE crisis in particular exposed a number of shortcomings in the food safety and health policies of several member states, especially in the United Kingdom, France, and Germany as well as at the EC. As a consequence, consumer trust in national and EU institutions responsible for risk assessment and risk management sharply declined. A number of far-reaching changes in food safety policy were made to restore consumer trust. These included national and EU institutional arrangements which were proposed in the White Paper on Food Safety[50] and swiftly implemented. A major intent was to separate scientific advice (risk assessment) more clearly from policymaking (risk management).

The crisis of trust in food policy has to be viewed in the context of a more general loss of trust in institutions and politics, particularly acute at the level of the EU. Loss of citizen confidence in the EU is certainly not helpful in dealing with the challenges of European integration. The White Paper on European Governance[51] proposed to improve policy delivery and to enhance legitimacy and accountability by opening the policymaking process to stakeholders and the general public and to make it more transparent. As with the White Paper on Food Safety[52] the relationship of expert advice to policymaking again came under scrutiny in the White Paper on European Governance.[53]

Both White Papers are proposals for long-term structural and procedural reforms at the EU and national levels. Several reforms have already been implemented, such as establishing of the EFSA as the independent scientific body responsible for risk assessment in food policy.

Both, the reforms that have been implemented so far and their underlying drivers, the malfunction of food safety policy and the apparent lack of trust in and legitimacy of EU policy and institutions, have been and still are affecting agricultural biotechnology policy and are key to understanding some of the current difficulties and peculiarities.

THE NEW REGULATORY REGIME

The extension and revision of EU GEO legislation has been especially dynamic in recent years. Following the introduction of horizontal legislation for deliberate release and marketing of GEOs in 1990 and revisions in the second half of the 1990s, additional sectoral legislation was introduced for GE food (1997), GE feed (2003), labeling and traceability (2003),[54] and transboundary movements of GEOs (2003). Furthermore, a Commission Recommendation was issued on the coexistence of GEPs, conventional and organic agriculture (2003).

The EU is also party to the UNEP Cartagena Protocol on Biosafety Convention on Biological Diversity, which came into force on September 11, 2003. The main EU legislation covering both experimental releases and marketing of GEOs is Directive 2001/18/EC (which replaced Directive 90/220/EEC in 2001).[55] This probably marks the watershed between the old and new regulatory regimes on GEOs. The scope of this Directive includes cultivation, introduction into the environment, storage, import, and seed production of GEOs. Prior to any marketing of GEOs an EU-wide consent is needed. The authorization procedure largely follows the Directive 90/220/EEC with national CAs playing a major role as risk assessors and rapporteurs to EC committees. A submission for market authorization must include an environmental risk assessment (including risks for human and animal health) that identifies and evaluates potential adverse effects of the GEO. This Directive not only refers explicitly to the precautionary approach (see note 43) but also widens the scope of risk assessment beyond the preceding Directive to include direct or indirect, immediate, or delayed effects and taking into account any cumulative and long-term effects on human health and the environment. According to the original wording of the Directive 2001/18/EC: C2, "it is important not to discount any potential adverse effect on the basis that it is unlikely to occur."

Transparency of the decision-making process gained more importance than in the preceding legislative framework. For instance, a public consultation is mandatory for experimental releases (see Part B of the Directive 2001/18/EC), and member states are required to establish and maintain a public register of the particular locations of authorized release experiments and of GEPs authorized for marketing. A summary of the application for multisite experimental releases and for marketing of GEOs and GE food and feed products, as well as the opinion of CAs will be made available to the public. Comments can be submitted within thirty days and the final decision will be included in a public register.[56]

Regulation 1829/2003 on genetically modified food and feed recently replaced the Novel Food Regulation.[57] Besides dealing with potential adverse affects on human and animal health or the environment, it stipulates that GE food/feed must not mislead the consumer and must not differ from food/feed it is intended to replace to an extent that normal consumption would be nutritionally disadvantageous for the consumer/animal.

A premarket assessment remains mandatory for all GE food and feed. The application procedure is, however, different from the one in Directive 2001/18/EC because it provides for a centralized risk assessment. National Competent Authorities serve to forward applications to the EFSA, which is then responsible for the scientific risk assessment. EFSA's opinions will be made available to the public and the public will have the opportunity to make comments.

Within three months of receiving the EFSA opinion, the EC will draft a proposal to grant or refuse market authorization. This proposal is then subjected to vote within the Standing Committee on the Food Chain and Animal Health composed of member-state representatives. Again the comitology procedure will apply. Both the Regulation and the regulatory procedure have two important features. First, applicants need not request separate authorizations for food and feed applications. Instead, a single risk assessment is provided for a GEO and its possible uses. GEOs that are to be used as food and feed can only be authorized for both purposes. Thus, cases such as the Starlink maize, which was authorized only for feed use in the United States but eventually appeared in food, would be avoided. Second, the regulation is based on the so-called *one door–one key* principle; hence it is possible to file a single application for both authorizations, for marketing under the statutory requirements of Directive 2001/18/EC and for food and/or feed use under the criteria laid down in Regulation 1829/2003.

Market authorizations are generally granted for a ten-year period and where appropriate, are subject to a monitoring plan according to both Directive 2001/18/EC and to Regulation 1829/2003. Postmarket monitoring becomes mandatory, if it is aimed at cultivation of GE crops.

Directive 2001/18/EC came into force in September 2002 and Regulation 1829/2003 in 2003. So far, twenty-four applications for marketing of GE crops have been submitted under Directive 2001/18/EC, eleven of which are for import and processing only whereas the remainder include cultivation. Some of these applications had already been submitted under Directive 90/220/EEC and were "upgraded" to Directive 2001/18/EC, and about half of the applications are presently being changed to Regulation 1829/2003 applications.[58]

Approval of GEPs for cultivation according to Directive 90/220/EEC (now Directive 2001/18/EC) does not allow unlimited commercial cultivation of a GE crop. This requires inclusion of each commercial variety in the Common Catalogue of varieties of agricultural plant species.[59] A prerequisite of inclusion in the Common Catalogue is the inclusion in a national list. To be added to a national list, a variety must be distinct, sufficiently uniform and stable and, for agricultural crops, have satisfactory value for cultivation and use. Testing prior to inclusion in a national list does not include either environmental or health-related safety aspects. Rather it focuses on agronomic criteria. By January 2004, a total of twenty-three GE varieties were included into national lists (Spain, France, the Netherlands); however, none has yet been included in the Common Catalogue.[60]

Regulation 1829/2003 only recently came into force and no applications have yet been received by the EC. There are nine applications for GE food still pending under the preceding Novel Food Regulation.[61] Along with Regulation 1829/2003, the EC introduced a Regulation 1830/2003 on labeling and traceability of GEOs.[62] In contrast to the previous provisions under the Novel Food Regulations, labeling is no longer triggered by the appearance of measurable traces of recombinant DNA or proteins derived therefrom. Instead all food and food ingredients produced from GEO must be labeled. The EC is thereby meeting a long-standing demand from consumer organizations to enable freedom of consumer choice. Conventional food may contain up to 0.9 percent of GE material without being subject to labeling if it can be shown to be adventitious and technically unavoidable. The same rule applies to animal feed. In case of traces of nonauthorized GEOs that already have been granted a favorable ruling by the EC's Scientific Committees, labeling is not enforced below 0.5 percent.[63] These labeling provisions would only make sense if a strict system for traceability can be established and enforced; hence this Regulation also includes statutory requirements covering this issue. Business operators must transmit and retain information about products that contain or are produced from GEOs at each stage of marketing. As far as GEOs intended as food and feed are concerned, this means that operators may either provide the specified information or give a declaration that the product shall only be used as food or feed or for processing, together with the identity of the GEOs that were in the original mixture from which this product arose. In case of food and feed from GEOs, operators shall inform the next operator in the chain that the product is produced from GEO(s).

Given this labeling regime, the statutory requirements for GE residues, and the possibility of pollen dispersal, outcrossing, and contamination of the processing chain after plant harvest, the cultivation of GE crops is likely to have an impact on farmers who wish to produce non-labeled food or feed. This is a particularly important issue for organic farmers and organic producers because EU-harmonized legislation, and guidelines of their international associations, do not allow the use of GEPs, and largely exclude the presence of GEP-derived ingredients in food and feed. To ensure that farmers can choose their preferred production type, the EC must also deal with the question of coexistence. So far, the EC has the nonbinding Recommendation 2003/556/EC on coexistence that sets out guidelines for the development of national strategies and best practices to ensure coexistence.[64] Hence, it is up to the individual member states to develop and implement management measures for coexistence. This Recommendation states that concepts for coexistence need to be developed in a transparent way, based on scientific evidence and in cooperation with all stakeholders. Measures should be crop specific and not go beyond what is necessary to comply with EU threshold levels for GEO labeling. Regional restrictions on the cultivation of certain types of GE crop should be pursued as individual cases and only if it can be demonstrated that farm-level measures and coordination between farms cannot ensure coexistence.

Moves to a More Centralized System

Given the experiences in decision making on GE crops and food market authorizations, it was no surprise when the EC publicly agreed to reexamine, if not eliminate, the regulatory committees at the same time as it discussed possible options to refer routine decisions to more or less autonomous agencies. The White Paper on Governance proposed to grant the powers to decide individual applications to EU regulatory agencies and even questioned the present comitology procedure.[65] Given that the constraints of the Treaty restricts the delegation of power and due to fears of threats to democratic legitimacy,[66] the EC eventually (and at least temporarily) dropped the idea of autonomous decision making. There were similar concerns regarding the EC's proposal to abolish the regulatory committees,[67] and recent proposals retain member-state representatives. The EC is apparently still willing to pursue this route and it has pointed out that proposed changes to comitology would be "transitional in nature, pending a new system of delegation of implementing powers based on a revision of the treaty."[68]

This is reflected by the final design of the EFSA, in which only the risk assessment task was shifted to that agency while the risk management, including the decision making (e.g., on market authorizations), remained with the EC.[69] Establishment of EFSA and the enacting of the Regulations 178/2002[70] on general food law and 1829/2003 on genetically modified food and feed brought about a partial break with the comitology procedure. EFSA's undertaking of all risk-assessment and risk-evaluation activities for GE food and crops[71] that were previously conducted by national CAs is an opportunity to streamline the authorization process[72] by providing a more consistent risk evaluation. EFSA was also envisaged, to speak with a "single voice" that should not be "second-guessed" by member-state agencies.[73]

The new procedure for market authorization according to Regulation 1829/2003 assumes that national CAs are acting as "letterboxes" so that the evaluation of the dossiers is carried out by EFSA, which forwards its opinion to the EC. In the absence of a modified decision-making procedure, the comitology procedure still applies and the EC has to base its decision on EFSA's risk assessment to the regulatory committee. Hence, even if there is no formal mechanism that requires the views of the national CAs *in the course* of the risk assessment itself, their views will eventually be considered *afterward* when the decision is being put to the vote within the regulatory committee. As the Regulatory Committee will hear doubts or objections to the risk assessment, it is appropriate to establish some mechanisms for (informal) consultation with national CAs in order to better prepare the decision-making process.[74] However, such informal mechanisms will most likely not add to the transparency of the whole process.

Improving the Risk-Assessment Process

The EC and member states acknowledged early on that promoting biosafety research was important to improve the scientific database. Research has been supported in successive EU Framework Programs from 1985 to the present day. Over this fifteen-year period eighty-one projects have been supported at a cost of approximately €70 million to the Community.[75] Some of the projects, such as ENTRANSFOOD, were aimed directly at improving approaches and methods of GEO risk assessment. Under the Sixth Framework Programme (FP6), the Commission intends to invest a total of €685 million to promote and fund research efforts to enhance food quality and safety including safety of GE food.[76] This action is also reflected at the national level. Germany, for instance, supported a total of seventy projects before 2001 and forty-two new projects were added in 2001. Biosafety research costs in Germany presently amount to a total of €7.9 million per year.[77]

These projects delivered very interesting insights into various aspects of biosafety, but it is clear that they are raising additional safety relevant questions. On the other hand, relevant initiatives to improve practical aspects of GEO and GE food risk assessment by providing guidance were not commonly funded at EU level. More detailed Guidance documents became available in 2003 and 2004.[78] About 460 comments on the most recent EFSA Draft Guidance Document[79] and subsequent discussion at the EFSA Stakeholder Consultation clearly showed that there are still a number of issues to be resolved. For example, it is still not clear in what particular cases the Scientific Panel on Genetically Modified Organisms (this is the official name of the panel) would require a toxicological ninety-day whole plant/food study . . . The GEO Panel knows that particular approaches and methods of risk assessment are more contested than others; for instance, the methods to collect evidence on the allergenic potential of a GEO and newly introduced proteins. The Draft Guidance was thus presented as a "snapshot" (which emphasized the issues that were most topical). While it is clear to the Panel that revisions of the Guidance will become necessary in the future, this will not mean that there will be a continuous revision.[80]

The contentiousness of certain requirements for risk assessment is also reflected by an apparent lack of agreement between Guidance Documents. A comparison of Guidance Documents issued by international organizations and scientific committees revealed that OECD and the Scientific Committee on Plants are criticizing in vitro digestibility studies (required as indicators of allergenicity), whereas FAO and EC's Scientific Steering Committee favor these tests. Proposed requirements for toxicity endpoints also vary: acute toxicity is advocated by the European biotechnology industry association, Europa-Bio, and to a certain extent by the Belgium Institute of Public Health and the Biosafety Council. The latter, which is the EC's Scientific Steering Committee and the EFSA Draft Guidance, also advocated twenty-eight-day repeated-dose toxicity whereas the International Life Sciences Institute (ILSI) and the Royal Society of Canada proposed a ninety-day sub-chronic toxicity.[81] These examples clearly reflect uncertainty with regard to a methodological issue (in vitro digestibility) and to issues of possible endpoints (which in turn indicate uncertainties due to a rather poor scientific database about protein toxicity). It is unclear whether recommendations are simply based on a different set of scientific evidence or (probably more likely) on different interpretations of the same evidence and different ideas about appropriate level of protection based on different grades of precaution. Whatever the case, the aforementioned examples illustrate that problems with different views on risk assessment are not likely to disappear soon.

CONCLUSIONS

Biotechnology is perceived as key to the EU's knowledge-based economy, and ever-increasing efforts are being made to harness the economic benefits and to catch up with the United States, which is its major competitor. While this goal is being pursued in principle at both the EU and national levels, economic stakes between EU member states differ a great deal. Furthermore, promotion of Community R&D only amounts to a small share of total R&D expenditures. In this policy area, the national level is still the most important.

While economic prospects of agricultural biotechnology have always been a major driver, EU harmonization has largely been preoccupied with issues of environmental and health risks. Biotechnology governance in the EU is, therefore, largely dominated by risks and regulatory issues.

Early discussions on risks associated with GEOs and GE food revealed uncertainty and ignorance in the underlying scientific basis and eventually led to the establishing of a precautionary-based legislative framework in 1990. Following a period of granting market authorizations for both GEPs and GE food, steps to reduce legislative hurdles to market entrance seemed to get closer and closer despite an overall unfavorable public perception of GE crops and food.

Intergovernmental deliberations of the legislative regime for GEOs were affected by a number of food scandals and scares, which shook consumer trust in governments and institutions on the national as well as on the EU levels. While these incidents were not directly related to GE crops or food, they happened at about the same time as when the first shipments of genetically modified soybeans arrived from the United States and when a number of illegal releases of GEPs occurred. These incidents diminished trust in institutions and policymakers responsible for risks management as well as in the underlying science, and the scandals most likely contributed to the rather negative perception of agricultural biotechnology in the EU.

Consequently policymakers found themselves in a difficult position that did not allow relaxation of the legislative framework and a new regime that emerged after intense deliberations was even more strict and precautionary. Statutory provisions of the new framework provided increased transparency and public participation associated with risk assessment and market authorizations, and better separated science from policy intrusions.

Linking increased precaution and reluctance in market authorizations to public opinion may, however, only partly explain the dynamics and design of biotechnology policy. A more comprehensive picture would emerge from additional analysis of underlying structural constraints specific to the EU, including differences between EU member states in cultural, legislative, political, and economic terms as well as in public support of agricultural biotechnology, and on the dynamics and constraints of European integration and harmonization. Clearly, harmonization and integration of an ever-increasing number of policy fields are possibly equally important to the understanding of the dynamics and the characteristics of biotechnology governance in the EU. In fact, reasons for European harmonization seem to have been crucial for establishing a separate, process-oriented legislative framework in the first place, which allows for European diverse views to unfold.

Commentators who are more inclined to economic views or who tend to argue that the apparent divergence between EU members is a naïve positivist stance are simply disregarding the science as being unsound. The extent of public support and the economic stakes are in essence driving the divergent views in decision making on GEPs and GE food in the context of market authorizations. Insights from science policy studies revealed considerable divergence and discretion in interpreting and drawing conclusions from scientific evidence in risk-assessment and risk-management tasks. Others argue that national policies may, at least in part, be attributed to preexisting policy focuses in the different member states. For instance, Denmark had already introduced a policy of reducing herbicide use, partly to protect ground water for drinking purposes.[82] Austria has the world's largest area of organic cropland, in absolute numbers as well as in percentage of overall arable land.[83] The Italian Parliament considers GEPs as threatening to traditional cultivars and high-quality products through both environmental effects and economic competition.[84] The key conclusion from these studies is that risk assessment in general and, particularly in the GEO issue, is characterized by scientific uncertainty and ignorance. The more a particular field of risk assessment is dominated by uncertainties and ignorance of its scientific base the more important it becomes to consider the normative baselines very carefully (this is essentially the job of risk management). The broad adoption of the precautionary principle in EU legislation and in risk management (in our case especially in the context of the new regulatory regime under Directive 2001/18/EC and Regulation 1829/2003) as a guiding tool for both designing the scope and interpreting the findings of risk assessments prompts risk managers to favor particular interpretations in order to be "on the safe side." It also extends their interpretative discretion.

The EU has been supporting research to reduce uncertainties in the scientific basis of GEO risk assessment since the mid-1980s. More recently it also resumed work on the guidance principles and more detailed requirements for risk assessment in terms of scope, base-lines, endpoints, approaches, methods, and so on. Results of this kind of research and experience with this kind of guidance clearly failed to meet expectations in the short term. This is not to say that pursuing both tasks will fail to improve and facilitate risk assessment and deci-sion making in biotechnology policy. The EU's difficulties of work-ing with scientific uncertainty in a dynamic political system, which is preoccupied with achieving economic and political integration, are likely to continue because of fundamental cultural, political, and eco-nomic differences between member states. Ongoing deliberations on normative baselines and scopes for and interpretations of risk assess-ment are essentially shaped by the institutional and legislative con-straints of the EU, especially the comitology procedure.

It is, therefore, not surprising that the EC has attempted to reform the risk assessment and decision-making process with regard to GEO products. Efforts to centralize risk assessment within one EU author-ity and to insulate the scientific assessment task from "policy intru-sions" are possible starting points. The last rounds of EU expansion and problems with the comitology procedure have led the EC to press for major reforms that are likely to alter, if not replace, the procedure and might even lead to delegation of decision-making power to inde-pendent EU agencies. A first round of proposed transitional amend-ments to the comitology is already under review by the European Parliament and the Council.

Given the problems of "getting the science right" in risk assess-ments, it might be tempting for the EC to focus on "engineering" the decision-making process to run more smoothly and efficiently and to reduce possibilities of different criteria in GEO risk assessment and decision making. This may conflict with other EU policy goals such as enhancing legitimacy, transparency, accountability, and participa-tion of a broader range of stakeholders in EU governance. If this path is pursued it will certainly not improve the assessments made and de-cisions taken, because it treats the symptoms only, while leaving the underlying causes unaddressed.

The biggest challenge in the near future is to maintain the balance between improving the efficiency of the regulatory procedures for GEO risk assessment and market authorization and addressing scien-tific uncertainty and the choice of normative baselines in a transparent and participatory way. The outcomes will be difficult to predict in the dynamic policy environment of an expanding EU.

NOTES

1. European Commission. 2004. *Questions and answers on the regulation of GMOs in the EU*. Memo/04/102. Brussels, April 30, 2004.

2. http://www.agbios.com/dbase.php. 2004. According to http://www.isb.vt.edu/cfdocs/biopetitions1.cfm 11 GMPs are authorized by EPA http://www.epa.gov/pesticides/biopesticides/pips/pip_list.htm. The FDA had completed consultations on a total of fifty-eight GMPs by July 29th 2004. http://www.cfsan.fda.gov/~lrd/biocon.html. URLs last accessed in July 2004.

3. International Service for the Acquisition of Agri-Biotech Applications (ISAAA). 2003. *Preview. Global status of commercialized transgenic crops.* ISAAA Briefs 30.

4. Cantley, M. 2004. How should public policy respond to the challenges of modern biotechnology? *Current Opinion in Biotechnology* 15: 258-263; Miller, H., and Conko, G. 2004. EU is out of step over regulation of modified products. Op-Ed & Articles letter. *Financial Times,* May 24; The statutory requirements for labeling are perceived as strict even by Greenpeace. See Greenpeace. *Europe adopts the world's strictest legislation for GE labelling.* Toronto: Greenpeace Canada, November 28, 2002. http://action.web.ca/home/gpc/alerts.shtml?sh_itm=a5174ee079ce80 bbeb 9e2a0d1a1c6d36. Last accessed in July 2004.

5. Not to be confused with the Council of the European Union (sometimes referred to as the Council of Ministers), which is the EU's main decision-making institution. In contrast to the EU, which was primarily aimed at economic integration, the Council of Europe was intended as a political institution from the very start. By now, the Council of Europe comprises forty-five countries and about 800 million people.

6. European Commission. 2003. *Life sciences and biotechnology—A strategy for Europe. Progress report and future orientations.* Communication from the Commission to the European Parliament to the Council and to the European Economic and Social Committee. 5.3.2003, COM. 96 final; European Commission. 2002. *Life sciences and biotechnology—A strategy for Europe.* Communication from the Commission to the European Parliament, the Council, the European Economic and Social Committee and to the Committee of the Regions. Brussels, 23.1.2002, COM. 27 final.

7. Isaac, G.E. 2002. *Agricultural biotechnology and transatlantic trade. Regulatory barriers to GM crops.* Oxford/New York: CABI.

8. Ibid., p. 206.

9. European Commission, 2002. See note 6.

10. According to a survey among private companies and research institutes, 39 percent of respondents have cancelled R&D projects on GMOs in recent years. The main reasons given were the unclear regulatory framework and the uncertain market

situation. Furthermore, there was a sharp decline in notification of field trials after 1998; see Lheureux, K., Libeau-Dulos, M., Nilsagard, H., Cerezo, E.R., Menrad, K., Menrad, M., and Vorgrimler, D. 2003. *Review of GMOs under research and development and in the pipeline in Europe.* ESTO Report. Most recently, press reports about reduction in the EU biotech crop research and a shift to the United States have fueled long-standing concerns about draining the knowledge base in Europe. See Hirschler, B. 2004. Syngenta to end biotech crop research in UK. *Reuters News,* July 1; Mettler, A. 2004. Europe is steadily losing its scientific elite. *Financial Times,* July 15.

11. European Commission. 2003. See note 6.

12. Isaac, G.E. 2002. See note 7.

13. The numbers are from 1997.

14. The numbers are from sixteen member states reported in December 2000. See Allansdottir, A., Bonaccorsi, A., Gambardella, A., Mariani, M., Orsenigo, L., Pammolli, F., and Riccaboni, M. 2002. *Innovation and competitiveness in European biotechnology.* European Commission—DG Enterprise: Enterprise Papers 7.

15. European Commission. 2002. See note 6.

16. European Commission. 2003. See note 6. p. 22.

17. The OECD Report, *Recombinant DNA safety considerations,* states "that there is no scientific basis for specific legislation to regulate the use of recombinant DNA organisms." See OECD. 1986. *Recombinant DNA safety considerations. Safety considerations for industrial, agricultural and environmental applications of organisms derived by recombinant DNA techniques.* Report.

18. "The Commission had felt itself obliged to legislate on the technology, in spite of OECD advice and the US example, because it was faced with a proliferation of divergent national legislation, starting with the Danish Gene Technology Law of June 1986. The Catenhusen Commission on Opportunities and Risks of Gene Technology reported in January 1987, advocating German legislation, which followed in 1990. In both cases, the adverse consequences and absence of benefit from this restrictive legislation forced subsequent major amendment. It was also clear that the advent of gene technology presented a major opportunity to expand the authority of Environment Ministries, who encouraged and welcomed the initiative which was prepared within the Commission by the corresponding Directorate-General, DG XI (Environment)." See Cantley, M. 1999. *EC regulation of genetic modification in agriculture. Second report of the selected committee on the European communities.* Written Evidence. In House of Lords. 1999. Session 1998-99m, HL 11-l. London: The Stationary Office. http://www.parliament.the-stationery-office.co.uk/pa/ld199899/ldselect/ldeucom/11/11we01.htm. Last accessed in July 2004.

19. Galloux, J.-C., Prat, H.G., and Stevers, E. 1998. Europe. In *Biotechnology in the public sphere. A European sourcebook.* Edited by J. Durant, M.W.Bauer, and G. Gaskell. London: Science Museum, pp. 177-185; House of Lords. 1999. See note 18; Torgersen, H., Hampel, J., von Bergmann-Winberg, M-L, Bridgman, E., Durant, J., Einsiedel, E., Fjaestad, B., Gaskell, G., Grabner, P., et al. 2002. Promise, problems and proxies: Twenty-five years of debate and regulation in Europe. In *Biotechnology—The making of a global controversy.* Edited by M.W. Bauer and G. Gaskell. Cambridge: Cambridge University Press, pp. 21-94.

20. However, in deference to scientific opinion and the demands of existing regulatory authorities, the Commission's proposal acknowledged the logic of returning regulatory competence to the sectoral authorities concerned, even for the products of novel technology. It was clearly understood, and recorded by a Commission declaration in the Council minutes of April 23, 1990, that as modern biotechnology moved into application in various sectors, the sectoral legislation would add . . . necessary environmental safety requirements; and oversight of the placing on the market of genetically modified products would revert from Part C of 90/220 to the sectoral legislation concerned (cf. Paragraph 2(b) of Article 10). An obvious example is in the Novel Foods Regulation; Cantley, M. 1999. See note 18.

21. Torgersen, H. et al. 2002. See note 19.

22. Directive 90/220/EEC (*Official Journal L 117, 08/05/1990 P. 0015-0027*) used the term prevention as "action by the Community relating to the environment should be based on the principle that preventive action should be taken," whereas the amended Directive 2001/18/EC (*Official Journal L 106, 17/04/2001 P. 0001-0039*) in Preamble 8 explicitly refers to the precautionary principle: "The precautionary principle has been taken into account in the drafting of this Directive and must be taken into account when implementing it."

23. This practice was established under Directive 90/220/EEC but also applies to Directive 2001/18/EEC that replaced Directive 90/220/EEC. See note 22.

24. Assessments of national competent authorities (CAs) and of EU Scientific Committees cannot be considered as fully-fledged risk assessments; they should rather be seen as an evaluation of the applicant's risk assessment.

25. With establishment of the European Food Safety Authority (EFSA), the EC's Scientific Committees were also restructured. The Scientific Panel on Food and the GMO Panel, both located at the EFSA, took over the responsibilities of the Scientific Committee on Food and Scientific Committee on Plants.

26. This particular design of the authorization procedure allows member states to exert control over implementation measures of the EC. In case of health-related issues, the Commission can only adopt implementing measures if it obtains the approval by qualified majority of the member states meeting within the respective Regulatory Committee.

27. A slightly different shortcut procedure applied in case of GE food that was shown to be substantially equivalent to the conventional counterpart (food/plant) according to Article 5, Novel Food Regulation. In this case, national CAs focus mainly on the claim of substantial equivalence. In cases where objections were raised and justified, the application would have to undergo the normal procedure (Article 5) as described in the text.

28. One more important difference pertains to the basis upon which the national CAs must decide. Although there is only a statutory requirement to circulate summaries of the applications, whole dossiers were circulated during implementation of Directive 90/220/EEC. See Spök, A., Hofer, H., Valenta, R., Kienzl-Plochberger, K., Lehner, P., and Gaugitsch, H. 2002. *Toxikologie und Allergologie von GVO-Produkten. Empfehlungen zur Standardisierung der Sicherheitsbewertung von gentechnisch veränderten Produkten auf Basis der Richtlinie 90/220/EWG (2001/18/EG).* Monographien Band 109. Vienna: Umweltbundesamt Wien. In contrast, in

case of Novel Food applications only summaries—although varying in size and detail—were circulated. See Spök, A., Karner, S., Stirn, S., and Gaugitsch, H. 2003. *Toxikologie und Allergologie von GVO-Produkten – Teil 2B. Untersuchung von Regelungen zur Sicherheitsbewertung von gentechnisch veränderten Lebensmitteln in der EU und den USA.* Monographien Band 164B. Vienna: Umweltbundesamt Wien.

29. The comitology procedure also applies to the present Regulation 1829/2003 (*Official Journal L 268. 18/10/2003 P. 0001-0023*) on genetically modified food and feed. The new comitology procedure laid out in Decision 1999/468/EC (*Official Journal L 184, 17/07/1999 P. 0023-0026*) even granted more power to the member states by enabling the Council to reject the Commission's proposal by qualified majority.

30. Provided that the Council does not object by a qualified majority.

31. At the time of the transition into the new regulatory regime of Directive 2001/18/EC (see note 28) two applications for GE cotton were awaiting a decision of the Council and several applications were waiting for first consideration by the Standing Committee. Most of the pending applications were subsequently amended and considered further under Directive 2001/18/EC.

32. European Commission. 2004. *Information note in support of the Communication to the Commission (from the President in association with Mrs. Wallström, Mr. Byrne, Mr. Fischler, Mr. Lamy, Mr. Liikanen and Mr. Busquin) for an orientation debate on genetically modified organisms and related issues.* COM 2004.

33. Also including market authorizations for GEPs according to Directive 2001/18/EC (see note 28).

34. Norway is not a member state of the EU but, as a member of the European Free Trade Association and the European Economic Area, it is pursuing its policy goals in very close similarity to the EU.

35. Austria, for instance, introduced a criterion that products have to be socially acceptable ("sozialverträglich"). Otherwise they could be removed from the market.

36. Actually the German national CA is the Federal Office of Consumer Protection and Food Safety (BVL), which is a subordinate authority to the Federal Ministry of Consumer Protection, Food and Agriculture.

37. Von Schomberg, R. 1998. *An appraisal of the working in practice of Directive 90/220/EEC on the deliberate release of genetically modified organisms.* Draft Final Report to the Scientific and Technological Options Assessment Unit (STOA) of the European Parliament, January 2.

38. Levidow, L., Carr, S., and Wield, D. 2000. Genetically modified crops in the European Union: Regulatory conflicts as precautionary opportunities. *Journal of Risk Research* 3: 189-208.

39. The observations outlined in this paragraph pertain to herbicide-tolerant rape MS1× RF1 marketed by Plant Genetic Systems.

40. The observations outlined in this paragraph pertain to insect-resistant maize *Bt*176 marketed by Ciba-Geigy (Novartis).

41. Levidow, L. and Carr, S. 2000. UK: Precautionary commercialization? *Journal of Risk Research* 3: 261-270; Torgersen, H. and Seifert, F. 2000. Austria: Pre-

cautionary blockage of agricultural biotechnology. *Journal of Risk Research* 3: 209-217.

42. In fact, the inclusion of agronomic and socioeconomic criteria had been essential for achieving national consensus on GMO regulation. von Schomberg, R. 1998. See note 37.

43. Spök, A. et al. 2002. See note 28; Spök, A., Hofer, H., Valenta, R., Kienzl-Plochberger, K., Lehner, P., and Gaugitsch, H. 2003. *Toxikologie und Allergologie von GVO-Produkten-Teil 2A. Untersuchungen zur Praxis und Empfehlungen zur Standardisierung der Sicherheitsbewertung von gentechnisch veränderten Lebensmitteln.* Monographien Band 164A. Vienna: Umweltbundesamt Wien.

44. Spök et al., 2003. See note 43.

45. Spök, A., Hofer, H., Lehner, P., Stirn, S., Valenta, R., and Gaugitsch, H. 2004. Risk assessment of GMO products in the European Union. Toxicity assessment, allergenicity assessment and substantial equivalence in practice and proposals for improvement and standardisation. *Series Berichte* 253. Vienna: Umweltbundesamt. http://www.umweltbundesamt.at/fileadmin/site/publikationen/BE253.pdf. Also published as Volume 7-04 in the Series of the Bundesministerium für Gesundheit und Frauen, Vienna, available at http://www.bmgf.gv.at/cms/site/attachments/6/8/7/CH0255/CMS1090828056047/risk_assessment_of_gmo_products-bmgf-layout.pdf. URLs last accessed in July 2004.

46. Spök, A. et al. 2003. See note 43; Spok, A. et al. 2004. See note 45.

47. Gaskell, G., Allum, N., and Stares, S. 2003. *Europeans and biotechnology in 2002. Eurobarometer 58.0.* 2nd ed. Brussels: European Commission's Research Directorate-General.

48. Bauer, M.W., Durant, J., and Gaskell, G. 1998. Biology in the public sphere: A comparative view. In *Biotechnology in the public sphere. A European sourcebook.* Edited by J. Durant, M.W. Bauer, and G. Gaskell. London: Science Museum. pp. 217-227.

49. In Italy, the shift from a rather positive to a more critical stance was probably more heavily influenced by a concurrent debate on transgenic maize. Lassen, J., Allansdottir, A., Liakopoulos, M., Mortensen, A.T., and Olofson, A. 2002. Testing times—the reception of Roundup Ready soya in Europe. In *Biotechnology—The making of a global controversy.* Edited by M.W. Bauer and G. Gaskell. Cambridge: Cambridge University Press, pp. 279-312.

50. European Commission, 2002. See note 6.

51. European Commission, 2001. *European governance. A white paper.* COM. Brussels: European Commission.

52. European Commission, 2000. *White paper on food safety.* COM (1999). Brussels: European Commission.

53. European Commission. 2001. See note 51.

54. A labeling regime had already been in place since 1997 embedded in the Novel Food Regulation. This regime turned out to be a matter of constant dispute; so in 2003, a more comprehensive labeling system was introduced.

55. Directive 2001/18/EC repealing Council Directive 90/220/EEC. See note 22.

56. This procedure is largely similar for Directive 2001/18/EC (see note 22) and for Regulation 1829/2003 (see note 29).

57. In terms of GM food only, as the Novel Food Regulation is still valid for non-GM food.

58. European Commission 2004c. *Communication to the Commission.* For an orientation debate on Genetically Modified Organisms and related issues. No date of publication stated but issued for the first time in January 2004. http://europa. eu.int/comm/food/food/ biotechnology/gmfood/gmo_comm_en.pdf. Accessed July 26, 2004.

59. Seed legislation. Directive 98/95/EC. *Official Journal L 025, 01/02/1999 P. 0001-0026* requires GMO varieties to be approved under Directive 2001/18/EEC (see note 22) prior to inclusion in the Common Catalogue. If intended for food and/or feed purposes an authorization according to Regulation 1829/2003 (see note 29) is also required.

60. European Commission, 2004d. *Answer of the Commission to written question of MEP Caroline Lucas,* E-3542/03. March 22, 2004. http://www. carolinelucasmep.org.uk, accessed July 28, 2004. This might be one reason for the huge differences in areas of GEP cultivation between the EU and the United States.

61. According to provisional regulations applications that were originally submitted under the Novel Food Regulation must be considered under the preceding regulatory framework. Status: January 2004.

62. Regulation (EC) 1830/2003 concerning the traceability and labeling of genetically modified organisms and the traceability of food and feed products produced from genetically modified organisms and amending Directive 2001/18/EC. *Official Journal L 268, 18/10/2003 P. 0024-0028.*

63. Above 0.5 percent it is prohibited to put the product on the market.

64. Recommendation 2003/556/EC: Commission Recommendation of July 23, 2003 on guidelines for the development of national strategies and best practices to ensure the coexistence of genetically modified crops with conventional and organic farming. *Official Journal L 189, 29/07/2003 P. 0036-0047.*

65. European Commission, 2001. See note 51.

66. Reviewed in Spök, A. 2003. *"Uneasy Divorce" or "Joint Custody"? The separation of risk assessment and risk management in food policy.* MSc Dissertation, Science Policy Research, University of Sussex.

67. European Commission 2003. *Report from the Commission on European governance.* Luxembourg: European Commission.

68. Ibid.

69. Spök, A. 2003, See note 66.

70. Regulation (EC) No 178/2002 laying down the general principles and requirements of food law, establishing the European Food Safety Authority and laying down procedures in matters of food safety. *Official Journal L 031, 01/02/2002 P. 0001-0024.*

71. This is also likely to apply for Directive 2001/18/EC (see note 22) applications, in case they are sought in addition to a food or feed authorization according to Regulation 1829/2003 (see note 29) and the EFSA Stakeholder Consultation. 2004. *Draft guidance document for the risk assessment of genetically modified plants and derived foods and feed.* April 2004.

72. European Crop Protection Association. 2002. *Perspectives on crop protection and crop science.* A briefing for policymakers from the European Crop Protection Association. June 2002; Association of the European Self-Medication Industry. 2002. AESGP position paper providing comments to the discussion paper on the implementation of the Novel Foods Regulation (EC) No. 258/97 concerning novel foods and novel food ingredients. *Official Journal L 043, 14/02/1997 P. 0001-0006.*

73. Byrne. 2000. House of Lords report for May 24. p. 21, para 69.

74. Given the fact that Regulation 1829/2003 (see note 29) has applied since April 2004, there is very limited experience using the new procedure so far. Thus, it is not clear how the procedure will be pursued in practice and, for example, whether the EFSA Advisory Committee, which is composed of members of national CAs, will play an important role for informal negotiations between national CAs and EFSA.

75. Kessler, C. and Eonomides, I., eds. 2001. *EC sponsored research on safety of genetically modified organisms. A review of results.* EUR 19884. http://europa.eu. int/comm/research/quality-of-life/gmo. Accessed July 27, 2004.

76. Biosociety Research-Online. Food Quality and Safety. Basic Information. http://europa.eu.int/comm/research/biosociety/food_quality/basics_en.htm. Accessed July 27, 2004.

77. Informationsportal "bioSicherheit" eröffnet. 16.04.2002, press release 76/2002, http://www.bmbf.de/press/615.php. Accessed July 27, 2004.

78. Scientific Steering Committee. 2003. *Guidance document for the risk assessment of genetically modified plants and derived food and feed.*

79. See European Food Safety Authority, 2004. See note 71.

80. Ibid. The author of this chapter participated in this Consultation.

81. References in Spök, A. et al. 2004. See note 45.

82. Levidow, L., Carr, S., Wield, D., and van Schomberg, R. 1997. European biotechnology regulation: Framing the risk assessment of a herbicide tolerant crop. *Science, Technology and Human Values* 22: 472-505.

83. Austria has about 12 percent of total cropland under organic farming compared to Canada (1.3 percent) and the United States (0.2 percent). Willer, H. and Yussefi, M. 2004. *The world of organic agriculture—statistics and emerging trends.* 6th rev. ed. Bonn: International Federation of Organic Agriculture Movements.

84. Levidow, L. et al. 2000. See note 38.

Chapter 13

Regulatory Regimes for GE Crops in Africa

Jennifer A. Thomson

This chapter discusses the most developed African biotechnology biosafety regulatory regimes, in particular those in South Africa, Kenya, and Egypt, and reports on the current status in a number of other countries where regulations are more preliminary. The Conference of the Parties to the Convention on Biological Diversity adopted an agreement known as the Cartagena Protocol on Biosafety on January 29, 2000. The Protocol seeks to protect biological diversity from the potential risks posed by living modified organisms resulting from modern biotechnology. It established a procedure for ensuring that countries are provided with the information necessary to make informed decisions before agreeing to the import of such organisms into their territory.

A problem facing Africa in particular is the lack of a dynamic private sector to take agricultural technologies to the farmer. It has also been estimated that regulatory costs might exceed the costs of research and experimentation to develop a given GE crop. This is a major problem in releasing such products to the market. The regulatory costs of generating food and environmental safety data may be reduced by establishing regional "centers of excellence" that are responsible for reliable, regionally relevant, and culturally sensitive testing and enforcement.

SOUTH AFRICA

South Africa's Genetically Modified Organisms (GMO) Act was passed on May 23, 1997 but the necessary regulations for implementation were not published until November 26, 1999. The Act is man-

Genetically Engineered Crops
© 2007 by The Haworth Press, Inc. All rights reserved.
doi:10.1300/5880_13

aged by the Registrar of Genetic Resources within the National Department of Agriculture. She or he receives applications for trial or commercial releases and passes these on to the Scientific Advisory Committee, a body of up to eight members knowledgeable in the field and two persons from the public sector with knowledge of ecology and GEOs. This Committee sends the applications out for scientific review, coordinates the reviews, and sends a report to the Registrar, who is also responsible for obtaining public input. The report is then sent to the Executive Council, which consists of officers from the Departments of Agriculture, Science and Technology, Environment, Health, Labour, and Trade and Industry, with knowledge of the implications of GEOs for their departments. This body makes the decision on whether approval is given and under which conditions, taking into account socioeconomic impacts on the communities that will be affected. They pass this information on to the applicants via the Registrar. The Registrar issues permits, appoints inspectors, and ensures that the conditions of permits are complied with. Inspectors ensure that trials are carried out in accordance with the GMO Act and issue warrants for any violations. They also conduct routine inspections of trials.

It is the responsibility of the applicant to notify the public of trial and commercial releases. Notices of releases must be placed in three different newspapers in the affected areas and copies must accompany applications for releases. The Council takes into account public response when deciding whether or not to allow the releases.

All facilities that import, export, develop, and use GEOs require permits. However, academic and research institutes are exempt from these as long as their facilities are registered with the Registrar. These registrations must be renewed annually to take into account changes in GEO-related activities. Users are also required to submit risk assessments for all projects.

Permits issued for import, export, field trials, commodity clearances, and commercial planting in 2002 and up until July 2003 are shown in Table 13.1. The countries involved as either suppliers or receivers were Argentina, Austria, Brazil, Chile, China, Colombia, Egypt, France, Germany, Indonesia, Israel, Philippines, Romania, Spain, Uruguay, and the United States.

One example of field trials and subsequent commercial releases, among small-scale farmers in the Kwa-Zulu Natal province, is of in-

TABLE 13.1. Permits for GE crops issued during 2002 and until July 2003.

Year	Import/export	Field trials	Commercial plantings	Commodity clearances
2002	136	50	11	19
2003	64	1	7	4

sect-resistant cotton, expressing a *Bacillus thuringiensis* (*Bt*) toxin. In 1997 four farmers took part in the trials. Their crop yield was increased so successfully, presumably due to the engineered resistance, that the following year seventy-five farmers planted *Bt* cotton and by 2001, 1,184 farmers were planting *Bt* cotton. Independent assessments of the commercial effects of these plantings were carried out by Ismael et al.[1,2] who found that 60 percent of the producers in the region had adopted the new technology. It was found that the average yield per hectare and per kilogram was higher among adopters of insect-resistant cotton and that the increased yield and reduction in chemical costs outweighed the higher cost of the seed. Adopters were also found to be more efficient (62 percent compared with 46 percent). Adopters had a gain (yield per kg seed) of 12 percent in the 1998/1999 season, 45 percent in the 1999/2000 season, and 42 percent in the 2000/2001 season. In addition the adopters suffered far less crop losses than the non-adopters during the wet 1999/2000 season.

Bt yellow maize has been in commercial use for a number of years and accounts for about 20 percent of the crop, which is mainly used for animal and chicken fodder. *Bt* white maize has only been commercialized for the past few seasons and accounts for about 2 to 3 percent of the crop.

KENYA

Biotechnology regulations are being developed in Kenya. The Kenya Agricultural Research Institute (KARI) in Nairobi is conducting field trials on sweetpotatoes engineered to be resistant against several viruses, including Sweetpotato feathery mottled virus. In 2002, yields in Kenya were 7 tons per hectares compared with 18 in China and 33 in the United States. Part of this poor performance is due to infection by Sweetpotato feathery mottled virus present in Kenya and infecting farmers' fields in the Association for Strengthening Agricultural Research in Eastern and Central Africa (ASARECA) region.

The Monsanto Corporation, and, more recently, the Donald Danforth Plant Science Center in St. Louis, Missouri, United States, have given technical assistance. Monsanto and KARI signed a nonexclusive, royalty-free licensing agreement in 1998, which allows KARI to use and further develop the transgenic virus resistance technology in sweetpotato. KARI is also permitted to protect the resulting transgenic varieties under the plant breeders' rights convention or similar regulations that are effective in Kenya, and the technology may be transferred to any other country in Africa. It has been estimated that these varieties could produce an aggregate annual benefit equivalent to US$5.4 million.[3]

Trials took place in 2001 and 2002 but no significant differences were seen between control plants and transgenics. In retrospect this is not surprising as the coat protein (CP) gene that provided resistance was taken from an American strain of the virus. However, the trials were considered a success in that the transgenic plants thrived in the field, scoring for disease symptoms did not increase, and the trials were managed effectively and in compliance with regulations at four different field trial sites. This places Kenya second only to South Africa among sub-Saharan African countries able to undertake field trials of transgenic crops, and this was a valuable contribution to capacity building.

Scientists at the Donald Danforth Center are currently analyzing transgenic plants of the Kenyan sweetpotato cultivar CPT60 that have been transformed with the CP and replicase genes from the severe (based on visual symptoms ranked on a five-point scale, where five is the most severe) Kenyan Muguga strain.

EGYPT

The Agricultural Genetic Engineering Research Institute (AGERI, http://www.agri.gov.eg/gene.htm) in Cairo is carrying out most of Egypt's GE research and field trials. Through collaboration with Monsanto, they have developed an insect-resistant long-staple GE cotton strain by crossing elite Egyptian germplasm with Monsanto's Bollgard II. Egyptian cotton is one of the world's finest and the country's most important agricultural export. Field trials are pending. Other GE crops that have been in field trials for more than one season include potato tuber moth-resistant potatoes, virus-resistant squash and tomatoes, corn borer-resistant maize, and drought-tolerant wheat.

Problems cited for the slow passage from experimental, to trial, to commercial stage include the lack of capacity to negotiate licenses to use genes and research techniques patented by others, especially for crops with export potential. In addition there are difficulties in meeting regulatory requirements and a lack of effective public commercialization modalities and working extension networks.

OTHER COUNTRIES

Turning to biosafety in other African countries, GE crops are being tested in Malawi, Zambia, and Zimbabwe. In June 2001, the United Nations Environmental Programme (UNEP) and the Global Environment Facility (GEF) have started a worldwide biosafety project (http://www.unep.ch/biosafety) to assist more countries wishing to embark on field trials and commercial releases. This is designed to help up to 100 eligible countries to prepare their own National Biosafety Frameworks (NBF) and to promote regional and subregional collaboration in the field of biosafety. Completion is anticipated within three years.

The cost is estimated at $38.4 million with $26.1 million coming from the GEF and $12.3 from UNEP and participating countries. On a national level, the process must be country driven; hence the requirement for national investment either financial or in kind. The outcome will be that participating countries set up NBFs for the management of Living Modified Organisms (LMOs), hence allowing them to meet the requirements of the Cartagena Protocol (for executive summary see http://www.biodiv.org/biosafety/background.asp).

The aims of regional and subregional collaboration include the exchange of experience on issues of relevance to the NBF to make efficient use of financial and human resources. It also aims to establish networks and promote the harmonization of risk assessment procedures and regulatory instruments.

The process for the development of NBFs at the national level is as follows:

- Information gathering.
- Analysis, consultation, and training.
- Preparation of a draft NBF, which will require draft legal instruments, administrative systems, risk-assessment procedures, and systems for public participation and information.

At the regional level, steps include the following:

- Mechanisms for sharing of risk assessment and management experiences.
- Ways to coordinate capacity-building activities.
- Networking to share lessons and experiences.

UNEP/GEF eligibility requirements are the following:

- The country must be a signatory to or have ratified the Cartagena Protocol on Biosafety.
- The country must have the ability to borrow from the World Bank, or to receive technical assistance grants from the United Nations Development Programme.
- The country must not have received previous assistance from the UNEP-GEF Pilot Project on Biosafety.
- The national GEF Focal Point must have formally expressed the country's interest in participating in the project.

Ten African countries participated in the Pilot Phase during 2001-2002; eighteen are already participating in the project, eight only need GEF endorsement, and twenty-two countries still need to sign the Cartegena Protocol. Table 13.2 summarizes the status of biosafety in fifty-three African countries.

TABLE 13.2. Status of biosafety in Africa.

Status	Countries
GE legislation and functioning framework	South Africa, Zimbabwe
GE legislation and framework in development	Malawi, Cameroon
Draft legislation and interim framework	Egypt, Kenya
Draft legislation and framework not yet reviewed	Cote d'Ivoire, Mauritius, Zambia, Nigeria
UNEP-GEF biosafety development process	Forty-three countries

The institutional structure set up to manage the process is as follows:

1. National Executing Agency (NEA)
   - This is the legal entity responsible for executing the UNEP-GEF project in order to set up the NBF.
   - It must be a body with a central role in biosafety issues.
2. National Coordinating Committee (NCC)
   - This is established by the NEA to advise and guide the preparation of the NBF.
   - It needs to be multidisciplinary and multi-sectorial.
   - It needs to be flexible and respond to changing circumstances.
   - It needs to represent government departments, private and public sectors.
3. National Project Coordinator (NPC)
   - The coordinator must be appointed by the NEA.
   - She or he will act as secretary of the NCC and its link to the NEA.
   - Her or his major task is to ensure that the NBF is completed within the required three years.

In conclusion, at the end of the UNDP/GFP project it is hoped that 100 countries worldwide will have National Biosafety Frameworks in place and be in compliance with the Cartagena Protocol. They will also have the ability to develop, use, and exchange LMOs.

NOTES

1. Ismael, Y., Bennett, R., and Morse, S. 2001. Can farmers in the developing countries benefit from the modern biotechnology? Experience from Makhathini Flats, The Republic of South Africa. *International Service for the Acquisition of AgriBiotech Applications publication. Crop Biotechnology.* Brief Vol. 1 (no. 5). www.isaaa.org/kc.

2. Ismael, Y., Bennett, R., and Morse, S. 2002. Do small-scale *Bt* cotton adopters in South Africa gain an economic advantage? *Proceedings of the 6th International Conference on Agricultural Biotechnology, Ravello, Italy.*

3. Qaim, M. 1999. The economic effects of genetically modified orphan commodities: projections for sweetpotato in Kenya. *International Service for the Acquisition of AgriBiotech Applications (ISAAA).* Brief No. 13.

Chapter 14

GEO Research and Agribusiness in Brazil: Impact of the Regulatory Framework

Marília Regini Nutti
Maria José Amstalden Sampaio
Edson Watanabe

INTRODUCTION

Humans have cultivated plants for thousands of years, during which time crop plants have been continually selected for improved yield, growth, disease resistance, or other useful characteristics. Plant breeding is an exceptionally successful enterprise that has generated modern high-yielding crops on which we now depend. Until recently, plant breeders had to depend on empirical methods to reach their goals.[1]

With the advent of molecular biology and biotechnology it became possible not only to identify a desirable phenotypic trait but also to identify the precise genetic material responsible for that genetic trait. Recombinant DNA and plant transformation techniques have made it possible to alter the composition of individual plant components (lipids, carbohydrates, proteins) beyond what is possible through traditional breeding practices.[2]

Improvements in agronomic traits such as yield and disease resistance continue to be driving forces behind today's seed industry but, increasingly, attention is focused on speciality traits, including high oilseed grains and delayed ripening fruits and vegetables, which command premium values in the marketplace.[3]

Genetically Engineered Crops
doi:10.1300/5880_14

The use of biotechnology in agriculture is an important tool not only to increase productivity, but also to increase the income of the rural poor, who depend primarily on agriculture for income generation. Questions have been raised on consumer acceptance, environmental impacts, public versus private sector role, biosafety and intellectual property rights (IPR) issues, and so on. These factors have contributed to a general apathy in developing country governments in promoting adoption of this technology for their agricultural systems.

Biotechnology is a complex topic that embodies difficult technical, social, and economic issues played out against a backdrop of human hunger, economic marginalization, and environmental degradation.[4] Taken as a major tool to help the development of agribusiness as a whole, biotechnology has not enjoyed the same acceptance by consumers in Brazil as it has in the United States or Argentina. Legal, ethical, environmental, social and economic issues have become entangled and are associated with other complicated items such as labeling, market barriers, and global commerce, making the situation very difficult for the consumer to understand and to trust any given governmental decision.

The current scenario indicates that, due to its intimate relation with the need for safety, the development of agricultural biotechnology will follow a different route when compared to other industrial sectors. The development of any novel industry usually emphasizes market issues. Agricultural biotechnology must also consider another aspect, the delivery to the consumer of precise information about this new technology, using the most trustable scientific evidence.

The current discussion is concentrated on some genetically engineered (GE) crops, which were released to the market in the last five to ten years, as a result of the "first wave" of genetic engineering (GE). These products, which were tolerant to herbicides and resistant to insects, presented a small benefit to final consumers. Most of the described benefit went to the farmer or to the environment, far away from the table. The "second wave" will bring products with new characteristics to the market, for example, soybeans with higher concentration of oleic acid, which offer improved health to consumers. However, the major revolution in agricultural production is expected with the "third wave," with crops engineered to contain medicines and other important components for the human health and animal production, which could result in a "health revolution."[5]

Although much experience has been acquired with products developed and commercialized in other countries, safety protocols should be developed and/or adapted to local conditions. This is imperative for environmental safety since impacts may differ from those in other regions of the world. Food safety assessment must be complemented in cases where no data is available (e.g., where a new transgene is used) or if there are other applications to the food chain.

This chapter analyzes the governance of agricultural biotechnology in Brazil, one of the countries where GM plants were just approved for commercial use. Only soybean has been specially approved and has been released for the 2002/2003, 2003/2004, and 2004/2005 planting seasons.

The recently modified body of biosafety legislation and the requirements for research and commercial release of GE crops in Brazil is discussed along with the current legislation for GEO labeling, and finally a brief review on food safety assessment and environmental impacts of GE organisms is presented. All the listed studies are being considered for the safety evaluation of GEO developed by the Brazilian Agricultural Research Corporation (Embrapa) through its multidisciplinary Biosafety Network.

THE BRAZILIAN LEGAL FRAMEWORK FOR GM CROPS

To set the playing field for readers, Brazil is one of the largest countries in the world, with an area of 8,500,000 square kilometers, a large supply of fresh water, abundant solar energy, and a rich biodiversity. Besides being home to the world's largest tropical forest, the country has over 200 million hectares of savannah (known as "cerrado") with immense agriculture and livestock production potential. The cerrado is the main agricultural frontier in Brazil and one of the largest in the world, not only due to its considerable areas still available for farming expansion, but also due to its potential of productivity increase through technology improvements. It has been estimated that Brazil could increase its total cultivated area by 420 million acres or more if key legal, technical, and financial developments occur in the near future (USDA, 2003).[6]

The Brazilian Biosafety Law

The first Brazilian Biosafety Law[7] for genetically modified organisms (Law no. 8974/95) was approved in 1995 and provided for a horizontal type of regulation that interfaced with other regulatory frameworks in the areas of agriculture, health, and environment. Several ministries and federal regulatory departments and agencies were involved in the implementation of biosafety regulations. Due to the conflicting and overlapping regulations, many of the GE projects had to be submitted to duplicate, and sometimes triplicate, analysis by different ministries with great loss of time and opportunities for developers to advance in the development of transgenic agricultural products. In 2002/2003, Congress began to discuss a new law to define better the mandate for biosafety risk analysis. Finally, on March 28, 2005, the new law was published (Law no. 11.105/05)[8] putting an end to a ten-year period of legal uncertainty, which gave rise to many legal disputes and almost stopped public investment on the development of GM products. The new Law reedits the National Technical Biosafety Committee (CTNBio) with a wider membership of scientists and other representatives, all PhDs, in the areas that have an interface with biotechnology and biosafety. CTNBio continues to be linked to the Ministry of Science and Technology and is the only designated authority to analyze petitions and issue binding authorizations for laboratory, greenhouse, and field experimental releases.

The Law innovates with the introduction of a Ministerial Council, which is in place to take care of socioeconomic discussions, relieving the National Committee from the political burden and assuring that the Committee will only take care of biosafety issues.

Plant Incorporated Protectants (PIPs)

In the case of *Bt* plants or other plants containing any biopesticide characteristic (PIPs), a specific law for pesticides used to apply. Hence, the use of virus coat protein could not be deregulated as an exception as happened in the United States. Authorization for laboratory, greenhouse, and field studies of these products had to be obtained from the

SDA (Plant Protection Secretariat—Ministry of Agriculture), IBAMA (Brazilian Institute of Environment—Ministry of Environment), and ANVISA (Ministry of Health). Since the publication of the new biosafety Law in 2005, all PIPs will be analyzed like any other genetically modified organisms by CTNBio unless the product is itself to be used as a pesticide, when the Pesticide Law also applies. This change comes as a blessing for the *Bt* products because the procedural simplification will allow for experiments to be done with the sizes and designs one needs to prove environmental safety. Under the Pesticide Law many more restrictions would apply for other reasons than GE biosafety.

Environmental Law

Further regulations contained in Resolution 305, approved by the National Environmental Council (CONAMA)[9] in June 2002, that were applied to GE crops in general, including pest-resistant and herbicide-tolerant varieties, before the approval of Law no. 11.105/05, will now apply only to those products considered by CTNBio as having potential environmental impact. In this case, the Ministry of Environment and IBAMA authorities will continue to require these GEOs to go through an elaborate battery of environmental impact studies and tests.

EMBRAPA'S BIOSAFETY NETWORK

Although in 2000 so much was required by so many different agencies, all private, university, and Embrapa laboratories had stopped research with GE biopesticide plants. The reason was that IDEC[10] and Greenpeace appealed, alleging that GEO-specific regulations did not exist within the Pesticide Law. The petition was approved and several public meetings and extensive consultations were required before implementation of these regulations was complete in November 2002.

Between September 2001 and early 2002, several interested institutions submitted petitions to restart research in Brazil. However, due to the change of government, in January 2003, and accompanying reorganization in the various ministries, final regulations were only published in December 2003, and Embrapa received the first license to plant a small field in the state of Bahia with GE papaya.

Brazilian farmers are beginning to demonstrate in favor of the new technology even more stridently than Greenpeace and other environmental NGOs that fiercely oppose the use of GE organisms in Brazil. In the 2002/2003, 2003/2004, and 2004/2005 seasons, farmers were permitted by a special law to plant "stealth" soybean seeds; seeds obtained and multiplied without the IP owner's permission.[11]

All of this led to confusion among farmers[12] who are used to keeping their own seeds for three to four years, and could not understand why the legality of GE soybeans could change from year to year. Indeed, the policy changes seemed to be political and ideological. This is now past history because the new biosafety Law has clarified the situation by giving legal and definite approval for the planting and commercialization of glyphosate-resistant soybean. At least for this GE crop, all pending issues were decided and it should now be treated as any other conventional plant.

THE SOYBEAN SAGA

For a number of different reasons, including the alleged lack of specific regulations, Brazil was not able to develop a unique scientifically based approval process for transgenic crops to be commercialized. The proposal for transgenic glyphosate-tolerant soybeans, which was approved for commercial release by the CTNBio, has been in the hands of the judiciary since 1998. The present government has passed three Provisional Laws since 2003 to legalize the commercial production of transgenic glyphosate-tolerant soybean planted mostly in the south of the country by farmers who brought the technology from neighboring Argentina during the late 1990s.

"Farmers will always seek the use of the technology that gives them the best return"—and should receive due support. This was one of the reasons why Congressmen, helped by scientists and policymakers, put a lot of their efforts to approve a Law that would give the country its needed clear legal framework, allowing for an organized process of risk analysis to be followed by applicants. From March 2005 onward, the planting of glyphosate-tolerant soybeans is completely legal and hopefully the same will happen to varieties of *Bt-*

cotton *and Bt*-corn, which are already adapted to Brazilian conditions, and which should follow the normal course of approval by the new CTNBio.

THE LABELING DECREE NO. 4680

Decree no. 4680,[13] published in April 2003, requires labeling of GEO products and derivatives. It also requires GE products to stamp a visible stamp of a capital "T" (for transgenics). The Decree made an exception only for the 2002/2003 soybean crop, its products and derivatives. There are difficulties for its implementation in the short term due to the huge size of the country, and the lack of specialized infrastructure to harvest, segregate, and transport the crops. Labeling issues are discussed later in this chapter.

The billions of dollars spent all around the world on very uncertain compliance and enforcement of biosafety rules, regulations, and laws might be better used in other ways; indeed the cost may not be worth the potential benefit.[14] The development of accepted international standards for the environmental protocols would at least harmonize the requirements for similar cases and develop a better understanding that risk analysis, risk communication, and risk management of GEOs should differ from the more commonly known environmental impact analysis demanded before the construction of roads, dams or buildings.

THE CARTAGENA PROTOCOL

Brazil has a new regulatory framework that can be used to promote scientific research and at the same time assure society that the necessary care will be taken to ensure environmental and food/feed safety.[15] However, Brazil has ratified the Cartagena Protocol on Biosafety of GEOs in 2004 and has designated its provisional Focal Point but it has not set the biosafety clearing house mechanism (CHM). With the approval of the new Law in March of 2005 and the reedition of CTNBio, it will be possible to implement the CHM, which should be managed by the CTNBio Secretariat. Quite a lot of compromise is expected from the government, with a major responsibility falling on the Ministry of Science and Technology because of its link with CTNBio to implement both the new Law and the Protocol. The latter could became a major barrier for exporters to reach markets in countries that are already making demands based on the rules of the Cartagena Protocol and, therefore, the matter needs careful attention and implementation.

CURRENT GEO LABELING LEGISLATION IN BRAZIL

As the discussions of international trade and biotechnology issues are intensified, there is a growing need for differentiation between genetically modified foods and those that have not been genetically modified. In several countries, food-labeling legislation has established threshold levels, which are the maximum levels allowed for the presence of a GEO in a non-GEO product. Foods that contain levels of genetically modified ingredients above the established threshold should be labeled as "genetically modified."

As an example, the Japanese legislation has established a 5 percent threshold level for soybeans. Due to the possibility of cross pollination, no threshold has been established for corn. Australia and New Zealand had set a 1 percent threshold level for genetically modified foods that have been evaluated and approved in these countries, and amounts above this level require mandatory labeling.

It is agreed that labeling should not be required for foods containing highly refined ingredients (e.g., sucrose and refined vegetable oils), but do not contain measurable amounts of a novel protein or DNA, since any engineered genetic material and protein, which could be present in those products, are destroyed or removed during the refining process. The food end product is, therefore, not itself genetically modified and cannot be distinguished from that produced by conventional means.

In Brazil, Decree No. 4680[16] established that all processed or raw packed foods, feeds, or bulk products, which contain more than 1 percent GEO material, should be labeled with the list of ingredients that informs the consumer of the species from which the introduced gene was derived. This decree also required that food and food ingredients produced from animals that were fed GEO feed should be labeled with the following expression: "(name of the animal) fed with feed containing transgenic ingredient." This decree mandates consumers' right to information, in accordance with the Brazilian Consumer Defense Code (CDC),[17] which can be considered one of the most important achievements in the consumer' defense history in Brazil.

Decree No. 4680 also raises several other issues:

- As the Decree applies to all prepacked foods and feeds (bulk, processed, or raw), this law introduced new criteria to Brazilian food-labeling standards, which previously applied only to prepacked food. Implementation required information for bulk, segregation[18] and identity preservation[19] of the food chain.

- Although so far the decree has not been implemented, it can be presumed that all foods, feeds, or food ingredients where the detected protein or DNA amounts to or is below 1 percent shall not be labeled. It can also be presumed that when protein or DNA cannot be detected, the food or feed need not be labeled. It is important to point out that Brazil is the first country that has established the threshold level for the final product, and not for each separate ingredient as did Australia, New Zealand, Japan, and the European Union. Such a threshold should be taken into account by food manufacturers, as they will need to control the origin of each ingredient in order to determine whether or not the final product should be labeled. As the GEO detection and quantification should rely on internationally validated methodology, this issue has been discussed in several international fora but a consensus has not been reached. Realizing that most currently available methodology has not been internationally validated; the Codex Alimentarius Commission has referred this matter with high priority to the Codex Committee on Methods of Analysis and Sampling (CCMAS).

- Another interesting innovation in the Brazilian Decree is the mandatory declaration of the gene-donor species, which should be displayed on the label near the list of ingredients. There are still doubts about how this declaration will be presented, for example, in the case of a product with a *Bt* maize variety as ingredient, the label might have to include "donor species: *Bacillus thuringiensis*" in the list of ingredients. This issue has raised different concerns because *"Bacillus thuringiensis"* is not an ingredient of the product and the information may be misleading. According to the Brazilian Consumer Defense Code, Articles 6 and 31, the information displayed on the label should be clear and not misleading; hence this point should considered by the regulators prior to implementation of the decree. The publication of a list of approved GEOs, the inserted genes, and their respective donor species might help in this matter, as there may well be cases where the GEO will have more than one inserted gene or donor species.

- The decree also establishes the use of the term "transgenic," highlighted at the front panel, followed by a symbol or a logo, with one of the following expressions: "transgenic *(name of the product)*," "contain transgenic *(name of the ingredient(s))*" or "product produced from transgenic *(name of the product)*." The logo was established by Ministerial Order 2658 published by the Ministry of Justice and gave food producers sixty days to implement it. Use of the term "transgenic" may cause some difficulties for its implementation and regulation, because the Brazilian Biosafety Law[21] defines and uses the expression "genetically modified organism," as it is done in most of the components of the GMO biosafety legislation in Brazil.

- The Decree also established that labels of food and food ingredients produced from animals that were fed with feed containing GEOs should declare this condition highlighted in the front panel, with one the following expressions: "*(name of the animal)* fed with feed which contains transgenic ingredient" or "*(name of the ingredient)* produced from animal fed with feed which contains transgenic ingredient." A very important fact related to this issue is that several studies have shown that DNA or protein from the genetic modification are not detected in the meat, milk, or eggs of animals fed with feeds containing GEO. Therefore, it is not possible to determine whether animals fed with GEO feed were used in the production of a food or food ingredient. For the enforcement of this matter, traceability[22] of the food chain will be required.

- For food and food ingredients, which are not produced or derived from GEO and when the corresponding GEO has been approved for the Brazilian market, the use of the expression "*(name of the product)* transgenic-free" is voluntary. As far as the use of the term transgenic-free is concerned, it will be very important for the enforcement of the labeling decree that the absence of the DNA or protein should be determined by detection analysis or certification of the food chain.

- The Decree has a different terminology for the 2002/2003 soybeans crop, its products and derivatives, which were approved by the Brazilian government, and allows the use of the expressions *"may contain transgenic soybeans"* or *"may contain ingredients produced from transgenic soybeans"* on their label. In this case, the use of stickers or any other printing form is allowed. Soybeans that are certified GEO-free[23] or that have been produced in regions where GEO presence has not been verified will be exempt from using the label.

FOOD SAFETY ASSESSMENT
OF GENETICALLY MODIFIED FOODS

The use of toxicological animal models for the safety assessment of compounds such as pesticides and food additives is well established. In these cases, animals are directly fed such compounds at a range of doses some several orders of magnitude greater than the expected dosage for human consumption, and safe intake levels are estimated by the application of the appropriate safety factors. Whole foods, however, due to their variation in composition and nutritional value, and their bulk and effect on satiety, can usually only be fed to animals at low multiples of the amounts that might be present in the human diet.[24] Thus, the difficulties of applying traditional toxicological tests to whole foods required the development of an alternative approach for the safety evaluation of genetically modified foods.[25]

In 1993, the OECD formulated the concept of substantial equivalence as a guide in the safety assessment of genetically modified foods. The concept of substantial equivalence, which has been clarified over the years,[26] is part of a safety-assessment structure based on the idea that already existing foods can be used for comparison of the genetically modified food with the appropriate conventional counterpart.[27] Such an approach is not intended to establish absolute safety, which is impossible to achieve for any food, but does warrant that the food, and any substance introduced in it as a result of genetic modification, is as safe as its conventional counterparts.[28]

Nowadays, the concept of substantial equivalence is normally limited to genetically modified plants, but it could also be expanded to other genetically modified organisms and organisms modified by other biotechniques or traditional breeding.[29] The application of the concept is not by itself the safety evaluation, but helps to identify similarities and possible differences between the conventional food and the new product, which can then be submitted to additional toxicological evaluation.[30]

Factors in the safety assessment include identity, source, composition, effects of processing and cooking, transformation process, protein expression product of the novel DNA (effects on function, potential toxicity, and potential allergenicity), possible secondary effects from gene expression (composition of critical macro-nutrients, micro-nutrients, anti-nutrients, endogenous toxicants, allergens, and physiologically active substances), and potential intake and dietary impact of the introduction of the genetically modified food. The type and extent of further studies depend on the nature of the differences and whether or not they are well characterized.[31]

Studies should be carried out in accordance with good laboratory practices, which require all laboratories to use validated methodology and analytical conditions that yield internationally acceptable results and can be compared with results from other laboratories. Questions have been raised concerning the safety of genetically modified foods, which are being introduced into the food chain. The United States, Japan, and Australia, as well as countries of the European Union, already have established rigorous safety-assessment protocols for cultivating and importing genetically modified foods.[32]

ENVIRONMENTAL IMPACTS

The basic characteristics of general risk assessment of genetically modified organisms are understandably different from those associated with chemicals. GEOs are living organisms and, unlike chemicals that may become diluted, have the potential to disperse to new habitats, colonize these sites, and multiply. Their novel activities, including the production of metabolites, enzymes, and toxins, will occur as long as the GEOs remain metabolically active. Once established in the environment, living organisms cannot be recalled.

The potential impacts of genetically modified organisms on the environment have been extensively reviewed.[33] Additionally, Dale et al.[34] studied the issue of direct and indirect impacts of genetically modified crops.

The main topics of concern related to the release into the environment of GEO are as follows:

1. Risks to human and animal health, which include toxicity and food quality, allergenicity, and resistance to pathogens and drugs (antibiotic resistance);
2. Risks to the environment: persistence of the gene or transgenes (voluntary plants, increase in fitness, invasiveness) or transgenic products (cumulative effects), susceptibility of nontarget organisms, increase in the use of chemicals in agriculture (herbicides), and unpredictable gene expression or transgene instability;
3. Risks to agriculture: resistance/tolerance of nontarget organisms, weeds or superweeds, alteration in the nutritional value (attractiveness of the organism to pests), and reduction of cultivars (increased susceptibility), and loss of biodiversity.

FINAL CONSIDERATIONS

Certainly, biotechnology has been established worldwide. With every new discovery comes improved control of its development, and an array of possibilities is emerging from the use of functional genomics. Its impact on the sustainability of agricultural production could indeed be enormous. As more developing countries begin to experiment and see the benefits of the local applications of the technology, society may well begin to accept the new varieties as the product of a novel and different biological process. This is already the case with many of the medicines produced through biotechnology that are currently available in the market.

It is unfortunate that a diversity of other issues, such as proprietary matters, the buildup of gigantic transnational companies with enormous economic influence around the world, unresolved ethical issues, a variety of fear campaigns against the initial products, the influence of precautionary decisions taken by rich nations whose populations are largely overfed, and many overarching political issues, have become entangled with the use of biotechnology in such a way that clear scientific discussion can seldom take place.

As observed in several fora, biosafety regulatory systems around the world are becoming increasingly complex. The coming years will be difficult for developing countries as they address issues such as the need for better-trained personnel (including those responsible for policing the experiments) and the need for local investment to continue developing their capacity in potentially beneficial areas of biotechnology and biosafety, in spite of the negative campaigns.

One can expect that the coming years will also give scientists a higher confidence in terms of developing the appropriate methods to predict, on a case-by-case basis, a possible negative environmental impact. With such efforts, they will be able to better address the apprehensions and the doubts about the beneficial use of transgenic plants and other genetically modified products.

Finally, it seems appropriate to comment on the potential return of Embrapa investments and the prohibition of the commercialization of GEO crops in Brazil. According to Avila et al.,[35] the organization has been spending around US$10 million annually on agricultural biotechnology research over the last ten years. It has already developed some transgenic lines of soybean, potato, papaya, and common beans, which will be released for commercial use after completion of biosafety tests, pending public acceptance. The Brazilian agribusiness has been so far undermined as a result of the precautionary policy pursued by Brazil as influenced by the European and Japanese markets.

The labeling issue is still far from an international consensus. Progress in this discussion requires not only maturity from all the stakeholders involved, but also a clarification of the weight attributed to the different factors such as the right of consumers to make a choice based on information, cost impacts in the final product, and availability of internationally validated methodologies for the detection of GEOs that may interfere with governments' ability to make the final rules.

NOTES

1. Atherton, K.T., ed. 2002. *Genetically modified crops—Assessing safety.* New York: Taylor & Francis.

2. Thomas, J.A. and Fuchs, R.L., eds. 2002. *Biotechnology and safety assessment* (3rd ed.). San Diego: Academic Press.

3. Atherton, K.T. 2002. See note 1.

4. Traynor, P.L., Frederick, R.J., and Koch, M. 2002. *Biosafety and risk assessment in agricultural biotechnology: A workbook for technical training.* East Lansing: The Agricultural Biotechnology Support Project, Institute of International Agriculture, Michigan State University.

5. Portugal, A.D., Sampaio, M.J., Contini, E., and Ávila, F. 2001. Agricultural biotechnology in Brasil—Institutionality and implications of genetically modified organisms. *Proceedings of 5th International Conference of the International Consortium on Agricultural Biotechnology Research (ICABR) on Biotechnology, Science and Modern Agriculture: A New Industry at the Dawn of the Century.* Ravello, Italy.

6. USDA/FAS. Foreign Agricultural Service. 2003. *Brazil: Future Agricultural Expansion Potential Underrated.* www.wto.org/english/res_e/statis_e/its2001_e/stats2001_e.pdf (Last accessed September 5, 2006).

7. Ministry of Science and Technology. http://www.mct.gov.br/legis/leis/8974_95.htm.

8. Casa Civil da Presidencia da República–Governo Digital. www.planalto.gov.br/legislação.

9. Ministry of Environment. http://www.mma.gov.br/port/conama/index.cfm.

10. IDEC—Instituto Brasileiro de Defesa do Consumidor—Brazilian Consumers Association. http://www.idec.org.br/paginas/english.asp.

11. Herring, R. 2003. Underground seeds: The lessons of India's *Bt* cotton episode for representations of the poor, property claims and biosafety regimes. *Proceedings of the workshop, Transgenics and the poor: Science, regulation and development strategy.* Cornell University, November 7-8, 2003.

12. Bonalume Neto, R. 2003. GM confusion in Brazil. *Nature Biotechnology* 21 (11): 1257-1258.

13. Decree No. 4680/2003 issued by the Ministry of Justice. www.mj.gov.br.

14. Brody, J. 2002. In a world of hazards, worries are often misplaced. *New York Times,* August 20.

15. Convention of Biological Diversity. www.biodiv.org.

16. Ministry of Justice, www.justica.gov.br/DPDC/servicos/legislacao/pdf/Decreto4680.pdf. Accessed April 24, 2003.

17. Law 8.078/90. December 12, 1990. http://www.justica.gov.br/DPDC/servicos/legislacao/cdc.htm.

18. Segregation: Refers to a system of crop or raw material management that allows one batch or crop to be separated from another. It implies that specific crops and products are kept apart, but does not necessarily require traceability along the production chain.

19. Identity Preservation: A system of crop or raw material management and trade which preserves the identity of the source or nature of the materials.

20. Ministerial Order 2658. December 22, 2003, defines the logo which should be printed on the labels of transgenic foods and food ingredients. http://www.justica. gov.br/DPDC/servicos/legislacao.htm.

21. See note 10.

22. Traceability: Measures covering feed, food and their ingredients. It includes the obligation for feed and food businesses to ensure that adequate procedures are in place to withdraw feed and food from the market where a risk to be the health of the consumer is posed. Operators should keep adequate records of suppliers of raw materials and ingredients so that the source of the problem can be identified.

23. Article 4, Law 10688. June 13, 2003. Ministry of Agriculture, http://extranet. agricultura.gov.br/agrolegis/do/consultaLei?op=viewTextual&codigo=3076.

24. Donaldson, L. and May, R. 1999. *Health implications of genetically modified foods*. Site do Department of Health. http://www.doh.gov.uk/gmfood.htm. Accessed July 13, 2000.

25. World Health Organization (WHO). 2000. *Safety aspects of genetically modified foods of plant origin*. Report of a Joint FAO/WHO Expert Consultation on Foods Derived from Biotechnology. Geneva: WHO, p. 29.

26. Erickson, B.E. 2000. Detecting genetically modified products in food. *Analytical Chemistry* 454A-459A, July 1.

27. Kuiper, H.A., Kleter, G.A., Noteborn, H.P.J., and Kok, E.J. 2001. Assessment of the food safety issues related to genetically modified foods. *The Plant Journal* 27: 503-528.

28. World Health Organization. 2000. See note 25.

29. Pedersen, J. 2000. *Application of substantial equivalence data collection and analysis*. Joint FAO/WHO Expert Consultation on Foods Derived from Biotechnology. Topic 2.

30. Kuiper, H.A. et al. 2001. See note 27.

31. World Health Organization. 2000. See note 25.

32. Lajolo, F.M. and Nutti, M.R. 2003. *Transgênicos—Bases Científicas da sua Segurança*. São Paulo: SBAN.

33. Shelton, A.M., Zhao, J.-Z., and Roush, R.T. 2002. Economic, ecological, food safety, and social consequences of the deployment of *Bt* transgenic plants. *Annual Reviews of Entomology* 47: 845-881; Snow, A.A., and Palma, P.M. 1997. Commercialization of transgenic plants: Potential ecological risks. *Bioscience* 47: 86-96; Traynor, P.L., and Westwood, J.H., eds. 1999. *Proceedings of a workshop on ecological effects of pest resistance genes in managed ecosystems*. Bethesda, MD. Blaksburg: Information Systems for Biotechnology; Wolfenbarger, L.L., and Phifer, P.R. 2000. The ecological risks and benefits of genetically engineered plants. *Science* 290: 2088-2093.

34. Dale, P.J., Clarke, B., and Fontes, E.M.G. 2002. Potential for the environmental impact of transgenic crops. *Nature Biotechnology* 20: 567-574.

35. Avila, A.F.D, Quirino, T, Contini, E., and Rech, E.L. 2002. Social and economic impact. Ex-ante evaluation of Embrapa's biotechnology research products. In *Economic and social issues in agricultural biotechnology*. Edited by R.E. Evenson, V. Santaniello, and D. Zilberman. Wallingford: CABI Publishing, pp. 287-308.

Chapter 15

Toward a Liability
and Compensation Regime
under the Biosafety Protocol

Kristin Dawkins
Josh DuBois

INTRODUCTION

In February 2004, the world's governments began to craft liability provisions to compensate and redress damages that may be caused by imports of genetically engineered grains and other organisms with altered DNA. These liability provisions will supplement the Cartagena Protocol on Biosafety (CPB), which was adopted on January 29, 2000. Liability provisions were not included in the original draft of the CPB, but CPB Article 29 required signatories to begin crafting such rules at their first meeting and to complete the process in four years.

Liability provisions for the CPB were to be crafted to provide maximum environmental protection. Other environmental treaties have incorporated features such as strict liability,[1] recovery for both direct and indirect damages, and statutes of limitations[2] which toll[3] until latent harms[4] are discovered. These features are important in environmental liability regimes to ensure that contracting states will benefit fully from international agreements. The international community is experienced in creating liability provisions for environmental trea-

Genetically Engineered Crops
© 2007 by The Haworth Press, Inc. All rights reserved.
doi:10.1300/5880_15

ties. By borrowing principles from similar agreements the signatories should be able to develop liability rules for the CPB quickly.

The CPB is not the first environmental agreement of its kind. Three previous international environmental agreements provide useful guidance for the development of a liability regime for Cartagena, including the following:

- The Convention on International Liability for Damage Caused by Space Objects (hereinafter "Space Objects Convention"), which deals with damage from spacecraft,
- The Basel Protocol on Liability and Compensation Resulting from Transboundary Movement of Hazardous Wastes (hereinafter "Basel Protocol"), which deals with international trade in hazardous wastes, and
- Convention on Civil Liability for Oil Pollution Damage (hereinafter "CLC"), which deals with damage caused by oil spills.[5]

The Space Objects Convention addresses liability for damage caused by the launching of spacecraft. The liability problems surrounding spacecraft are similar to those that surround genetically engineered organisms (GEOs): in both instances, new technology gives rise to the potential for widespread, unanticipated environmental damage.[6] A further similarity is that both spacecraft and GEOs are produced by relatively few countries, but create potential for harm throughout the world. International trade in hazardous wastes is another area in which developed nations (waste exporters) may benefit at the expense of environmental damage to less developed countries (waste importers). The Basel Protocol's regulation of the hazardous waste trade is particularly important because many developing nations lack robust domestic environmental regulations, and prior to the Basel Protocol waste exporters could take unfair advantage of this regulatory gap.[7] The CLC, addressing environmental damage from traffic in oil, is a good example of a liability regime for a lucrative commodity trade that poses severe environmental threats. All three agreements are instructive for the development of a liability regime for the CPB.

SUGGESTIONS FOR THE DEVELOPMENT
OF BIOSAFETY LIABILITY RULES

A Strict Liability Regime Is Appropriate

Strict liability is an appropriate and commonplace liability regime for "ultrahazardous"[8] activities such as transboundary movement of GEOs.[9] Several other multilateral environmental treaties adopt a strict liability regime with regard to harm resulting from new, untested technology with the potential for large-scale environmental disruption.[10]

There is growing evidence that the harm resulting from a GEO accident could have catastrophic results, causing irreparable harm to agricultural ecosystems, crops, export earnings, indigenous knowledge systems, and threatening food security.[11] This qualifies any activity relating to GEOs as ultrahazardous.[12] Strict liability will deter reckless behavior in the development and marketing of GEOs.

Some agreements adopt a two-tiered approach to liability.[13] Liability for ultrahazardous activity that is *non volenti,* where the victim has not agreed to risk the injury by his or her own conduct, should be strict. For example, farmers whose fields and crops are contaminated through horizontal gene transfer have not voluntarily assumed a risk. Thus, liability for horizontal gene transfer should be strict. For activity that is less hazardous, or where the injured party has voluntarily accepted risk, liability may be fault based.

A two-tier approach to liability is appropriate for the CPB. Strict liability should apply to all "ultrahazardous" activity and to all *nonvolenti* harms, such as horizontal contamination. Fault-based liability is acceptable where activity is less dangerous or where the injured party has assumed risk.

DAMAGE THAT SHOULD BE RECOVERABLE

Environmental Damage

There are two facets of liability for environmental damage. In one class of cases, environmental damage will result directly in lost profits to private parties. This kind of case might arise, for example, when GEO contamination injures an individual farmer's ability to earn a living from his or her land. When environmental damage can be shown to have caused direct economic harm, the damage should clearly be recoverable.[14]

In other cases, a country's environment or natural resources may suffer damage without direct, immediate economic harm. Examples of this second kind of damage might include damage to publicly owned or communally held environmental resources; damage to resources that are valuable but not currently being exploited; and damage such as loss of biodiversity, which will adversely affect a country's natural resources in the long run but which may not have immediate, direct economic impact. These kinds of damages to the long-term health of a country's environment must also be recoverable.

Recovery for environmental damage must include the reasonable cost of environmental restoration.[15] The cost of preventative measures taken to prevent future harm must also be recoverable.[16] In some cases environmental restoration may be impossible, when for example damage done by GEOs is irreversible. Calculating damages in such cases will be difficult. Various national laws may provide some guidance in calculating damages for both kinds of situations.

Socioeconomic Damages

Article 26 of the CPB allows parties to take into account the socioeconomic impacts of GEOs when making decisions or enacting domestic laws related to the CPB. Article 26(2) specifies that GEO impact on indigenous and local communities merits special attention. Introduction of GEOs may have an adverse effect on traditional agricultural systems and practices of such communities.[17] A liability regime should allow recovery of socioeconomic damage, particularly that suffered by local and indigenous peoples. Calculating a dollar amount for cultural damage may be difficult. Damage to traditional agricultural practices, however, is a type of economic damage that should be amenable to calculation. Small farmers, fisher folk, and local populations and communities must be taken into account when liability rules are developed.

Life, Health, and Property

Damage to life and property, and personal injury to individuals, should all be recoverable losses under the CPB. These are fundamental damages, which any liability regime should address. The CPB also seeks to protect "human health." Article II(2) lists risks to human health alongside concerns for biodiversity as primary purposes of the agreement. Article IX(8) lists concern for human health again in its discussion of the Precautionary Principle. Given the CPB's repeated stress on human health concerns,[18] this language should be interpreted broadly to encompass nonphysical injury and illness, and also health impairments to the population that are indirectly caused by GEO accidents.

Illegal Transboundary Movements

Article 25(1) of the CPB defines "illegal" transboundary movements to be those that do not conform to domestic measures passed to implement the Protocol. Article 25(2) provides that recovery of damages will be allowed for such illegal transboundary movements. Damages in the case of an illegal movement include costs of repatriation of the GEOs to the exporting country, or destruction of the imported GEOs. Allowing for recovery of these costs is an explicit requirement of the CPB, so any set of liability rules must allow for this kind of recovery. Additionally, the fact of an illegal transfer should strengthen a plaintiff's case for environmental, health, or socioeconomic damages.

REMEDIES

The injured party should be restored to its position before the harm. The CPB liability provisions should allow an injured party to be restored to its position before the harm caused by transboundary movement of GEOs. As discussed earlier, this may include repatriation or destruction of GEOs,[19] recovery of lost profits,[20] the cost of preventative measures[21] and compensation for long-term environmental damage.

There should not be liability caps. Harms caused by GEOs could be colossal in magnitude.[22] Particularly for *non volenti* harms, harms where the injured party has not assumed risk, it is unfair to allow the injured party less than a full recovery. *Non volenti* harms are precisely those for which strict liability is imposed; therefore, strict liability should not be accompanied by recovery caps.[23] The CPB must ensure that the handful of GEO-producing countries are responsible for whatever damage their products cause abroad,[24] by allowing full recovery for injured parties.

There must be a compensation fund. An international fund for payment of damages, which may otherwise not be recoverable, should be part of the CPB.[25] The fund should be accessible when the responsible party is insolvent, and, if liability caps are imposed, when such caps are reached but do not fully compensate for the damage sustained. A biosafety compensation fund should be funded by both the biotechnology industry and countries that benefit from the export of GEOs.

There must be compulsory insurance for trade in GEOs. Financial guarantees must be in place to assure that injured parties will be able to recover from harms caused by GEOs. Compulsory insurance or bonds for those who trade in GEOs should be made part of the Cartagena liability rules.[26] The Cartagena Protocol should require GEO traders to offer proof of insurance as a prerequisite for trading activity.[27]

Breaches of Duty under the Protocol

Because a strict liability regime should be adopted for all or most harm due to transboundary movement of GEOs, parties will often only need to show damage and causation, and not breach of duty, in order to recover.

For transboundary movements that are illegal under the CPB, breach of duty will be an issue. Under the Biosafety Protocol, exporting states have an obligation to notify an importing state and obtain advance consent for introduction of GEOs. According to Article 8.1, the Party of Export must notify the importer state or require the exporter to undertake that notification. The domestic laws of a country will most certainly reflect these provisions in the CPB. Any transboundary movement in violation of the notification principles will then be illegal under Article 25(1) of the Protocol. The first duty of a state in breach then is to make reparation. The content of this duty of reparation was stated by the Permanent Court of Justice in the *Chorzow Factory* Case as follows:[28] "The reparation must, as far as possible, wipe out all the consequences of the illegal act and re-establish the situation which would, in all probability, have existed if that act had not been committed."

So the parties must be restored to the position before the breach. The affected party (the state) can request the state of origin to dispose of the GEO by repatriation or destruction as appropriate [Article 25(2)]. The cost must be borne by the latter state.

Parties

Liability of States

The state of export is liable for illegal transboundary movements, according to Article 25 of the CPB. Illegal transboundary movements will occur primarily as a result of a state's breach of its obligation to notify (or to require notification by the exporter) under Article 8, so state liability in these cases make sense.

Liability of Private Parties

Several private parties will come in contact with GEOs during their life cycle. In most circumstances, joint and several liability should be imposed on a variety of parties.[29] Private parties should be free to enter into agreements by which they contractually assume risk and apportion liability.[30]

Joint and several liability will result in a just apportionment of risks. For example, in cases where the GEO is inherently flawed, so that it will cause damage even when properly used and contained, common products liability theory says the manufacturer should be liable.[31] When damage results from an incident during transportation (a "spill" or similar event), then liability should be imposed on the transporter. Simple apportionments of liability such as these fail to consider the chain of actors, including producers as well as exporters, importers and brokers, who will benefit from a trade in GEOs: those who benefit from the trade in GEOs must all share the risks contemplated by the Protocol. It is important to note that imposition of liability on the end user (e.g., a small farmer) will often be unfair. These users will have relied on information supplied by a producer or broker to make safety assessments, they will benefit less from GEO use than will GEO producers, and they will have less economic power.

The CPB should allow for claims against any one or more of the following for damage resulting from a GEO as applicable:

- The country of export if damage results from the deliberate introduction of a GEO into the environment (following the formulation of the Biosafety Protocol) of the country of import, in breach of the obligation to notify, or require the exporter to notify in accordance with Article 8 of the Biosafety Protocol.
- The manufacturer of the GEO if the harm is caused by the properties of the GEO, the genetic modification, and the conditions under which the GEO is introduced, and continues to remain in the received environment.
- If the damage results from the failure to provide an adequate system of safety, such as physical barriers, then the operator responsible for this default.
- In all other cases, any one or more of the following: the manufacturer of the GEO, the exporter, the country of export, and anyone else responsible for putting the GEO in circulation into the environment.

Parties Who May Act As Plaintiffs

Any injured party should have standing to bring a claim.[32] Plaintiff parties could include individuals, businesses, and other public or private bodies, including states, that have suffered injury.[33] There is no reason to limit the standing of any injured party to bring a claim, and without such a reason all injured parties should be allowed to bring suit and recover.

Statute of Limitations

The statute of limitations for bringing the claims under the CPB must not be an undue barrier to recovery by a plaintiff.[34] The statute of limitations for GEO pollution or damage should be long enough to allow at least two generations of the GEO to reproduce, so that the long-term effect of a GEO can be ascertained before a claim is brought. The statute of limitations should also "toll"[35] until damage is discovered, since some harms may not be apparent at the moment of an incident of contamination.

Structural and Procedural Provisions

Claims made under the CPB should be brought in national courts. Substantive law in each signatory country must be enacted, which will allow for recovery under the CPB.[36] The alternative to use of domestic courts is the creation of an international body to try cases that arise under the agreement, but use of such an international body is expensive and procedurally complex. Jurisdiction should exist:[37]

- where the damage is suffered, or
- where the incident giving rise to damage occurred, or
- where the defendant resides, or
- where the defendant's principal place of business is located.

Since substantive provisions for recovery under the CPB will be created in the national courts of each signatory state, motivation for forum shopping will be minimized.[38]

CONCLUSION

The CPB is a strong environmental treaty, and liability rules should be developed with environmental protection in mind. The CPB is guided by a need to protect the environment from the unknown effects that GEOs may have. Furthermore, the CPB is designed to impose restrictions on exporters of GEOs so that the few parties who produce GEOs, and stand to benefit from their widespread use, will also bear the responsibility of potential harm. A final important feature of the CPB is that it recognizes the need to protect not only the natural environment, but also the social and economic fabric of local and indigenous communities that may be affected by the introduction of foreign GEOs.

Strict liability is appropriate for transboundary GEO trade because of its ultrahazardous nature. Costs recoverable under the CPB must include direct and indirect damage resulting from harm to the environment (including environmental restoration costs and the costs of measures to prevent future harm); socioeconomic damage (with special focus on damage to indigenous and local communities); damage to individual life, health, or property, as well as general damage to the

health of a population as a whole; and, particularly in the case of an illegal transboundary movement, the cost of repatriation or destruction
of the GEO at issue. To ensure full recovery for these damages, there
must be no liability caps that would limit damage awards, and there
must be mandatory insurance for GEO traders and an international
compensation fund to compensate victims when recovery is otherwise
unavailable. All injured parties should be able to bring claims, and
the statute of limitations should not be an unreasonable barrier to suit.

The other international agreements discussed herein serve as guides
during the formation of liability rules for the CPB. This should allow
the Parties to quickly develop a regime that serves to protect the environment and serves to protect parties, who are non-exporters of GEOs,
from potential harm.

NOTES

1. This chapter is an edited version of a longer work that discusses precedent
agreements in more depth—more information about the Space Objects Convention,
the CLC and the Basel Protocol (and their implications for the liability rules for the
CPB) can be found at http://www.twnside.org.sg/title/blp.pdf. Accessed February
13, 2004. By building on the successes of these previous agreements, a strong set of
liability rules for Cartagena can be developed soon.

2. *Black's Law Dictionary* defines *statute of limitations* as: "A statute establishing a time limit for suing in a civil case, based in the date when the claim accrued (as
when the injury occurred *or was discovered*)" (emphasis added). Ibid. at 1422.

3. A statute of limitations *tolls* when it is stopped temporarily from running, for
example, because the potential plaintiff does not yet know of the harm that has occurred. *Black's Law Dictionary* defines *toll* as follows: "2. (Of a time period, esp. a
statutory one) to stop the running of; to abate <toll the limitations period>." Ibid. at
1495.

4. This chapter is intended for a broad audience, so some common legal terms
will be defined throughout. According to *Black's Law Dictionary* (the preeminent
English-language legal dictionary), *strict liability* means: "Liability that does not
depend on actual negligence or intent to harm, but that is based on the breach of an
absolute duty to make something safe. Strict liability *most often applies either to
ultrahazardous activities* or in products-liability cases" (emphasis added). *Black's
Law Dictionary.* 1999. (7th ed.). New York: West Publishing Company, p. 417.

5. For readers unfamiliar with locating international agreements: The United Nations publishes *The United Nations Treaty Series,* a compendium of international
treaties. Treaties are submitted to the United Nations for inclusion in the *Series,* and
the full text of each treaty is published for reference. Citations to treaties in this document take the form NN U.N.T.S. PPP, where U.N.T.S. stands for *United Nations*

Treaty Series, NN is the volume number in which the treaty is found, and PPP is the page number within the volume on which the treaty begins. The U.N.T.S. is available in hard copy at many libraries, or may be searched online at http://untreaty. un.org/English/access.asp. Accessed February 12, 2004.

Citations for the three treaties discussed most often in this document are as follows:

* Convention on the international liability for damage caused by space objects, March 3, 1972, 961 U.N.T.S. 187 (hereinafter "Space Objects Convention").
* Protocol on Liability and Compensation for Damage Resulting from Transboundary Movements of Hazardous Wastes and Their Disposal at http:// www.basel.int/meetings/cop/cop5/docs/prot-e.pdf. Accessed February 13, 2004) (hereinafter "Basel Protocol". [Note: this protocol has not yet been published in the U.N.T.S. It is available on-line.].
* International Convention on Civil Liability for Oil Pollution Damage, November 29, 1969, 973 U.N.T.S. 3 (hereinafter "CLC").

6. See, for example, *Scientific Findings and Elements of a Protocol: Report of the Independent Group of Scientific and Legal Experts on Biosafety* (1996) at http://www.twnside.org.sg/title/bios-cn.htm. Accessed February 13, 2004 (more information regarding this source may be obtained from: Third World Network, 228 Macalister Road, 10400 Penang, Malaysia, Tel: 6-04-2266159, Fax: 6-04-2264505).

7. That is to say that waste producers in developed countries would not be able to dispose of their waste cheaply at home, because developed countries may have environmental regulations that make waste disposal an expensive business. If waste producers can sell their waste to less developed countries that have less stringent environmental regulations, they essentially circumvent their own domestic environmental regulations at the expense of the waste-importing country. Environmental regulations are difficult and expensive to develop, and may not be top domestic priorities for smaller or less developed nations. International environmental treaties play a valuable role in that they can stand in place of domestic environmental regulations that small countries may not be able to develop easily on their own. For a discussion of the global trade in hazardous waste, see UNEP, Environmental Data Report 345 (1991), pp. 335-336 documenting that 20 percent of the global trade in hazardous waste goes from developed to developing countries.

8. *Black's Law Dictionary* defines ultrahazardous activity as follows: An undertaking that cannot be reasonably performed safely even if reasonable care is used while performing it, and for which the actor may face strict liability for any harm caused; especially an activity (such as dynamiting) for which the actor is held strictly liable because the activity (1) involves the risk of serious harm to persons or property, (2) cannot be performed without the risk, regardless of the precautions taken, and (3) does not ordinarily occur in the community. Under the Restatement (Second) of Torts, determining whether an activity is abnormally dangerous includes analyzing whether there is a high degree of risk of harm, whether any harm caused will be substantial, whether the exercise of reasonable care will eliminate the risk, whether the activity is a matter of common usage, whether the activity is appropriate to the place in which it occurs, and whether the activity's value to society out-

weighs its dangerousness. Restatement (Second) of Torts § 520 (1977). Black's Law Dictionary (7th ed.) 5: 1524.

9. Strict liability is featured in the CLC, the Space Objects Convention, and the Basel Protocol. See CLC, note 5, art. III, 973 U.N.T.S. at 5; Space Objects Convention, note 5, art. II, 961 U.N.T.S. at 189; Basel Protocol, note 5, art. 4. Strict liability is not out of place in a multilateral international treaty—strict liability regimes exist in countries around the globe, within a wide variety of legal systems. See Gurdial Singh Nair. 1997. *Liability and compensation in a biosafety protocol.* ISBN 83-9747-26-6. Available for purchase from the publisher, Third World Network, at http://www.twnside.org.sg/title/prot-cn.htm. Accessed February 13, 2004.

10. Other examples of strict liability for technology-based, hazardous activity include the following—for air transportation: 1929 Warsaw Convention for the Unification of Certain Rules relating to International Transportation by Air; for nuclear power: The Vienna Convention on Civil Liability for Nuclear Damage, 1963, Brussels Convention on the Liability of Operators of Nuclear Ships, 1962, Paris Convention on Third Party Liability in the Field of Nuclear Energy, 1960, The Convention Supplementary to the Paris Convention of 1960, 1963; and for space flight of course the Space Objects Convention.

11. See note 6.

12. See also note 8. The factors listed from the Restatement of Torts in determining whether an activity is "ultrahazardous" indicate that trade in GMOs should qualify.

13. The Space Objects Convention uses a two-tier liability approach. See Space Objects Treaty, note 5, art. II & III, 961 U.N.T.S. at 189-90. The Basel Protocol also adopts a two-tiered approach. See Basel Protocol, note 5, Art. 4-7. The CLC adopts a generally strict liability regime, with some provision made for exoneration based on lack of fault. See CLC, note 5, art. III, 973 U.N.T.S. at 5.

14. The Basel Protocol, for example, covers damages that include "Loss of income directly deriving from an economic interest in any use of the environment. . . ." Basel Protocol, note 5, art. 2(c)(iii): See generally Bruce Hurwitz. *State liability for outer space activities.* 1972. Springer, pp. 14-18.

15. Basel Protocol, note 5, art. 2(c)(iv).

16. Ibid., art. 2(c)(vi). The CLC also allows the costs of preventative measures to be recovered. CLC, note 5, art. IX, 973 U.N.T.S. at 8.

17. Article 8(j) of the Convention on Biological Diversity acknowledges that these practices have a salutary beneficial effect on the conservation and sustainable use of biodiversity.

18. The Cartagena Protocol is not the first treaty to focus on the broad issue of human health as an environmental concern: the Space Objects Convention is also concerned with threats to human health. See Space Objects Convention, note 5, art. I(a), 961 U.N.T.S. at 189.

19. CBP, art. 25.

20. See note 14 and accompanying text.

21. See note 15 and accompanying text.

22. See note 11 and accompanying text.

23. Some treaties allow liability caps for strict liability, but no caps where fault can be shown. The Basel protocol and the CLC take this approach. Under the CLC, there is no limit for liability based on fault. However, for claims based on strict liability, the owner of a ship is entitled to limit his or her liability but the amount of the limitation is fixed. CLC, note 5, art. IV(1),(2), 973 U.N.T.S. at 5-6. The Basel Protocol uses a similar system: no limit for fault-based liability; for strict liability, the financial limits are determined by domestic law (the Protocol does, however, impose minimum bounds for liability limits). Basel Protocol, note 5, art. 12 and annex B. There is no limit to the amount of compensation recoverable under the Space Objects Liability Convention. See Space Objects Convention, note 5, 961 U.N.T.S. at 187.

24. See note 7 and accompanying text (discussing other international environmental treaties that protect against environmental exploitation). The position is that a mere handful of manufacturers will be the primary beneficiaries of the GMO trade, and that it is unjust for them to reap the profits of their trade without bearing the risk as well.

25. The liability cap provisions in the CLC are only available when a compensation fund exists to cover damages. CLC, note 5, 973 U.N.T.S. at 6. This structure is a strong incentive for the creation of such funds.

26. The CLC makes insurance mandatory for all ships carrying more than 2,000 tons of oil as cargo. CLC, note 5, art. VII, 973 U.N.T.S. at 7. The Basel Protocol mandates insurance coverage and also sets minimum coverage limits. Basel Protocol, note 5, art. 14.

27. The CLC contains detailed provisions on the insurance requirement, including rules about proof of insurance and the right of each state that is a party to the treaty to ensure that insurance given by other states is valid. See CLC, note 5, art. VII, 973 U.N.T.S. at 7-8.

28. Case Concerning the Chorzow Factory (Germany v. Pol.), 1928 P.C.I.J. (ser A) No. 17, at 47.

29. The CLC imposes liability jointly and severally when an accident involving two or more tankers occurs. See CLC, note 5, art. IV, 973 U.N.T.S. at 5. The Space Objects Convention imposes joint and several liability when two or more parties act together to launch a space object. Space Objects Convention, note 5, art. V, 961 U.N.T.S. at 190. The strict liability article in the Basel Protocol provides joint and several liability in the case that two or more parties are responsible for damage. Basel Protocol, note 5, art. 4(6).

30. The Space Objects Convention makes an explicit allowance for parties that may be jointly and severally liable to negotiate for apportionment of liability. Space Objects Convention, note 5, art. V(2), 961 U.N.T.S. at 190.

31. See note 1 (discussing that strict liability often applies in product liability cases).

32. The Basel Protocol, for example, does not place limits on who may bring a claim. See Basel Protocol, note 5.

33. The CLC contemplates claims by any of these parties. See CLC, note 5, art. I(2) (defining "person") and art. VI(1)(a) (contemplating that "person[s]" will have claims for damage under the CLC), 973 U.N.T.S. at 4, 7.

34. The CLC and the Space Objects Convention both provide explicit statute of limitations provisions. Ibid., art. VIII at 8; Space Objects Convention, note 5, art. X, 961 U.N.T.S. at 191.

35. The Space Objects Convention uses the rule that the statute of limitations tolls until the harmed party knew or should have known about the damage. See Ibid.; See *generally* note 3 (explaining what it means for a statute of limitations to *toll*).

36. The CLC requires member states to ensure that their courts can entertain claims under the treaty. CLC, note 5, art. IX, 973 U.N.T.S. at 8. The Basel Protocol also requires that member states allow their courts to hear claims, and further stipulates that domestic courts shall hear claims in a nondiscriminatory manner. Basel Protocol, note 5, art. 10.

37. Ibid., art. 17.

38. *Forum shopping* is the practice of strategically choosing the most favorable court in which to file a lawsuit. Standardization of laws among signatory parties will simplify plaintiffs' decisions about where to file a case and may prevent defense lawyers from convincing a court to dismiss a case so that it can be brought in a different (and possibly more defense-friendly) forum. *Black's Law Dictionary* defines *forum shopping* as "[t]he practice of choosing the most favorable jurisdiction or court in which a claim might be heard" (1999) *Black's Law Dictionary.* (7th ed.). p. 666.

PART III:
CHALLENGES TO CIVIL SOCIETY

Chapter 16

Public Spheres Pushing for Change: Public Participation in the Governance of GE Crops

Simon Joss

INTRODUCTION

Switzerland, June 7, 1998: Swiss citizens were called to the ballot box to vote in the first national referendum on genetically modified organisms.[1] The "popular initiative" (a frequently used tool of Swiss direct democracy) had been prompted by well over 100,000 citizens' signatures having been collected by a coalition of campaigners critical of GE (genetic engineering) technology. The "gene protection initiative" led to an intensive, public debate that lasted many years, in which citizens, politicians, regulators, scientists, environmental campaigners, social commentators and the media took part in activities ranging from local debates, media broadcasts, and parliamentary hearings, to national conferences and street demonstrations. Both the government and parliament recommended rejection of the popular initiative, but nevertheless agreed to amend existing legislation in response to criticism raised by the initiative. In the event, a surprisingly large majority (66 percent) of the voting population rejected the popular initiative in the hotly awaited and widely reported ballot.

B G Kere, Karnataka state, south-eastern India, March 2000: Fourteen local farmers met on a farm in this small village over five days to debate GE crops.[2] In what was billed as a "citizens' jury," the seven women and men, aged between thirty and seventy, heard evidence from eight different experts on various aspects of agricultural biotechnology research, development and regulation. Asked to consider whether they would grow GE crops—the proposed policy of the Indian government and multinational companies—a clear majority said they would not and that they favored a

Genetically Engineered Crops

doi:10.1300/5880_16

moratorium on the commercialization of GE crops. According to the charity ActionAid, which organized the citizens' jury, this initiative showed how poor and marginalized farmers can not only be involved in the GE crops debate but can also have a voice in decision making on matters that affect their existence.

Forest Row, East Sussex, south-east England, June 27, 2003: A couple of hundred local people filed into the old village hall of this rural village on a warm summer evening to participate in a lively debate about the proposed commercialization of GE crops.[3] A pro-GE scientist, a representative of the anti-GE environmental interest group Friends of the Earth, and the local Member of Parliament were invited to explain their positions on GE crops, and to discuss questions raised by the villagers. The event, hosted by a group of concerned citizens, formed part of the GM Nation? The Public Debate initiative organized by the Agriculture and Environment Biotechnology Commission (AEBC), a regulatory body in charge of assessing biotechnology and advising the U.K. government on GE policy. It was one of several hundred local activities organized "from below" across the country by citizens groups, research organizations, and various interest groups; these complemented the "top-down" national and regional events hosted by the AEBC itself. After the debate, participants were asked to complete questionnaires asking for the public's views and attitudes; these questionnaires provided formal feedback to the AEBC, which published the results of the various debates and of the over 35,000 returned questionnaires in autumn 2003.[4] The overwhelming majority of those consulted did not want the commercialization of GE crops to go ahead at this point.

These are three examples of how, in recent years and in different national and cultural contexts, social actors have increasingly become involved in publicly discussing, scrutinizing, and contesting issues of modern biotechnology. These examples show how the new technology of genetic modification has become the subject of sometimes controversial debates in the public sphere. These debates raise scientific-technical, political and ethical questions, such as what the risks are of releasing GE organisms into the environment, how GE technology might help alleviate world hunger, and whether it is morally right to manipulate genetic material. As such, they point to the increasing politicization of GE technology. This politicization has been prompted, on the one hand, by the maturation of the technology (GE processes and products have left the laboratory and entered the market) and, on the other, by a growing skepticism toward the technology on the part of a broad coalition of interest groups, nongovernmental organizations (NGOs), grassroots activists, critical experts, media representatives, and sections of the "general" public. These diverse social actors not only critically discuss the technology in terms of its scientific, economic, social, and ethical viability, but also use it as a focus for considering wider political issues, such as globalization, the role of multinationals, and the accountability of scientific and technological decision making.

The public questioning of GE technology has manifested itself to a varying extent in different parts of the world. In China and the United States, for example, where the application of GE technology is widespread,[5] there seems to have been relatively little overt public debate to date (although, interestingly, some of the earliest debates about the emerging recombinant DNA technology in the early 1970s took place in the United States). In contrast, in many European countries, and more recently in a growing number of countries in Latin America (for example, Brazil), sub-Saharan Africa (Zambia, South Africa), and Australasia (India, Japan, New Zealand), the technology has become subject to often prolonged and intensive public debate and scrutiny. In response to such public resistance and volatile public opinion, regulators and policymakers have begun to address more proactively the political and social dimensions of GE technology, and to include public concerns in their decision making. The Swiss parliament and government, for example, in countering the "gene-protection" initiative, strengthened the regulatory system. The British government, stung by an unprecedented broad public campaign in the late 1990s against its GE technology policy, overhauled the regulatory system and introduced greater transparency and public involvement in the decision-making process by setting up, among other bodies, the AEBC, and by funding the *GM Nation? The Public Debate* initiative.

In this context of an increased politicization of GE technology, and of the apparent failure of policymakers to address sufficiently the technology's social dimensions, "public participation" has become a fashionable catchphrase to denote a supposedly new style of public engagement with this emerging technology: such participation is seen by growing numbers of policymakers, politicians, scientists, and NGOs as a way of addressing the cognitive, normative, and practical uncertainties surrounding the technology. Numerous initiatives and activities have mushroomed in the public sphere in recent decades, involving members of the public, stakeholders, scientists, and politicians in policy deliberation and public debates. In short, public participation has become an important characteristic of GE technology in the public sphere in significant parts of both the developed and the developing world. The significance of public participation in the governance of GE technology is, however, not immediately obvious. First, public participation comes in many institutional, political, and cultural shapes and forms, from public information campaigns in the media and formal public consultations in policymaking to Internet chat rooms, town hall meetings, and local protest initiatives. Second, the contribution of public participation to the governance of GE technology may vary greatly: it includes raising public awareness, analyzing "public opinion," scrutinizing policymaking, and involving citizens in decision making. Third, public participation has a strong normative dimension: different social actors thus judge its value for policy- and decision making and its legitimacy differently, depending on their standpoints. Finally, participatory governance is not just a group of specific methods and procedures, but also a sociopolitical phenomenon set in a wider context of politics, public discourse, and social mobilization.

In this chapter I analyze the relationships between various manifestations of public participation and the governance of GE crops. Drawing on relevant literature on public policy and governance as well as on empirical studies of participatory practice, I consider different forms of public participation both in institutional settings and in the wider public sphere, and I discuss the sociopolitical relationships between these procedures and decision making on the development, use, and regulation of GE crops.

GENEALOGY OF PUBLIC DEBATE

In trying to understand the significance of public participation in GE technology, it is useful to look at the technology's historical development. Taking the first successful experiments to create recombinant DNA in the early 1970s as the starting point of GE research and development, the following three phases can be distinguished:

1970s-Early 1980s: Early Signs of an Emerging Public Debate

GE technology emerged in the early 1970s against the backdrop of the environmental and consumer movements as well as of the participatory experimentation in planning and social policy that had developed in the United States and Northern Europe since the early 1960s. Recombinant DNA (rDNA) technology, as it was first known, caused a lot of excitement and speculation in the scientific community at the time. However, as a nascent technology confined to the laboratory and scientific discourse, it did not enter the realms of policy, politics, and wider public debate as prominently as other environmental and health issues.

Nevertheless, there were early signs of the public debate and participatory engagement that followed a decade or so later. It was leading scientists themselves, most prominent among them a molecular biologist, Paul Berg of Stanford University, who were concerned about potential risks of rDNA technology; in 1974 they imposed a two-year moratorium to allow for further risk assessment.[6] At the famous "Asilomar" conference in 1975, researchers agreed on safety guidelines (ethical issues were deliberately excluded from consideration), with the effect that research was subsequently resumed and the U.S. Congress did not enact restrictive legislation. While the U.S. government seemed to welcome this self-governance by the research community, it also issued governmental guidelines. At least one authority sought to further scrutinize the implications of rDNA research in public. In 1976, the City Council of Cambridge, Massachusetts, temporarily voted to ban rDNA research at Harvard University and the Massachusetts Institute of Technology for six months, and held several public hearings into the possible risks of GE technology.[7] As a result of these hearings, the City Council set up the public Cambridge Biohazards Committee to oversee rDNA research and to report on safety violations.

Overall, however, these examples were the exception, not the rule, in the handling of GE technology at the time: its governance was characterized by behind-closed-doors policymaking, which largely centered on self-regulation and voluntary codes of practice, with minimal involvement of legislators. The assessment of the technology was predominantly an expert affair, and as such typical of how more generally scientific and technological issues were dealt with by leading institutions, such as national academies of sciences, technology assessment offices, and national health institutes. The public dimension of the technology hardly featured in regulation and policymaking, and, where it was addressed, was usually limited to the one-way provision of scientific information for the education of the general public.

Mid 1980s-Mid 1990s: Experimentation with Institutionalized Participation

From the 1980s, governments and legislators increasingly recognized that modern biotechnology was of significant strategic importance for national research and development (R&D) and, thus, for economic competitiveness. Biotechnology was heralded as the third strategic technology of the postwar period, following nuclear and information technology.[8] As a consequence, large R&D support programs were launched through national funding agencies. Regulatory frameworks were put in place to guide and monitor research and technological development. This was done in some countries by amending existing laws and regulation, and in others by specifically introducing new legislation. In 1985, for example, Denmark was the first country to enact a gene technology law.[9] The legislative and regulatory process was accelerated by a number of controversies about reproductive medicine at Danish hospitals in the early 1980s, and the announcement by two leading Danish biotechnology companies (Novo and Nordisk Gentofte) of their plans to manufacture human insulin and growth hormone using GE technology. In 1990 the European Commission began to introduce regulation concerning the contained use of GE organisms, releases to the environment, and novel foods and products.

NGOs, such as environmental campaigners, consumer groups, and patients' organizations, used campaigns and media work to help raise critical awareness about developments in biotechnology amongst their membership and the wider public. They were also important in prompting policymakers to regulate the technology, and to scrutinize proposed policy. In Switzerland, for example, the Swiss Action Group on Gene Technology (SAG), a heterogeneous coalition of more than a dozen civil society organisations, began to exert considerable influence in the 1980s, as a *quasi* "extra-parliamentary" political actor group, on the parliamentary and governmental policymaking process.[10]

Scientific organizations also attempted to improve the public's knowledge of gene technology by providing scientific information, thus hoping to achieve positive public attitudes toward modern biotechnology. Various "public understanding of science" programs were put in place, targeting the public directly through events, such as "science festivals" and exhibitions, and indirectly through the media.

More and more, legislators and regulators began to address the wider public dimension of the technology, including its legal, ethical, and social aspects.[11] This was done partly in response to growing public concerns, and partly as a proactive strategy of assessing the potential future social and political repercussions of GE technology. This approach reflected the concerns among policy and decision makers that the technology could meet significant public resistance, not unlike what had happened in the nuclear energy debates of the 1970s and 1980s.

Characteristic of this assessment of GE technology, particularly in Europe, was the growing involvement of various stakeholders, for example, environmental interest groups, patients' organizations, and "laypeople," alongside scientists and other experts.[12] The institutionalization of "technology assessment" (TA) led to new forms of what became known as "constructive" and "participatory" TA. This aimed to analyze a broad range of issues, such as the possible adverse impacts of GE technology on the environment, animal husbandry, and human health, and the social and ethical implications of manipulating genetic information.

In 1987 the newly established Danish parliamentary Board of Technology, for example, received funding through the first national biotechnology R&D program to support public assessment and debate activities. The Board of Technology then developed the so-called consensus conference (also known as "citizens conference") as a new, participatory form of TA. At the center of a consensus conference is a fourteen-to-thirty strong panel of citizens that discusses socially relevant issues with various experts, assesses the evidence, and makes written recommendations to policymakers. The aims are to identify— through in-depth discussion and assessment—the concerns, expectations, and interest of members of the public vis-à-vis scientific–technological issues, and to encourage wider public debate. Consensus conferences are typically held in public, with media representatives, interested experts, and stakeholders and other members of the public invited to attend. In Denmark, they often take place in the parliament building so that members of parliament can attend the proceedings and respond to the publication of the citizens' panel's report. The media have widely reported some consensus conferences on topical issues.

Early Danish consensus conferences looked at gene technology in agriculture, industry, and medicine (1987), and human genome mapping (1989). Members of the Danish parliament who were surveyed and interviewed stated that consensus conferences provided them with a better understanding of citizens' views and concerns about socially sensitive issues of science and technology, and helped them to identify priorities for policymaking.[13] For example, the 1987 conference clearly showed that transgenic animal research was unacceptable to the citizens' panel. The members of parliament present used this evidence, together with similar evidence in opinion surveys and from public debates, to support exclusion of transgenic animal research from the first national biotechnology R&D program. The citizens' panel at the 1989 conference expressed strong concerns about the possibility of information gained from human genomic research being used by employers and insurers. Shortly after the 1989 conference, a group of members of parliament from different political parties used the citizens' panel report to initiate new legislation on the use of genetic information in employment and insurance. The law was enacted in 1996.

Other forms of technology assessment have focused on the interaction between researchers and technology developers, on the one hand, and interest group representatives, on the other. In Denmark, for example, researchers at universities and in industry invited environmentalists, consumer organizations, and local residents to "stakeholder dialogue" meetings; these meetings aimed to improve understanding of different viewpoints. In the Netherlands, as part of "constructive" TA, (prospective) technology users were involved in discussions with technology developers in the upstream phases of technological development—for example, in a series of stakeholder forums on "novel protein foods."[14] If social aspects of technology were addressed early in the development process, it was argued, technology could be designed to be more directly responsive to users' needs. Similar stakeholder TA initiatives took place in other countries; for example, in Germany in the early 1990s, when the Berlin Science Centre (*Wissenschaftszentrum Berlin,* WZB) conducted a participatory TA project on herbicide-resistant plants that involved geneticists, social scientists, ecologists, and various other stakeholders.[15]

Throughout the 1990s, the experience of participatory TA grew, as is evidenced by its diversification in the form of the "citizens dialogue," "voting conferences," "deliberative opinion polls," and "interactive" TA methods, as well as by its expansion into new institutional and national settings. By 2002, more than sixty-five citizens conferences and consensus conferences had been held (of which over thirty were on biotechnological issues) in eighteen different countries, including Argentina, Brazil, Canada, France, Japan, Norway, South Korea, and the United States.[16]

Late 1990s: Widening and Intensification of Participation in the Public Sphere

The second half of the 1990s saw a significant, and in some instances dramatic, intensification of public debate and involvement. This triggered renewed policymaking activities, both nationally and internationally, such as the 2000 Cartagena Protocol on biosafety.

In Europe, critical public engagement became a major feature of biotechnology governance, with significant social and political repercussions.[17] In the United Kingdom, for example, a series of grassroots protests, high-profile media campaigns, and the intervention of leading figures from science, politics, and public life from 1998 onward led to what has been described as "the great GM debate."[18] In Germany, Greenpeace and other environmental campaigners succeeded in raising a broad media debate in 1998 about food products containing GE ingredients. This led to consumer boycotts which, in turn, led to retailers and food manufacturers withdrawing GE products from the market. In Denmark, public controversy erupted once again in 1996, as a result of the import of GE soy. This surprised some observers who had thought that Danish policy on GE technology with its tradition of participatory TA had made public debate less controversial. With the shipment of GE corn to Lisbon harbor in 1997, public controversy erupted in Portugal, where up to that point there had virtually been no public debate about GE technology. (This lack of debate was a consequence in part of the country not having a significant national biotechnology industry, in part of the government's strategy of trying to keep GE regulation out of political debate, and in part of the lack of a coordinated strategy by NGOs to raise the issue in the political and public spheres.)

In much of Europe, GE crops and food became hot topics of public debate and led to various forms of public involvement: local protests, direct action (which in a few cases led to the destruction of fields of GE crops by anti-GE campaigners), letters to newspapers, radio phone-ins, petitions, Internet discussions, demonstrations, public conferences, media campaigns, and consumer boycotts. The main trigger of this intensive public debate was the arrival on the market of GE soy and corn from the United States. There were also other contributing factors, such as the BSE (bovine spongiform encephalopathy, or "mad cow") epidemic, which shattered public confidence in agriculture and food production in the United Kingdom and Europe, the successful cloning of "Dolly the Sheep," and Dr. Arpad Pusztai's controversial research on the effect of GE potatoes on the immune system of rats.[19]

The intensity and suddenness of the public debate took many regulators and decision makers by surprise. Some decision makers at first reacted by accusing campaigners, the media, and the wider public of being irrational and misinformed, and by insisting on the effectiveness of the regulatory systems in place. The British prime minister, Tony Blair, at the height of public controversy in 1999, expressed his frustration in the House of Commons at the outcry over genetic modification, complaining about the "hysteria of public reaction" and the "extraordinary campaign of distortion" by "parts of the media," and he pledged to "resist the tyranny of pressure groups."[20] However, with the debate showing little sign of abating, and confronted with a broad coalition of social actors, decision makers changed tactics and began to insist that they were taking public concerns seriously and responding to them. Prime Minister Blair, writing in a British broadsheet newspaper in February 2000 (one year after his critical comments), said that there was "cause for legitimate public concern," which had been reflected in the media and which his government understood well.[21] In what was described as one of the biggest U-turns of government policy to date, he said that his government had recognized that "consumers and environmental groups" had "an important role to play" in finding answers to the questions raised. He explained that the U.K. government had "radically overhauled the regulatory and advisory processes so that consumers have a real say on GE foods," and that, by making the whole regulatory system "open, transparent and inclusive," confidence in it would be increased.

This regulatory streamlining in the United Kingdom came as two new advisory bodies were set up: the Agriculture and Environment Biotechnology Commission (AEBC), and the Human Genetics Commission. Their task was to provide strategic advice on biotechnology in the agricultural and human domains, including ethical and social issues. The new *Food Standards Agency* (FSA) was given statutory powers to regulate and monitor food-safety issues. In line with the pledge to design more open and transparent regulatory processes, the AEBC's meetings were held in public and its minutes posted on the Internet, and the FSA has used newspaper advertisements to invite the public to attend its meetings in different parts of the country.

Similar adjustments of policymaking and regulatory structures and processes in the face of critical public debate took place in other European countries. In Germany, the government of Chancellor Gerhard Schröder in 1998 transformed the Ministry of Agriculture into the Ministry of Consumer Protection, Food and Agriculture, and promoted transparency and consumer information as its goals.[22] The government also set up the new Federal Institute for Risk Assessment and the Federal Office for Consumer Protection and Food Safety. The minister in charge launched the so-called green biotechnology dialogue (Dialog Grüne Gentechnik) initiative in 2001/2002, aimed at involving all interested stakeholders in helping to clarify issues related to GE crops and food stuffs. In Denmark, reflecting the marketization of GE technology, the government emphasized, as a way of ensuring consumer choice, the labeling of GE products. The Ministry of Industry and Trade set up the so-called BioTik expert commission (a fusion of the words "biotechnology" and "ethics") in 1997, the recommendations of which were presented to parliament in 2000. The commission's task was to consider how to incorporate ethical criteria into formal decision making on GE technology. The Centre for Biotechnology and Risk Assessment was set up, and together with other research institutions, undertook research into various aspects of GE crops and transgenic animals. In Portugal, the main arena of public debate turned out to be parliament, which, prompted by the environmental campaigns against GE food products, in 1999 launched a series of debates and legislative procedures, leading to an effective moratorium of GE crops. In 1999, the Ministry of Agriculture revoked two licenses granted earlier for the commercialization of GE seeds.

These national reactions to and adjustments of the regulatory systems governing GE technology from the late 1990s influenced policymaking at the European level; in 1999 a majority of European Union (EU) countries established a de facto five-year moratorium on the commercialization of GE crops and food products.[23] This was a direct response to critical public debate across much of Europe, and gave the EU's regulators and policymakers breathing space to overhaul its regulatory system. The European Food Safety Authority came into being in 2002, and Council Directive 90/200/CE was revised.[24] The moratorium ended in 2004 when, under growing pressure mainly from the United States and other countries and from pending WTO rulings, the European Commission approved new GE crop products.

Significantly, various participatory activities in the public sphere have spread to other countries in Latin America, Asia, and Africa, where increasingly vociferous groups of farmers, local business people, environmentalists, and consumer groups have voiced their criticism of the GE policies promoted by governments and multinational industry. In Brazil, for example, NGOs were instrumental in pushing for a moratorium on the cultivation of GE crops (this was later lifted in autumn 2003). In Africa, a network of NGOs in 2003 challenged the policy of the Bush administration to introduce GE crops in Africa, stating that "Washington is not entitled to speak on behalf of African states on the matter."[25] The Southern African Catholic Bishops Conference added their voice, stating that GE could destroy the knowledge and sustainable agricultural systems of local farmers. Earlier, in 2002, several South African countries, including Malawi, Mozambique, Zambia, and Zimbabwe, refused GE crops during the food crisis, insisting on milled products to prevent the spread of GE organisms. Zambia rejected GE products outright.

In India, farmers, environmental groups (such as Gene Campaign and Greenpeace India), and grassroots movements (such as Jaiv Panchayat, which represents fifteen village groups in Orissa) took to the streets on several occasions in 2001, demonstrating against attempts by multinational companies, such as Monsanto, to penetrate the Indian market.[26] Demonstrations were held outside the Indian parliament in New Delhi, as well as near new experimental field sites run by Monsanto. The Research Foundation for Science, Technology and Ecology (RFSTE, New Delhi), which was founded by Dr. Vandana Shiva, one of the most prominent leaders of the growing movement against GE crops in India, has challenged both government and industry on the legality of official policymaking about GE crops. In an effort to counter mounting public criticism, Monsanto launched "farmer awareness programs" to educate farmers about *Bt* (*Bacillus thuringiensis*) cotton and other GE crops. The Indian media increasingly report on the GE controversy.[27]

In New Zealand, the issue of GE crops became a key issue in the 2002 general election campaign. This was as a result of a scandal ("Corngate") that gripped New Zealand's politics over the government's alleged cover-up of GE contamination during the mandatory moratorium on GE crops and foods in place since 2001.[28] The government was forced to release internal documents allegedly showing a close relationship between the biotechnology industry and government departments. Support for the incumbent government party (Labour) slumped to the benefit of the Green Party, which campaigned to extend the moratorium.

Even in countries where agricultural biotechnology was until recently assumed rarely to be a cause of public debate and controversy, there have been signs of a growing engagement with the issue by policymakers and significant sections of the public. In the United States, for example, a wine-growing region in California recently declared itself a GE-free zone, and legislative measures to introduce tighter regulation on GE crops and foods in several states (such as Vermont and North Dakota) have been under consideration.[29] Native American communities (such as the Navajo, Apache, and Hopi peoples) have spoken out against GE foods; specifically, they have criticized related NAFTA trade policy under which, in 2001, the United States exported GE corn to Mexico even though the Mexican Congress had passed a resolution against GE crops three years earlier.[30] In June 2004, an activist group, calling itself "Reclaim the Commons," staged a protest in response to Bio 2004, a large biotechnology conference being held in San Francisco.[31] One academic from Georgetown University, Washington DC, wondered aloud at the meeting whether the United States was "catching the European disease," with recent opinion surveys reportedly showing that a majority of Americans opposed GE fruit and vegetables.

In Australia, moratoria on GE crops and related legislation were announced by New South Wales, Victoria, South Australia, and Western Australia in spring 2004, drawing swift criticism from the Australian biotechnology industry.[32] China's widely reported "love affair" with biotechnology cooled considerably when it curtailed R&D of GE crops by international companies in 2002, reflecting growing concerns about food safety as well as trade protection.[33] The Chinese government introduced new GE labeling rules, in response to growing consumer concerns about GE food stuffs.

PARTICIPATION AS GOVERNANCE

What is the connection between this diversity of public participation and the governance of GE crops? What are the dynamics between the different (formal and informal, structured, and unstructured, small- and large-scale) participatory processes and the decision making about the development, use, and regulation of GE crops and food stuffs? What challenges does participatory governance pose for scientists, regulators, and politicians?

The burgeoning literature on governance provides several explanations, normative and empirical, of the characteristics and role of governance processes in contemporary society.[34] While the term "governance" is broad and encompasses a variety of definitions and concepts (Kooiman, for example, lists a dozen or so broad classifications[35]), it, nevertheless, has a number of common, core characteristics. One is the relatively nonhierarchical (or "heterarchical"[36]) and relatively self-organizing networking relationship between state and social actors involved in decision-making processes; these contrast with "traditional" hierarchical decision making in which the state steers and controls social actors. Linked to this is the multilateral, ongoing co-operation (including co-regulation, co-production, and cooperative management) between social actors in which they share tasks and responsibilities, and jointly solve problems, and create political, social, and economic opportunities. The boundaries between state and society shift and become more diffuse and diverse.

The rationale for this new mode of governance is that it will achieve more effective public policymaking through the active cooperation of social actors who have special knowledge, expertise, and/or interests in the areas concerned. The particular (often local) knowledge and expertise of "stakeholders"—for example, farmers and conservationists in the case of environmental protection, or patient groups in health care development—can be mobilized and their special interests considered in the drawing up of policies and implementation of decisions. This is considered particularly relevant where social actors are critical for the effective implementation of public policy and the state's influence is limited, or considered ineffective.

Such "participatory governance" is also seen as a means of tackling new, socially contested policy areas. These include the "not-in-my-backyard" (NIMBY) phenomenon apparent in many environmental issues and the introduction of "risky" technology, where significant opposition by special interest groups, or the wider public, may jeopardize public policies and undermine the legitimacy of the decision-making process and its institutions. The direct and regular involvement of social actors representing different types of expertise and special interests, as well as actors representing the public interest is supposed to strengthen the legitimacy of decision making. Increasing the opportunities for the mutual accommodation of interests should help ensure the quality and effectiveness of decision-making processes and their outcomes, as well as generate trust and accountability among those who participate.[37] The interest in sociopolitical governance thus stems in part from a desire to come to grips with new sociopolitical and scientific–technological developments and structural changes in modern societies characterized by increasing complexity, diversity and dynamics.[38] With regard to the issue of GE crops, several distinct spaces, or arenas, of participatory governance can be identified. The following sketches the main characteristics of these arenas, gives some examples, and discusses the dynamics between them in terms of how these different forms and processes of public participation have contributed to the governance of GE crops.

INSTITUTIONAL ARENAS
OF PARTICIPATORY GOVERNANCE

The arenas of participatory governance that have emerged from institutional settings have two main features: they *incorporate* various social actors (different types of experts, representatives of interest groups, NGOs, consumers, members of the public, etc.) and their respective contextual perspectives in the intra-institutional policy- and decision-making processes; and they are *open* to wider public sphere discourses and monitoring. This generally aims to:

1. broaden the cognitive basis of analysis and deliberation informing the policy- and decision-making processes by mobilizing the knowledge, expertise, and relevant "life experience" of social actors;
2. take into account (different) social values, expectations held, and needs felt in relation to the issues at stake;
3. consider both normatively and in practical terms the "public interest";
4. improve transparency and accountability by providing public information on, ensuring public access to, and allowing the monitoring of, policy- and decision-making processes; and
5. connect policy deliberations and decision-making more closely to related social discourses taking place within the public sphere.

Depending on the particular institutional or organizational settings, and the nature of the related policy- and decision-making processes, some of these aims are accentuated more than others, resulting in differences in the methodology, structures, and procedures of the participatory governance forms used.

Research Organizations

Research organizations have developed participatory procedures with several applications in mind, including: (1) involving (potential) users, such as patients and consumers, early in the technological development trajectory with a view to integrating social considerations into technological design; (2) involving various stakeholders and "lay people" in scientific and/or social-scientific research, in order to assess the potential economic, environmental, social, and ethical repercussions of specific scientific–technological issues; and (3) cooperating with people and organizations seeking expert advice in order to address issues and developments that affect them.

The "constructive TA" school,[39] developed in the 1980s by Dutch social scientists and TA experts in conjunction with research scientists and industry representatives is based on the premise that technologies are socially constructed and shaped. It seeks to anticipate the possible regulatory and social difficulties early in the innovation phase and to steer technological design accordingly; it also aims, through various interactive procedures, to make technology development more directly responsive to the needs of potential users. Consumers (and consumer representatives) participate alongside technical experts, industry representatives, and representatives of NGOs in the assessment of consumer perspectives on, in this case, GE crops, with the aim of informing the R&D agenda.[40] However, consumers often get involved only at the beginning of these procedures, for example, through in-depth interviews that are then fed into public opinion surveys on consumers' motives, preferences, and values. Consumers may not, therefore, directly interact with other participants in the assessment process. Furthermore, constructive TA is only indirectly oriented toward the wider public and political spheres; it functions more as an intra- and inter-institutional governance process. As such, it may have an impact on research developments, for example, by helping to define new research programs.[41]

A practical example of a participatory research network was the German discourse project on GE herbicide-resistant crops—referred to earlier—undertaken in the early 1990s by the Berlin Science Centre (*Wissenschaftszentrum Berlin;* WZB), Germany's largest public social science research institute, in collaboration with the Institute of Genetics of Bielefeld University and the Institute of Ecology of the Technical University Berlin.[42] The raison d'être of the project was to enlarge the technology assessment process so that it would not just entail scientific evaluation by technical experts but also be a "discourse arena," in which the social conflicts arising from the issue of GE crops could be articulated and assessed. The fifty or so partici- pants included proponents (industry representatives) and opponents (environmentalists) of GE herbicide-tolerant plant technology; repre- sentatives of regulatory authorities, agricultural associations, con- sumer organizations; and researchers. In a series of conferences and workshops over two years, they defined the scope of the research to be undertaken, evaluated some twenty specially commissioned re- search reports, and drew up conclusions. While the participants iden- tified a broad range of issues for assessment—from environmental risks and impacts on agriculture to the ethics of plant manipulation— in the course of the process the focus was narrowed to environmental risk assessment as a pre-regulatory strategy for considering herbi- cide-tolerant plants. Partly as a result of this particular framing, the environmental organizations involved withdrew from the process, complaining of a lack of comprehensive assessment (including a con- sideration of alternatives to herbicide-tolerant plants), and of a bias in the organizers' conclusions—namely, that there were no particular risks associated with GE crops. According to one external analysis, the project failed in the sense that the confrontation between two incompatible modes of assessment (one based on a scientific argu- mentation, the other based on a strategic-political argumentation) could not be resolved.[43]

"Science shops" are another model of cooperation between re- search institutions with people and organizations seeking specific expert advice.[44] If, for example, a group of neighbors are concerned about the possible impact of GE crops being cultivated in nearby fields, they can seek expert information about the nature of these crops as well as advice on how to analyze environmental assessments published by official agencies. While science shops primarily act as knowledge resource centers available to people, groups, and commu- nities who want to conduct their own assessments of a matter of con- cern, they also—by helping researchers better understand people's needs and concerns—aim to inform the R&D process within univer- sities. Science shops were first developed by Dutch universities in the 1970s but have since spread to many other European countries and further afield. The range of advice given is not limited to scientific–technical issues, but includes housing, tourism, and other areas of public policy.

Policy Institutions

Historically, the main application area of participatory TA in Europe has been in formal policy arenas at national and regional levels, including in parliamentary offices of technology assessment (such as the Danish Board of Technology and the French Parliamentary Office for the Study of Scientific and Technological Choices, OPECST), departments of the executive (for example, the Centre of Technology Assessment at the Swiss Science Council), various statutory and non-statutory advisory bodies and ethics councils (such as the Norwegian National Committees for Research Ethics), and national research councils (such as the U.K. Biotechnology and Biological Sciences Research Council).

Most participatory TA, in the form of, for example, citizens' conferences, consensus conferences, and round tables, has dealt with agricultural and medical biotechnology; this reflects critical public discourse on related policymaking.[45] The key objective of these procedures is to consider—as a form of open intelligence gathering—different kinds of expert and lay knowledge-framing and "worldviews." Other objectives are often given similar importance: for example, attempting to increase "public understanding" and to render policymaking more transparent, open, and inclusive with the aim of legitimizing decision-making processes and their outcomes. Hence, the output of these procedures typically takes the form both of statements and reports aimed at policymakers for intra-institutional use, as well as of reports and public events targeted at civil society organizations, the media, and the wider public.

However, the resulting intertwined character of these procedures often makes them and their outcomes somewhat ambiguous: their role as *quasi* policy advisory tools (delivering social, legal, and ethical assessments from a multi-actor perspective) may be unfavorably compared with established expert procedures; their use as *quasi* courts of public opinion (defining the public interest) also suffers when compared with, for example, representative public opinion surveys. The measurement of impacts on policymaking and public debate has also proven methodologically difficult; and evidence of tangible outcomes can be hard to come by. Hence, their value is often seen as more symbolic: an indicator of a more inclusive and publicly accountable style of policymaking.

Though their closeness to political decision-making process may increase the chances of their influencing these processes, participatory procedures can also suffer undue political instrumentalization by being used to deflect institutional criticism or retrospectively to justify a predetermined policy decision. The 1998 French national citizens' conference on GEOs, for example, lost credibility when it was held after the French government had made the decision to allow limited cultivation of GE crops. Some of these procedures had little impact because they were not closely enough linked with policymaking. The 2003 British *GM Nation?* public debate initiative (see earlier), for example, was held while the results of national GE field trials were still outstanding, and was not linked to the statutory process for recommending commercialization of GE crops that followed the voluntary moratorium agreed upon in 1999.[46]

Despite these ambiguities and criticisms it is worth noting that policymakers, experts, and various stakeholders seem to take public consultations seriously, at least judging by their increasing use in policymaking. This has also triggered stimulating meta-level, conceptual discourses on governance; this is reflected in recent academic literature as well as in policy documents, such as the seminal 2000 House of Lords Science and Society report and the European Commission's 2001 White Paper on Governance.[47]

Museums

Museums and science centers have been an important partner in developing participatory governance initiatives at the interface between institutional policy deliberation and wider social discourse.[48] For example, the first U.K. national consensus conference on plant biotechnology in 1994 was organized by the Science Museum, London, with funding from one of the national research councils. The Australian Museum in Canberra in 1999 was host to the first Australian consensus conference (initiated by the Australian Consumers' Association). The *Deutsches Hygiene Museum* in Dresden organized the first German national citizens' conference on genetic testing in 2001, with financial support from the Federal Ministry of Research.

Museums often host these initiatives because they offer participants, especially members of the public, a relatively "neutral" public space for deliberation and assessment that is not "pre-framed" or dominated by a particular specialist discourse of experts, policymakers, or politicians. Museums themselves are interested in developing new ways—beyond traditional exhibitions and public lectures—of interacting with various publics. The increasing involvement of museums points to the expansion of formal participatory technology assessment procedures from policymaking settings to the wider public sphere, with museums playing a possible bridging role between the world of policymaking and wider public discourse. The status of these museums in relation to policymaking as well as to the wider public is an important determining factor concerning the seriousness with which these procedures are treated and the impacts they have on policymaking and wider public debate.

Industry

Industry's engagement with stakeholders and citizens has often been limited to "open days" and public information campaigns. Few companies have proactively engaged in "stakeholder dialogues" and "round tables." One such is the Danish biotechnology company, Novo Nordisk, which has held round table discussions with NGOs and local stakeholders since the 1990s.[49] As part of the company's environmental accountability and social responsibility strategy, the purpose of these discussions has been to explore the company's activities in biotechnology development and thus to help shape future strategies and to develop a rapport with various public actors.

EXTRAINSTITUTIONAL ARENAS
OF PARTICIPATORY GOVERNANCE

The diversity of participatory governance at the institutional level is matched by a plethora of extrainstitutional participatory social activities. These occur in various, differently shaped, often interconnected public spheres, including local grassroots movements, national media debates, Internet chat rooms, and international campaigns. These activities involve many actors, including environmental activists, consumer organizations, scientists, farmers, educators, church organizations, business organizations, political parties, and concerned citizens. They pursue several related goals, including:

1. raising public awareness of developments in modern biotechnology;
2. producing knowledge about the social, environmental, and economic repercussions of GE plants and food; and
3. scrutinizing and challenging the politics of GE technology and related research and agricultural policy.

These extra-institutional processes are driven by events, such as the announcements of scientific breakthroughs, or the arrival of new GE products on the market. They are themselves driving events: for example, if they involve galvanizing local opposition to the release of GE crops into the environment, or leading GE product boycotts in restaurants and supermarkets. A common feature is their attempt to mobilize public interest in GE technology in order to disseminate and scrutinize information, shore up public support, or raise public opposition. They can thus be understood as part of a social, political, and cultural "appropriation" of GE technology, in which knowledge and information about the technology transcend the realm of the scientific laboratory and regulatory agencies and are more widely publicized, discussed, interpreted, scrutinized, contested, complemented, and adapted. The technology is no longer predominantly in the hands of scientists and technological developers, but is shared by a broader spectrum of social actors who claim different interests and stakes, and seek to gain control over the technology and how it is embedded in society and in the political process.

Other features of extra-institutional participatory governance are networking and coalition building among social actors. The issue of GE crops has brought together a diverse range of actor groups, who may each have particular concerns about the issue (for example, environmental risks, food policy, intellectual property, and international trade) but at the same time have a common interest in scrutinizing, supporting, or opposing certain public policy. By building networks of interest and by cooperating in participatory activities, these actors seek to create a critical mass that will attract the attention of policy-makers and the wider public—and, thus, to influence the direction of GE technology and its applications.

NGOs and Other Civil Society Organizations

For many environmental NGOs and their members, gene technology has become a major issue on which to campaign. Typically, they criticize industry and regulators for pressing ahead with the development and commercialization of GEOs without first having taken the necessary precautions and thus acting against the public interest. Since the 1980s, NGOs have brought the GE crops issue to wider public attention and have galvanized resistance to biotechnology. By offering (counter) expert assessments and outlining alternative scenarios, for example of organic and local-based agriculture, NGOs have often succeeded in making scientific experts and policymakers put information (regarding field trials, for example) in the public domain and also making them open up policymaking to wider public scrutiny. For example, the Andhra Pradesh Coalition on Defence of Diversity—which comprises 140 civil society groups across this Indian state, and, with the University of Hyderabad, ran citizens' juries on GE crops—commissioned two agricultural scientists to carry out a three-year study assessing the performance of *Bt* cotton in over eighty villages in four major agricultural districts.[50] The research, reported in spring 2005 in an Indian daily newspaper, contradicted earlier studies conducted by Monsanto (which markets *Bt* cotton in India): it showed that *Bt* cotton produced lower crop yields, cost more to cultivate, and was less pest resistant than traditional crops. The AP Coalition thus called for a policy review and demanded a moratorium on *Bt* cotton across the state.

There is now a vast array of NGOs campaigning on GE technology, from large international organizations with offices in various continents, such as Greenpeace, to small grassroots organizations spearheaded by a few dedicated campaigners and volunteers, for example, British-based GeneWatch, Five-Year-Freeze, and Genetix Snowball, the French ATTAC (an antiglobalization network campaigning against GEOs) and the Grupo de Reflexion Rural in Argentina. Other civil society organizations, such as charities campaigning for the developing world, church organizations, consumer associations, and farmers' unions, have added their critical voices to the widening debate, commenting on various economic, social, ethical, and political aspects of GE technology.

The tactics and strategies of these organizations range from launching eye-catching actions (such as destroying fields, blockading shipments of GE crops, and organizing street demonstrations), publishing pamphlets, and organizing citizens' petitions, to hosting public conferences and commissioning research by (counter)experts. Some NGOs have adapted public engagement procedures to carry out their own participatory assessments. For example, the charity ActionAid organized citizens' juries on GE crops in India (see earlier) and Latin America. Others have become involved in formal policymaking through their membership of governmental advisory bodies.

Grassroots Movements

Direct actions by grassroots movements have had an influence on the governance of GE crops by drawing the attention of local communities and the (national) media to developments, for example, field trials. In Britain in the late 1990s, grassroots movements were a significant driving force in the public controversy about GE crops. Several ad hoc groups of activists destroyed field trials, which provoked the Crown Prosecution Service into launching prosecutions against them. One court case ended in their acquittal because they were judged to have acted in the "public interest." Other forms of direct actions against GE crops have included street demonstrations and festivals—for example, as part of "reclaim the streets" campaigns—offering alternative visions to mainstream politics and criticizing public policy on matters such as international monetary policy, globalization, and environmental pollution.

Media Campaigns

The media have not only been an important source of information about biotechnological issues; they have also actively participated in campaigns for or against public policy on GE crops. In Britain, some print media, claiming to speak on behalf of the public and to be acting against "the establishment," became directly involved in the GE crops controversy in the late 1990s: they published critical information under the banner of "GE watch" and commissioned public opinion surveys that suggested there was overwhelming public disagreement with government policy.

The Internet has also provided an important tool for disseminating information, publishing critical articles, canvassing public opinion, and orchestrating public action. Critical writers and campaigners have brought the issue of GE crops to wider public attention in developing countries with the help of the world wide web. The Indian journalist and campaigner Devinder Sharma, for example, runs a dedicated website on GE crops in India.[51]

Sociopolitical Dynamics

Participatory governance is best understood as a dynamic sociopolitical process involving various interconnected arenas of participatory activities, rather than as a narrowly defined set of arrangements within particular institutional or social contexts. Analyzing the relationship between institutional arrangements of decision making and wider processes of social mobilization and public discourse is important for two reasons: it boosts understanding of the nature of governance processes and outcomes (particularly where the issues at stake are socially and politically complex and contested); and it helps with the design of new institutional governance procedures that relate more closely to their wider sociopolitical contexts.

In the case of GE crops, the interaction between extra- and intra-institutional arenas of participatory governance has tended to follow certain patterns. Extra-institutional spaces for participatory activities have often emerged as manifestations of social mobilization against developments in GE technology. Such mobilization is often driven by a substantive critique of the mainstream rationale and direction of research and technological development promoted by the "establishment" policymaking system. The mobilization is concurrently driven by a procedural critique of the perceived shortcomings of the decision-making and regulatory systems, with their alleged lack of opportunities for considering alternative paths of development and involving social actors in assessment and policymaking.

Public engagement in extra-institutional arenas may for some time be limited to relatively small numbers of social actors and organizations, and thus at first go largely unnoticed by policymakers and the wider public. Events, however, can quickly trigger the substantial mobilization of broader sections of the public and the widening of participatory arenas. Across Europe in the late 1990s, the import of GE soy and maize products was such a triggering event. In Britain, this came on the back of other, only indirectly related events, including the BSE crisis, the arrival of Dolly the Sheep, and the Pusztai controversy (see earlier). The long-standing, but limited social mobilization that had begun in the late 1980s and 1990s in response to developments in GE technology suddenly grew much larger as new, unusually broad coalitions of individuals and organizations criticized government policy. The organizations included radical environmental interest groups (for example, Greenpeace and Genetix Snowball), more conservative organizations (the National Trust; the Women's Institute), and major supermarket chains and consumer organizations. Left-of-center broadsheet newspapers (the *Independent* and the *Guardian*) as well as right-of-center tabloids (for example, the *Daily Mail*) championed the anti-GE cause (for instance, by launching a "GE watch"), thus greatly extending the arena of social mobilization. Prince Charles's intervention, in the form of his ten critical questions to government published in the *Daily Mail,* contributed to the growing controversy.[52]

The main thrust of such wider public mobilization is scrutiny of government and industry policy, and a demand for changes in institutional governance processes. Campaigners use diverse means and tactics, including popular referenda, media campaigns, demonstrations, consumer boycotts, and even the courts. Extra-institutional governance helps develop alternative scenarios for technological development and propose alternative policy options: scientific analyses by counter-experts are commissioned, literature reviews and media digests circulated, conferences and public debates hosted, declarations and manifestos published, and new legislation and policies drafted. The Swiss Action Group on Gene Technology (see earlier) took on a quasi extra-parliamentary role by instigating—in the perceived absence of sufficient parliamentary and regulatory action on GE technology—a policy review and publishing a proposal for new legislation. In India, farmers and NGOs considered the repercussions of GE technology for local agricultural and environmental policy; this was intended to counterbalance a government policy that was perceived to serve mainly multinational companies.

When such extra-institutional participatory governance gains sufficient momentum, intra-institutional governance is forced to react. This may first take the form of public reassurances that the policies in place are working, or even denunciations of the motives and legitimacy of the social actors concerned, as was the first reaction of the British government in 1999. If social mobilization persists and critical public discourse shows no sign of abating, the formal decision-making system then often adjusts to sociopolitical developments. In Britain, the government was forced to reform its policymaking system by putting GE technology in a more comprehensive regulatory framework; this included setting up new regulatory agencies in charge of considering wider social, economic, and ethical aspects through more transparent and open procedures. The formal decision-making system was adjusted less reluctantly in Germany in 1998, where it coincided with a change of government: the incoming government of Chancellor Schröder (which included the Green party, which in opposition had been a key catalyst of social mobilization against GE technology), reorganized, as indicated earlier, the agriculture and food policymaking system, placing a new emphasis on consumer protection, food safety and transparency; launched new risk assessment research; started a "green biotechnology dialogue" with stakeholders; and set up a new national ethics council.

In Europe at least, the effect of these (procedural) adjustments to the formal governance system in the late 1990s has been to slow down, or temporarily suspend decision making. In Britain, government and industry agreed to a voluntary moratorium on the commercialization of GE products while comprehensive scientific, environmental, economic, and social-scientific reviews were carried out. In Germany, manufacturers withdrew GE products from sale. At the European level, a de facto moratorium came into force for several years and existing regulation was overhauled (see earlier).

Sometimes, adjustment takes place *in anticipation* of potential future social mobilization. This was to some extent the case in Denmark and the Netherlands in the 1980s, where there had previously been strong environmental movements involving a broad public, especially in relation to energy policy. These had decisively influenced the course of national decision-making (neither country developed civil nuclear technology). Decision makers tried to make GE technology more socially compatible by considering its social, ethical, and legal repercussions and by encouraging broader involvement by social actors in both research and public discourse. As a result, accompanying social science research, TA projects and wider public debate programs were launched. Hence, in what became a significant feature of Danish and Dutch policymaking in the late 1980s and early 1990s, governmental initiatives stimulated and enabled wider social participation in biotechnology research and policy in various public spheres. The interaction between intra- and extra-institutional governance processes was actively sought. In Denmark, this was made possible by a red-green majority of members of parliament, who in the mid-1980s forced new legislation on biotechnology and TA onto a reluctant conservative government. The effect of this kind of anticipatory adjustment was not so much a slowing down or suspension of decision-making, as a proactive broadening of the scope of analysis and development to take account of wider sociopolitical perspectives.

These adjustments are also meant to create more openness and transparency in intra-institutional decision making. In Britain, for example, the government decided to publish information about its field trials, which had hitherto been kept secret; and its newly established regulatory agencies began to hold their meetings in public and to advertise their agendas in newspapers and on the Internet (see earlier). This transparency can often have the further effect of prompting critical, normative discussions of the adequacy and merits of these "new" policy- and decision-making procedures themselves. In other words, a reflexive, meta-level discourse on these new forms of participatory governance often accompanies the more substantive, "technical" discourse on, for example, GE crops. This can further fuel social mobilization, especially where social actors view such intra-institutional participatory governance skeptically or where the more substantive policymaking fails to make progress.

The experience of these reflexive debates has exposed a number of problems concerning new modes of participatory governance. How can the discourses and assessments in various extra-institutional participatory arenas be "translated" into these intra-institutional governance procedures? Who is invited to participate, who represents which interests and expertise, and how is public opinion "materialized" in these procedures? Some of the methods have been criticized for excluding certain stakeholders, or for attracting "biased" members of the public who do not fully represent "the public interest." There is the related problem of "framing." Who defines what can, and what cannot, be discussed, and the scope of the assessment and the nature of policy recommendations? There is often disagreement among different actors as to the remit of governance procedures, with some demanding a broader treatment of the issue (to take account of various perspectives) and others criticizing a lack of focus.

There is also the difficulty of linking these participatory governance processes to actual political decision making. It is sometimes unclear exactly how participatory procedures relate to policymaking processes, and how political decision makers and bureaucrats use their outcomes. This may partly be due to the novelty of these procedures: they must first establish themselves before they can find their place in the decision-making process. However, more critical observers may see simply lukewarm commitment, or even deliberate strategic maneuvering, on the part of decision makers at work here. Finally, linked to all of this, there is the problem of how to ensure the accountability and achieve the legitimacy of participatory governance. The mixing of responsibilities between policymakers and various stakeholders within these participatory procedures, and the latter's unclear relationship with formal decision making, can blur the lines of accountability and thus undermine the legitimacy of participatory governance.

CONCLUSIONS

As modern biotechnology has developed from a research endeavor in the 1970s, through its first industrial applications in the 1980s, to agricultural production and marketization since the 1990s onward, public participation has increasingly become a central feature of public policymaking and related social discourse. This is the case both in many postindustrial "risk" societies, as well as in much of the developing world.

In the course of this development, the nature and dynamics of public participation have changed. In the beginning, when biotechnology exclusively belonged to the expert community, public participation was, to a limited extent, conceptualized and practised as the one-way provision of information to the public, on the often tacit assumption that information would lead to acceptance. Since then, public participation has developed into a more sophisticated and complex deliberative relationship between experts, policymakers, stakeholders, and citizens. This relationship is more closely based on an understanding of biotechnological innovation as a process that is as much shaped by sociopolitical dynamics as by scientific and technological advances. Thus, public participation has become more widely recognized as a necessary ingredient of policymaking on GE crops and food: as a way of both gaining a better understanding of public concerns, expectations, and preferences, and of mediating between different specialist and public interests in an open and transparent manner. This has resulted in what is by now a widely adapted practice in many organizations, countries, and continents: institutionalized "participatory technology assessment" and "public engagement" as a way of addressing the ethical, social, and legal issues raised by GE technology. Such intra-institutional governance does not go uncontested: some participants and observers criticize its normative and methodological grounding as well as its practical effectiveness.

However, participatory governance of GE crops is not limited to these formalized, institutional procedures, even if these have been the main focus of attention of policymakers as well as scholars interested in GE public policy. Rather, participatory governance should be understood as a wider, dynamic social process of interaction between various, differently constituted arenas in the political, specialist, and wider public spheres. It is a process that involves experts, stakeholders, and members of the public not just in formal, structured procedures within institutional policy settings, but also in a multitude of mediated extra-institutional arenas that reflect the diversity of social actors and their different interests, strategies, and political agendas.

To understand public participation in the governance of GE crops as a constituent part of biotechnology and an important factor conditioning policymaking, we, therefore, need a more fine-grained picture of the dynamic relationship between the various forms of intra- and inter- institutional decision making, on the one hand, and extra-institutional processes of public mobilization and discourses, on the other. Otherwise, we risk ending up with too reductionist a view of the multifaceted phenomenon of public participation in the governance of GE crops; we will have only a limited understanding of its social characteristics, its changing dynamics, and its political significance.

The lessons learnt more generally for technological innovation from the case of GE crops are that ignoring or dismissing the various participatory and social mobilization processes in wider public spheres comes at a cost of misunderstanding the social, economic, and cultural conditioning of technological development. Furthermore, failing to recognize the inherently political nature of participatory governance means that one of its key functions—namely, to offer spaces for publicly contesting, scrutinizing, and thus co-determining technological innovation—is either missed or given insufficient weight in policymaking, with the risk that the latter's effectiveness and legitimacy will be undermined. Scientists, policymakers, and politicians alike would, therefore, be well advised to engage with the contents and processes of participatory governance as a key constituent part of technological innovation and related policy- and decision-making.

NOTES

1. Bonfadelli, H., Hieber, P., Leonarz, M., Meier, W.A., Schanne, M., and Wessels, H.-P. 1998. Switzerland. In *Biotechnology in the public sphere. A European sourcebook*. Edited by J. Durant, M. Bauer, and G. Gaskell. London: Science Museum, pp. 144-161.

2. ActionAid. 2000. *Indian farmers judge GM crops. ActionAid citizens' jury initiative*. Report prepared by T. Wakeford and edited by A. Wijeratna. London: ActionAid. (November 2, 2006). http://www.actionaid.org.uk/doc_lib/16_1_citizens_jury_initiative.pdf.

3. GM crops in East Sussex? speak now or forever hold thy peas! 2003. *Press notice of public debate in Forest Row Village Hall*. (June 27). See also http://there.is/forestrow/GM/ (November 2, 2006).

4. Agriculture and Environment Biotechnology Commission. 2003. *GM Nation. The findings of the public debate*. London: Agriculture and Environment Biotechnology Commission. http//www.gmnation.org.uk.

5. Together, the USA and China accounted for over 60 percent of commercially grown GM crops, according to 2001 estimates (see http://www.isaaa.org.or http://www.checkbiotech.org) (August 2005). According to the US Department of Agri-

culture, 68 percent of soybeams and 26 percent of corn is genetically modified in the USA. See *Indian Country Today*, March 8, 2002. http://www.indiancountry.com.

6. Barinaga, M. 2000. Asilomar revisited: Lessons for today. *Science*, 287: 1584-1585.

7. Spotts, P. 1992. Science, technology, and the public. *Agriculture and the Undergraduate*. Edited by Board on Agriculture/National Research Council. Washington, DC: National Academies Press, pp. 113-120. Rifkin, J. 1977. "Have the corporations already grapped control of new life forms?" *Mother Jones*. Original article published 1977. http://www.motherjones.com/mother_jones/FM77/rifkin.html.

8. Durant, J. Bauer, M., and Gaskell, G. 1998. Introduction. The representation of biotechnology: Policy, media and public perception. In *Biotechnology in the public sphere. A European sourcebook*. Edited by J. Durant, M. Bauer, and G. Gaskell. London: Science Museum, pp. 3-12.

9. See for example: Jelsøe, E. 1998. Denmark. In *Biotechnology in the public sphere. A European source Book*. Edited by J. Durant, M. Bauer, and G. Gaskell. London: Science Museum, pp. 29-42.

10. Hieber, P. 1999. Gentechnologiepolitik in der Schweiz. In *Gentechnologie im Spannungsfeld von Politik, Medien und Oeffentlichkeit*. Edited by H. Bonfadelli. Zurich: University of Zurich (IPMZ), pp. 21-62.

11. Vig, N.J. and Paschen, H., eds. 2000. *Parliaments and technology: The development of technology assessment in Europe*. Albany: State University of New York Press.

12. See, for example, Banthien, H., Jaspers, M., and Renner, A. 2003. *Governance of the European research area. The role of civil society*. Bensheim: Institut für Organisationskommunikation. http://www.ifok.de; Joss, S. and Durant, J., eds. 1995. *Public participation in science. The role of consensus conferences in Europe*. London: Science Museum; Joss, S. 2000. Participation in parliamentary technology assessment: from theory to practice. In *Parliaments and technology: The development of technology assessment in Europe*. Edited by N.J. Vig and H. Paschen. Albany: State University of New York Press, pp. 325-362; Joss, S. and Bellucci, S., eds. 2002. *Participatory technology assessment. European perspectives*. London: Centre for the Study of Democracy, University of Westminster.

13. Joss, S. 1998. Danish consensus conferences as a model of participatory technology assessment: An impact study of consensus conferences on Danish Parliament and Danish public debate. *Science and Public Policy* 25: 2-22.

14. Hamstra, A. 1995. The role of the public in instruments of constructive technology assessment. In *Public participation in science. The role of consensus conferences in Europe*. Edited by S. Joss and J. Durant. London: Science Museum. pp. 53-66; Van Est, R. 2002. The Netherlands: Seeking to involve wider publics in technology assessment. In *Participatory technology assessment. European perspectives*. Edited by S. Joss and S. Bellucci. London: Centre for the Study of Democracy, University of Westminster, pp. 108-125.

15. Gloede, F. and Hennen, L. 2002. Germany: A difference that makes a difference? In *Participatory technology assessment. European perspectives*. Edited by

S. Joss and S. Bellucci. London: Centre for the Study of Democracy, University of Westminster, pp. 92-107.

16. Joss, S. 2003. Zwischen Politikberatung und Oeffentlichkeitsdiskurs— Efahrungen mit Burgerkonferenzen in Europa. In *Burgerkonferenz: Streitfall Gendiagnostik*. Edited by S. Schicktanz and J. Naumann, J. Opladen: Leske & Budrich, pp. 15-35.

17. For an in-depth overview and analysis, see Gaskell, G. and Bauer, M.W. 2002. *Biotechnology 1996-2000. The years of controversy*. London: Science Museum.

18. For a comprehensive analysis of the "great debate" see Weldon, S. and Wynne, B. 2001. The UK National Report. *Assessing debate and participative technology assessment* (ADAPTA Final Report). Brussels: European Commission (project Nr. BIO-CT98-318). www.inra.fr/Internet/Directions/SED/science gouvernance/pub/ADAPTA/.

19. Joss, S. 2005. Between policy and politics. Or: Whatever do weapons of mass destruction have to do with GM crops? The UK's GM Nation Public Debate as an example of participatory governance. In *Democratising of expertise? Exploring novel forms of scientific advice in political decision-making. Sociology of the sciences*. Edited by S. Maasen and P. Weingart. Dodrecht: Springer, 24: pp. 171-187.

20. Lean, G. 2000. Blair: GM may be a health risk. Editorial comment. *Independent on Sunday*. February 27 . The reported statement of Tony Blair about the "hysteria of public reaction" is cited in: Moore, J.A. 2001. More than a food fight. *Issues in Science and Technology online*. http://www.nap.edu/issues/17.4/p_moore.htm.

21. Blair, Tony. 2000. The key to GM is its potential, both for harm and good. OpEd. *Independent on Sunday*. February 27.

22. *Bundesministerium für Ernährung, Landwirtschaft und Verbruacherschutz* (formerly *Bundesministerium für Verbraucherschutz*). http://www.bmelv.de/cln_045/ DE/00-Home/__Homepage__node.html__nnn=true (November 2, 2006).

23. For a more detailed analysis. See Grabner, P., Hampel, J., Lindsey, N., and Torgersen, H. 2002. Biopolitical diversity: The challenge of multilevel policy-making. In *Biotechnology 1996-2000. The years of controversy*. Edited by G. Gaskell and M.W. Bauer. London: Science Museum, pp. 15-34.

24. European Food Safety Authority. Http://efsa.eu.int/. (November 2, 2006).

25. Dauenhauer, K. 2003. Africans challenge Bush claim that GM food is good for them. Inter Press Service. June 23. http://www.cyberdyaryo.com/features/f2003_0623_03.html.

26. Jha, S. Seeds of death. Farmers in India are fighting to ban Monsanto's GM cotton. Article published at http://www.tompaine.com. Downloaded from http://www.organicconsumers.org/monsanto/indiacotton.cfm See also http://news.bbc.co.uk/1/hi/world/south_asia/843292.stm *India to test GM cotton*. July 20, 2000.

27. See, for example, Krishnakumar, A. Controversy. A lesson from the field. *India's National Magazine* 20 (11): May 2003. http://flonnet.com/fl2011/stories/20030606005912300.htm.

28. Rowell, A. and Burton B. 2002. Green reaper. New Zealand's PM could lose the election over GM scandal. *The Guardian (London)* July 2. http://www.guardian.co.uk/gmdebate/Story/02763,761959,00.html.

29. See for example, Reformer Staff and Associated Press. 2004. GMO bill passes Senate. *Battleboro Reformer.* April 17, 2004; and GMO ballot measure approved. *Associated Press*, April 16, 2004.

30. Taliman, V. 2002. Native Americans speak out against GE foods. *Indian Country Today,* March 8, 2002.

31. Collins, S. 2004. Anti-GM views growing in US. *New Zealand Herald,* June 8, 2004.

32. Ausbiotech. http://www.ausbiotech.org.

33. Kahn, J. 2002. The Science and politics of super rice. *New York Times,* October 22, 2002.

34. See for example, Kooiman, J., ed. 1993. *Modern governance. New government—society interactions.* London/Thousand Oaks/New Delhi: SAGE Publications; Pierre, J., and Peters, B.G. 2000. *Governance, politics and the state.* Basingstoke/London: Macmillan Press; Jessop, B. 2000. Governance failure. In *The new politics of British local governance.* Edited by G. Stoker. Basingstoke/London: Macmillan Press, pp. 11-32; Grote, J.R., and Gbikpi, B. 2002. *Participatory governance. Political and societal implications.* Opladen: Leske & Budrich.

35. Kooiman, J. 2002. Governance: A social-political perspective. In *Participatory governance. Political and societal implications.* Edited by J.R. Grote and B. Gbikpi. Opladen: Leske and Budrich, pp. 71-96.

36. Jessop, B. 2000. See note 34, p. 15.

37. See Schmitter. 2002. Participation in Governance Arrangements: Is There Any Reason to Expect it Will Achieve 'Sustainable and Innovative Policies in a Multilevel Context'? In Grote, J.R. and Gbikpi, B. 2002. See note 35, pp. 51-69.

38. "Social-political governing takes place in interactions between actors on micro, meso and macro levels of social-political aggregation. These interactions not only reflect the basic complexity, dynamics and diversity of our societies, they are themselves complex, dynamic and diverse. That is to say: they are connected in complex patterns, they move and change on the basis of dynamically working forces and they refer to different kinds of substances. In complexity the structural substances, in dynamics the changing substances and in diversity the different substances of mutual interdependences of social-political systems are expressed." (Kooiman, J., ed., 1993. Governance and governability: Using complexity, dynamics and diversity. In *Modern governance. New government—society interactions.* London/Thousand Oaks/New Delhi: SAGE Publications, p. 41). See also Pierre, J., and Peters, B.G. 2000. See note 34, pp. 65-66.

39. See, for example, Schot, J. and Rip, A. 1996. The past and future of constructive technology assessment. *Technological Forecasting and Social Change* 54: 251-268.

40. Vig, N.J. and Paschen, H., eds. 2000. See note 11.

41. Hieber, P. 1999. See note 10; Van Est. 2002. See note 14, p. 124.

42. Vig, N.J. and Paschen, H., eds. 2000. See note 11.

43. Ibid., pp. 124-125.

44. Ibid.

45. See for example, Irwin. A. 1995. *Citizen science. A study of people, expertise and sustainable development.* London: Routledge. A practical example is the

Science Shop for Biology, based at Utrecht University. For further information on the Science shop movement, see http://www.issnet.ac.nl.

46. See note 19.

47. House of Lords. 2000. *Science and society.* Paper 38. London: Stationary Office. European Communities. 2001. *European governance. A white paper.* Luxembourg: Office for Official Publications of the European Communities (ISBN 92-894-1061-2).

48. Weldon, S. and Wynne, B. 2001. See note 18; Joss, S. 2002. Toward the public sphere: reflections on the development of participatory technology assessment. *Bulletin of Science, Technology & Society* 22: 220-232.

49. Jelsøe, E. 1998. See note 9.

50. Akhileshwari, R. 2005. Genetically-modified Bt cotton a cropper. *The Deccan Herald,* Bangalore, Karnataka, April 13.

51. Link between biotechnology, trade and food security. Website of Devinder Sharma. http://www.dsharma.org.

52. HRH The Prince of Wales. 1999. My 10 fears for GM food. *The Daily Mail,* London, June 1.

Chapter 17

Risky Delusions:
Misunderstanding Science
and Misperforming Publics
in the GE Crops Issue

Brian Wynne

INTRODUCTION

Perhaps no other issue has more sharply exposed the differences between U.S. and European approaches to science and its techno-social deployments than that of GM crops and foods. The ramifications of this transatlantic dispute are thoroughly global, as reflected in the accusation leveled at the EU by U.S. President Bush in summer 2004, that EU opposition to GM was causing global starvation thanks to the refusal of U.S. GM food aid by developing countries such as Zambia, following Europe's stance. U.S. anxiety about winning the WTO dispute launched against the EU in May 2004 for its alleged six-year moratorium against GM crops and foods further reflects its concerns to signal the validity of its chosen scientific-technological trajectory for global agriculture to those world markets, than about European markets alone.

In such disputes, WTO decision rules and criteria (the Sanitary and Phytosanitary Agreements and the Technical Barriers to Trade rules) embody particularly narrow definitions of legitimate risk assessment

Genetically Engineered Crops
doi:10.1300/5880_17

and risk evidence for warranting intervention in free trade.[1] These definitions are said to constitute an objective, universal "sound science" for settling such major conflicts, which harbor complex cultural and social contingencies in their very scientific substance. The United States claims that the EU has ignored sound scientific principles and instead has allowed emotion and protectionist self-interest to prevail.

Yet despite these apparently sharply conflicting transatlantic perspectives on GEOs, their risks and the available science, there are less obvious but more substantial shared dimensions to a scientific knowledge culture that is allowed to define the issues. This culture shapes dominant global discourse-practices over GE crops and foods, including what is represented as the underpinning science, and the framing of "the public" embodied in this institutional science as presumed policy authority. I argue here that this common science-policy culture (not unique to GEOs issues) is deeply problematic in several unrecognized dimensions. In Europe, the majority public attitude is in deep tension with that science-policy culture as embodied in the technocratic policy culture of the European Commission, and especially its U.K. counterpart. These policy cultures are already thoroughly pro-GE anyway, since this is automatically (but falsely) equated with being "pro-science."[2]

The European experience of solid, mobilized, and lasting public opposition to GE foods and crops (reflected in retail and food industry rejection too, at least for the time being, on commercial grounds) has been ostensibly different from North American public reactions, though this apparent contrast may be less substantial than first appears.[3] European scientific and policy experts, and the political leaders they advise, have much more fundamentally in common with their U.S. equivalents than what the global symbolic action[4] of the WTO conflict might imply.

Thus despite differences of regulatory philosophy, with U.S. intervention based only on GE product risk assessment while EU intervention was from 1990 based in addition upon specific GE *process* factors,[5] the wider regulatory and public issue has also been defined almost exclusively by a fundamentally similar transatlantic institutional discourse of sound science and risk. More salient, and less obvious, this assumed discourse-practice of a universal rationality misunderstands and misrepresents its publics and the meanings that they invest in the issue. This means that the dominant scientific and policy bodies assume and impose a definition of the salient issue(s) to be addressed, which is problematic as it needs to be negotiated with other legitimate meanings that citizens invest in the issue, but which the dominant culture does not recognize, or if it does so, dismisses as rooted in misunderstanding. This includes their treatment of the key dimension of "uncertainty," on which unseen fundamental dislocations exist between the typical meanings of science-led institutions, and typical citizens. These unseen dislocations give rise to the sweeping institutional anxieties about what has been called a "crisis of public mistrust of science,"[6] GEOs being just one vivid recent example.

In this chapter I, therefore, outline the ways in which the GEOs issue has been presumptively and rigidly framed as one of risk and "sound science," and how this has in turn framed a particular, and deeply problematic, understanding of "uncertainty" as a crucial dimension of the public issue. I show how this also corresponds with—indeed embodies and projects back into the public domain—an imagined construct, "the public," its concerns and reactions. This imagined public is not only contrary to what evidence exists (including U.S. evidence) about public concerns over GEOs, but is itself emotively inspired—amongst institutions proclaiming their own science-based rationality.

For this task, I use recent theoretical perspectives that develop the following twin ideas:

- That objects like risk are not "given" but are intellectually constructed, a process that reflects not only objective realities but also implicitly selective human profiles of meaning and saliency. Imposed as if they were universal scientific givens, such definitions also involve unacknowledged moral, political acts—the tacit assertion of control and neglect of that which is uncontrolled, and a corresponding construction of human subjects as (inter alia) frightened of "uncertainty";[7] and
- That representational practices (like institutional scientific references to the GEOs issue as a "risk issue") are thus also tacitly performative[8] in the sense that they reflect, project, and impose on the public domain, implicit taken-for-granted but also nor-

mative assumptions about human subjects; here, that for them
the meaning of the issue at stake is indeed "risk."

This inadvertent form of projection and performance, of imagined
publics, their capacities, needs, concerns, *and meanings,* is taken by
its own nonself-aware authors to be only innocent representation.
This dominant scientific culture's profound misunderstanding of *itself* as well as its "other(s)" has further effects, which fuel a deeper
sense of public alienation, even if this remains unmobilized and unseen.

Much of this broader problem I suggest originates in unrecognized
shifts in the social-cultural role of science since the 1950s, along with
its rapid and unprecedented institutionalization as public policy authority. Simply stated, this constitutes a historical shift from the overt
role of *informing* policy issues, to in effect covertly *defining* them.
Science has thus become, by default, and with the instrumentalization of public life more generally, the presumptive definer of public
meanings. Along with this, its twin roles, of providing understanding, and generating instrumentally useful forms of manipulation of
nature (technological innovation), have been increasingly fused rather
than held in healthy explicit tension, and the instrumental ethic has
arguably been allowed to become *the* end of knowledge—its culture.
In the absence of any more transcendent human ethos of public policy and science, means have thus become ends; and as the GE case
exemplifies, a "hidden metaphysics"[9] haunts this institutional science-policy culture. The most significant marker of this abstract notion can be seen in the pervasive discourse of *risk* as a way of defining
public issues,[10] where the very limited commonalities of diverse substantive issues such as global terror, misadventure on school outdoors
trips, climate change, drug side effects, or GE crops are reduced to
common "risk issues"; thus public responses are "risk-attitudes"
(like "risk-aversion") based on "risk-perceptions."

As we are continually reminded, risk is a scientific issue. Thus the
common presumption that complex multidimensional issues like GE
crops are "(scientific) risk issues" reduces them *ab initio* only to sci-
entific-instrumental dimensions of meaning, as if the only salient
question is: "What harm [and only 'harm' as defined by institutional
science] do they inflict?"[11]

Risk, and risk assessment, has thus become the sovereign meaning of such issues, in threefold form: the supposed institutional-scientific representation of people's concerns and meanings; the correct means of addressing these; and the attempted reassurance of those concerns. From this unquestioned premise, continuing public refusal against repeated and escalating official assurances based on risk assessment science, is understood to be due to public deficit of ability (or willingness) "rationally" to understand that science (of risks), which is assumed to define the true public meaning of the issue. It is on such apparently unnoticed, taken-for-granted dimensions of science and its human context, including what is routinely defined as "science" and "the public," that I focus.

OBJECTS OF CONFUSION—AND PROVOCATION

It is a frequent scientific lament, that *"many of the public's concerns [on GE] have little to do with science,"*[12] as if this renders them automatically illegitimate, and beyond science's responsibility. In addition to showing why "nonscientific" public concerns may be salient, substantive, and legitimate, I want also to show how in crucial respects they are actually "scientific concerns," but in a fundamentally different and more challenging way than has been recognized. These are concerns not only about the impacts of science, but about its intellectual character as represented (what it does and does not claim to encompass reliably, which also takes on moral dimensions). Thus institutional science has embodied particular and contentious social assumptions and implications in its constitution of "science" for presumptive public authority, but then when faced with what are equally legitimate alternative social commitments and questions representing a cultural "other" (the typical public), it has rejected these a priori as "nothing to do with 'science'" therefore irrelevant, illegitimate and irrational.

As a result, the dominant culture has imposed a dictatorially preordained answer to what should be a continuing collective moral negotiation ultimately about the proper human ends of knowledge. It has done this in the name of "science" represented as innocent revelation. Thus it has also misunderstood public concerns *about* science, as if these concerns were not "scientific," therefore not the responsibility of science and scientists, when scientists themselves have imposed their own concerns *about* science (and their questionable normative vision of its proper role and authority in society) as part of what they have called just "sound science."

It is this implicit institutional-scientific confusion that comes across to the public as not only denial, but dishonest double standards. Thus if a central concern here is public confidence in science, scientific institutions need to examine themselves and this established post–World War II culture of science-policy as part of the question. Moreover I explain why this *self*-examination to replace the entrenched habit of *external* projection of responsibility has to include reconsidering the encultured imaginations of their public as embodied in their "science."

There has been a recently celebrated shift from discredited "public deficit" explanations of refusals to accept scientific assertions about new technologies and corresponding one-way "public education by science" policy responses toward two-way "public dialogue" or "public engagement with science." However I argue that these deeper and more problematic tacit dimensions of the institutional culture of science-in-policy threaten to derail this shift.

The escape route from this perverse effect of an otherwise progressive move out of the "deficit model" culture is to reflect the implicit demand seen in much *explicit* public concern about science, to address in accountable fashion the questions over the unaccountable social ends, and driving purposes shaping more upstream scientific commitments and trajectories;[13] that is, beyond the instrumental and downstream questions of risk alone. The exclusive and determined focus on risk in effect obscures this more constructive agenda and its hidden politics from wider access, leaving it to be shaped by unaccountable interests and untested imaginations of proper priorities and ends. This is the so-called upstream engagement move,[14] which has been misunderstood[15] as a task of engaging publics in earlier forms of anticipation of the possible effects of innovation so as to head off negative effects before they become entrenched and costly. This is not an adequate surrogate for what should be a genuinely democratic culture where science's driving purposes and ends are negotiated in such a way as to influence scientific practices indirectly and are consistent with respect for scientific specialist competences.

Strictly speaking, these further public questions, or objects of concern, are not scientific ones; but this is misleading. They may be deemed not scientific because they are not addressing the kind of propositional question that scientific work itself normally addresses, such as: "How big are the risks? What controls them?" However, they may be legitimate and necessary judgments of issues normally also left to science, thus "scientific" *in the sense typically meant (by scientists) when science is represented in public domains,* even though "non-scientific" in the narrow propositional sense referred to immediately above. Perhaps the key example is the way that "uncertainty" is taken to mean known uncertainties, usually reduced to being encompassed by "risk" and risk assessment, when for the typical public "uncertainties" mean unpredicted effects that are unknown to even the best scientific risk knowledge. The widespread public concern shown in the United States as well as Europe,[16] that in their repeated reference to risk assessment as response to public concerns, scientific institutions effectively *deny* such unknowns and the lack of scientific control they indicate, is instead treated as a naïve public concern that uncertainty *exists.* Given that this denial and corresponding exaggeration is performed in the name of "science," these public concerns about scientific exaggeration are "scientific concerns," even if they are not simple ones over propositional claims about risks.

Another example here is the official statement made about the 2000-2003 U.K. GE farm-scale evaluations (FSEs) of the effects on selective biodiversity indicators of growing herbicide-resistant GE crops with new pesticide-management regimes. The risks to "biodiversity" of these crops and their management regimes were thus defined, but the advisory body, the U.K. AEBC,[17] had already stated that this alone would not answer the question as to whether GE crops had acceptable environmental effects, let alone whether as a wider policy matter they should be commercialized. The question of GE commercialization had already been reduced by the science-policy institutional culture to a scientific question alone in official statements that the FSEs would provide the definitive scientific answer to the policy question of whether commercialization would be acceptable. Five such statements are listed in Annex C of the AEBC report;[18] but this reduction of many wider questions to one of science alone had been explicitly resisted by the AEBC. Moreover even *within* the science, the idea that all risk questions had been addressed just because a competent scientific risk assessment had been conducted was nonsense. As the scientific leader of the FSEs Les Firbank freely admitted to this author at a U.K. NERC meeting in 2001,[19] the trials could not practically examine the effects on soils, as these were too difficult to measure in a controlled and meaningful way. Soils were thus pragmatically ignored, even though scientists knew that soil microorganisms are a major element of biodiversity and soil quality.

Is this neglect of soil changes here a purely "scientific matter," or a policy one? Is the judgment that this neglect nullifies the role attributed to the risk assessment "science," of bearing the policy weight of commitment to GE commercialization, an exclusively scientific judgment, or a "nonscientific" one? Such policy-influential judgments when they are typically made privately by scientists on behalf of society, as Jasanoff has shown in detail,[20] have been routinely passed off as purely "scientific." Yet when alternative judgments are made on the same grounds, for example, here that the neglect of soil would justify a decision against commercialization at least until more is known about soils effects, this would be called "nonscientific," implying also, "unjustified." The boundary between "scientific" and "policy" is in this way always ambiguous, and open. These are judgments *about* the scientific knowledge and its role or meaning, rather than scientific-propositional judgments themselves, which is supposed to be the defining business of science. However, in practice because science has for long been routinely used more expansively to define and make these more ambiguous social–political determinations, when they are challenged this evokes the reaction that such concerns are merely "nonscientific" (implying "uninformed" or "irrelevant")— when *before* they were challenged, they were "scientific." This kind of systematic inconsistency *institutionalized* preemptively in favor of innovation and industry has been cumulatively exposed by science studies work, but so far in response no more consistent nor more robust intellectual position has been developed. Public mistrust, therefore, inevitably stalks the institutional dissembling that these issues suffer. This broader, more ambiguous, meaning of "science" in the public domain includes the institutional scientific presumptions, which are more than science, such as:

1. The conviction that even though scientific knowledge about, for example, the health and environmental effects of GE crops is uncertain, we know *enough* to be able to say with authority that the risks are sufficiently small to justify commercialization of this scientific and technological trajectory. Such a human commitment is represented as if the sole province of science, when it combines strict scientific knowledge (and uncertainty as always) with human judgments about the significance of ignorance, uncertainties, error costs, alternatives foregone, social implications, likelihood of being able to respond adequately to surprises, and so on—that is, *ethical as well as social and intellectual judgments, but made and represented (and defended) as if solely scientific-intellectual;*

2. The presumption that the objective, natural meaning of the issue is just scientific, that is, one of risk—over which science can and should be sovereign. In this sense, beyond merely informing policy, science now plays the much more pervasively potent role of providing the *meaning* of public issues involving science—as if, for example, the GE crops issue is, "What are the risks (as defined by institutional science)?" rather than, "What kind of agriculture do we want? Under what conditions could GE fit in, and are those conditions feasible and acceptable?" Or, in broader terms, "What kind of society do we want?"

I argue that a densely (and narrowly) *risk-centered* discourse has been allowed to provide the assumed meaning of the GE as well as many similar science-related issues. This also shapes what policy experts see as the problem of public opposition, misunderstanding, and mistrust of science. They do not see, however, that the problem is not primarily one with "other" actors like deficit-ridden public, sensationalist media, and misleading NGO propaganda, external to the institutional powers of modern policy-corporate-scientific-commercial actors.

Thus there is a dearth of what I call moral, as distinct from instrumental, learning about the character and salience of the social actors who are involved, and about their issues rather than what is dogmatically *presumed* to concern them. This translates into a need for not instrumental but reflexive institutional learning, about:

- themselves and their own habitual assumed ways of understanding and (mis)representing other actors;
- thus their own role in causing the problems like mistrust attributed to others; and
- their normal misunderstandings and misrepresentations of society and "the public," even after supposedly listening to them.

I suggest that the inability of institutional science to openly and
authentically recognize its own ignorance (as distinct from "uncertainty") is an inability to recognize "the other" in the form of real
surprise from unpredicted consequences. This betrays something
supposedly alien to modern science, namely a parochialism that con-
tradicts its *cultural* claims of open-mindedness.

I argue, further, that the problem of ignorance and unpredicted
consequences, which is largely concealed by the convenient reference to "uncertainty" as part of "the risk problem,"[21] is fundamentally an ontological problem, not only an intellectual or epistemic one
(though it is that too). This resonates with Latour,[22] and Stengers.[23]
Typical publics are more ready in practice to acknowledge the issue
of ignorance and lack of control, not of "uncertainty," which is different, than are scientific institutions. Yet publics responding to issues
such as the GE issue are still mistakenly defined as unable to recognize
and live with "uncertainty," and as mistrusting of science *because
they supposedly misunderstand that science cannot be expected to
deliver certainty.* This hints at some further cultural challenges facing
current scientific and technological institutions trying to cultivate a
more robust basis of public legitimacy.

RISK ASSESSMENT AND EXCLUDED DIMENSIONS
OF "UNCERTAINTY"

As I have noted elsewhere,[24] even when ethical issues have been
admitted to be part of the public issue over GE crops and foods, this
has been either subordinated to risk (utilitarian ethics—what are the
consequences?—which is returned to science); or, if deontological
(is it essentially right or wrong, like "playing God?"), defined as a
private matter only, thus resolved by labeling for individual con-

sumer, ethical choice. That people may have ethical problems with routine institutional demeanor (for example, the exaggeration of scientific control, thus entrenched hubris) is nowhere accommodated in this framing of "the ethical issues."

Thus I suggest, even when risk assessment is being honestly and diligently conducted and used, its institutional context has created a culture that implies that risk—and moreover a usually very reduced meaning of "risk" as defined by institutional risk-regulatory science—is the only meaning of the *public* issue. For example, when publics have rejected scientific accounts of the acceptability of GE crop commercialization, it has been unquestioningly assumed that this is on still-scientific grounds of risk—which the public must be misperceiving.[25]

A further major public concern indicated by social research[26] is about the institutional definitions of the quality and meaning of that scientific knowledge itself. Scientists and policy experts have assumed that science knows and handles uncertainty fully, but that publics are frightened of it. They allegedly misunderstand that in principle science cannot deliver certainty and zero-risk, which they supposedly crave. Sir Robert May, then chief-scientific adviser to the U.K. government, gave unequivocal voice to this dominant view:

> There is now an erroneous expectation that life can be "risk-free," and faith in the system tends to be further undermined every time this proves not to be the case. Science education in schools focuses too much on facts, rather than process, leading to the misleading impression that science deal(s) in certainties rather than, as is more often the case, conclusions based on the balance of probabilities after evaluation of the available evidence. Many policy decisions, for example on GM crops, have to be made while there is still a significant degree of uncertainty. Debate among scientists on these issues, which is an essential part of the process can be perceived as vacillation and weakness.[27]

Thus he invented a new, mark-2 version of the previous public deficit model in whose burial he had thrust himself into a prominent role.[28] This new version was based on a supposed public misunderstanding of scientific *process*, whereas the first version had attributed the agency in causing public rejection to misunderstood scientific *contents*. To see how this projection of the causes of public mistrust onto external agencies masked the scientific culture's own agency as cause of its own public mistrust problems, we need first to distinguish between different kinds of scientific uncertainty that are typically confused by the scientific institutions in policy debate. These are given in Exhibit 17.1.[29]

Ignorance, indeterminacy, and ambiguity are frequently mistaken as, or reduced to, "uncertainty." I include "disagreement" here because this too is often misrepresented as uncertainty as if it were capable of resolution by more research, which would inevitably produce agreement by eradicating what is assumed to be residual imprecision (within supposedly known questions) that allows such disagreement. Donald Schon first suggestively noted[30] how "uncertainty" tends to be reduced to "risk" in a constant institutional routine of presuming and projecting more (intellectual) control than is ever the case in reality. Moreover, crucially, he recognized this to be a cultural process, of habit and subconscious routine rather than a deliberate, strategically calculative one. However, Schon did not examine the further catego-

EXHIBIT 17.1. Different forms of scientific uncertainty.

RISK	Know the odds as well as the consequences
UNCERTAINTY	Know the possible consequences, but do not know the probabilities
IGNORANCE	Do not know the possible consequences (do not know whether we are asking the right questions)
INDETERMINACY	Processes subject to inconsistent outcomes from "same" (do we know?) initial conditions
AMBIGUITY	Differences of meaning, and thus of which questions, and which dimensions and variables, thus knowledge(s), are *salient*
DISAGREEMENT	These categories are not necessarily mutually exclusive

ries involved here. "Risk" is often used by scientists to describe their knowledge of relevant processes, as if they had knowledge of probabilities, when they should really describe it as uncertainty. This is so routine as to be unnoticed in science for policy.

Ambiguity is an important category because differences of meaning often pervade public issues involving science, as mentioned earlier where public divergence from scientific definitions of "uncertainty" is misunderstood as if framed within the same assumed meaning, then treated as if it were public ignorance of the scientific process, when it actually reflects a fundamentally different meaning, focussed on questions over ignorance, not *known* uncertainties.

Perhaps the most crucial distinction to be made here is of one that exists between uncertainty and ignorance. Scientific institutions, as in May's earlier assertion, typically allege a public inability to deal with "uncertainty," when they actually give a different meaning to the "uncertainty" which is at issue. For typical publics, uncertainty is about unknowns, as experienced in real-life effects of science and technology. For example, when in a five-country European project we put open-ended questions to people, to explain their problems with GEOs, typical spontaneous responses referred to "thalidomide," and chlorofluorocarbons (CFCs) and stratospheric ozone. Rather than treat these as irrational and irrelevant, we realized that they were attempting to explain an abstract concept through concrete analogy (both these were cases where unknowns at the time when the innovations were risk assessed and passed for commercial uses, later manifested as shocks unpredicted by the best scientific knowledge)—that ignorance, *and crucially, its effective denial,* was the reason for public disaffection.

In deep contrast to the entrenched institutional assumption, this is not at all an issue with the positive knowledge, or known uncertainty, which is the normal fare of scientific culture—except with its exaggeration. This assumed scientific definition of uncertainty, expressed in scientific responses to public concerns as continual reference to risk assessment, is experienced by ordinary citizens as scientific denial of the kind of uncertainty—that is, scientific ignorance, or lack of predictive control—which is of central meaning to them.

FROM "UNKNOWN UNKNOWNS"
TO KNOWN UNCERTAINTIES:
A TACIT CULTURE OF CONTROL

By definition, the object of ignorance is impossible to specify. So too is its extent. This forms a radical break with the more tractable condition of uncertainty. Although one can recognize ignorance in the abstract, it is impossible to do so in the concrete, which is a particular problem for an Anglo-American empiricist intellectual culture, as contrasted with continental European intellectual culture. This general problem was dealt with in a particular home-spun way in February 2004 by then U.S. Secretary for Defense Donald Rumsfeld, in trying to justify U.S. intervention in Iraq due to what was asserted to be unacceptable risk, by emphasizing the impossibility of proving the absence of weapons of mass destruction (WMDs) in Iraq:[31]

> Reports that say that something hasn't happened are always interesting to me, because as we know, there are known knowns; there are things we know [think?] we know. We also know there are known unknowns; that is to say we know there are some things we do not know. But there are also unknown unknowns— the ones we don't know we don't know. And if we look through the history of our country and other free countries, it is the latter category that tend to be the difficult ones [sic].

Rumsfeld was correct at least in his parting assertion, that "unknown unknowns" have been the most difficult categories, maybe especially for modern cultures whose identity has been founded on notions of scientific reason. In producing his philosophical apologia for the "evidence-free" supposition of an imminent risk from secret Iraqi WMD, Rumsfeld inadvertently identified himself and the U.S. government with a strong version of the precautionary principle, which is usually decried as irrational, illegitimate, and beyond the pale of serious consideration by that same government. Yet it is a much more common predicament than usually acknowledged. If we recognize that, even with the best foresight available, there is always the possibility of maybe drastic consequences simply unknown even as questions at the time of commitment, then basing policy on demonstrable evidence alone is problematic, especially as the speed of translation of immature scientific knowledge into commercial technologies is deliberately intensified. Whether we are talking about intervening in Iraq or in uncertain new technologies that one might logically need to open up for public consideration, the real driving purposes, ends, and interests behind such interventions in the face of this endemic (but variable) state of ignorance (not merely uncertainty) were conveniently omitted from Rumsfeld's "philosophical" rationalizations. They have suffered similar neglect in the precautionary principle and GEO debates.[32] The cultural framing of the issue as the instrumental one of risk automatically excludes any debate about appropriate human ends and purposes, and any recognition of the need to exercise explicit judgment about the quality of the knowledge in play, not only about risks but also about the innovations that precede them.

The same blind spot over the condition of ignorance as distinct from uncertainty, was vividly demonstrated at a public meeting of the U.K. advisory Agriculture and Environment Biotechnology Commission (AEBC) in 2001, when the scientific chair of the Advisory Committee on Releases to the Environment (ACRE), the UK regulatory scientific body for proposed GM crop releases, was interviewed by an AEBC member about dealing with uncertainties:

AEBC: Do you think people are reasonable to have concerns about possible "unknown unknowns' where GM plants are concerned?

ACRE CHAIR: Which unknowns?

AEBC: That's precisely the point. They aren't possible to specify in advance. Possibly they could be surprises arising from unforeseen synergistic effects, or from unanticipated social interventions. All people have to go on is analogous experience with other technologies. . . .

ACRE: I'm afraid it's impossible for me to respond unless you can give me a clear indication of the unknowns you are speaking about.

AEBC: In that case don't you think you should add health warnings to the advice you're giving ministers, indicating that there may be 'unknown unknowns' which you can't address?

ACRE: No, as scientists, we have to be specific. We can't proceed on the basis of imaginings from some fevered brow. . . .

Of course, one can recognize the ACRE scientist's problem, and, with qualifications as given later, his point. The allowed terms of regulatory decision making are very tightly defined, and if an advisory committee such as ACRE were to stray beyond them it would risk immediate legal and political embarrassment. However, the artificial necessity of such precise substantive criteria should not be equated by intelligent actors and their institutions, with the proper frames, bounds, and grounds of rational public policy debate. This was not a formal regulatory occasion, but an open discussion, but the ACRE scientist did not seem able to recognize any such distinction, nor the perhaps necessary but still real artificiality of his regulatory scientific frame of reference and responsibility. It is perfectly rational to ask who, if anyone, will be responsible for the effects of scientific ignorance as a predicament that always haunts even the best science, when these unpredicted effects issue as surprises somewhere and sometime in the future. A spontaneous and logical public question is, "Have they got a plan-B?" and "Can we trust that their response to such surprise will be driven by public interest?"

Hence the ubiquity of the trust issue, but for reasons that are not usually understood as rooted in the *logical,* empirically based public experience of the limits of science, which ironically science itself appears unable to acknowledge.[33] Institutionally, in law as well as regulatory habit, such effects arising from the realm of scientific ignorance (and which will only become concretely specifiable when they are no longer subjects of ignorance) are defined as "acts of God" even though they occur from deliberate *human* actions and decisions— they are deemed legitimate so long as the original decision to go ahead was made on the basis of then accepted scientific understanding.[34] Scientific ignorance, even though sometimes acknowledged as a general problem, is thus given no legal or moral status in collective deliberation of responsibility for intervening in nature for commercial purposes. If we did not even know to ask certain questions at the time of risk assessment, then human responsibility for any unpredicted effects, however drastic, is deemed nonexistent.

I suggest that this state of denial of the state of ignorance, and the implicit ethics of externalization that it involves, haunts science and its social-policy relations pervasively when science's social meaning is defined as commercial innovation.[35] It is not science's sole responsibility to rectify this, but neither can scientific institutions deny their responsibility altogether.

After having served as U.K. government chief scientific adviser through the public controversy over GEOs in the United Kingdom and EU, Bob May, in more recent guise as president of the Royal Society, addressed this issue in his November 2002 Anniversary Address. May is an enlightened scientist who acknowledges, more forthrightly than most, the provisionality especially of recently developed scientific knowledge, thus the endemic possibility of surprise:

> There are problems for scientists, and rather differently for policymakers and the general public, when possible—even remotely conceivable—adverse consequences cannot be accurately assessed. This is an issue in the present GM debate, with some opposed to GM crops arguing that you should not proceed with a new technology unless you can identify and quantify all potential risks. Some even argue that this stricture should cover both known and unknown consequences. Since by definition unknown consequences are difficult to identify and quantify, this—what might be called the Strong Limit of the Precautionary Principle—is a recipe for paralysis. On this basis no new technology could advance.[36]

However, a problem with May's statement is the default implied assertion, as in the ACRE scientific chair's "fevered brow" lament, that the impossible requirement to prove a negative (no adverse effects) has to be met before commitment can be allowed—which would as he says leave only paralysis—is a significant public attitude. This also reflects his view that the public erroneously expects certainty and zero-risk from science. Thus again, a wholly acceptable statement (provisionality of scientific knowledge) is married with a false universal (that the public will object to any innovation if we focus on that provisionality) to arrive at an ambiguous conclusion that appears to want its cake (provisionality) and eat it (innovation without debate over ends and purposes).

Here a central problem is that few, if any, scientists involved in public policy ever address the limitation of scientific knowledge in real-life cases. They might, like May, be ready to recognize it in the abstract, but not when commitments are being made in the name of scientific knowledge, such as "risk assessment," in real time. They thus effectively assert that scientific risk assessment can and does encompass all consequences and uncertainties, by pouring scorn on the other kinds of uncertainty as the ACRE chair did. When this fails, the impossibility of expecting all unknown effects to be known is immediately invoked—but never with recognition that if this is the case, then maybe the purposes and imagined societal meanings of doing innovation-oriented science (for which risk assessment is the limited clean-up work) are in need of upstream social debate as a key part of "doing science" robustly in the modern world.

Thus, for example, the U.K. government's assertion that "where there is scientific uncertainty the precautionary principle may be applied,"[37] and the EU's formal exaggeration of the power of scientific risk assessment by explicitly and formally placing it as a gate-keeper on whether the precautionary principle should be adopted for any such decision-making,[38] contradict any acknowledgment of endemic scientific ignorance. So too does the U.K. prime minister's science-advised major speech, "Science Matters," at the Royal Society, when he asserted: .

> The fundamental distinction is between a process where science tells us the facts and we make a judgement; and a process where a priori judgements effectively constrain scientific research . . . let us know the facts; then make the judgement as to how we use or act upon them.[39]

As a statement about scientific knowledge and public policy, this long-standing conventional pose again denies the reality of scientific ignorance, and the corresponding continual need to make judgments—which can never be scientific alone, but are also ethical, social, and political—about whether we have *enough* scientific knowledge, *for whichever purposes are involved*. This judgment rationally requires collective consideration of the driving interests, purposes, and the hoped-for benefits that may be achieved from the innovations being presumed. The statements of scientists like May about science's inherent limitations contradict the official assumption of scientific risk assessment's ability to define when "precaution" is needed, or to define proportionality of regulatory measures to the consequences. Yet when such serious exaggeration of scientific knowledge is regularly, as in such examples, performed in policy statements and commitments, *we find no institutional scientist standing up in public to correct such misrepresentations being performed in science's name.* If this exaggeration is performed in science's name, and lies uncorrected by scientists, it is small wonder that people take the resultant contradictions and inconsistencies to be science's responsibility.

CONFUSIONS OF MEANING: CONSTRUCTING THE PUBLIC BY "LISTENING" TO IT

There is a further element of this systematic institutional reduction of ignorance and thus lack of control to a pretence of "fully-known uncertainties" and thus of (only marginally incomplete) intellectual control. This is the correspondence between the institutional culture's misunderstanding of its own epistemic state, or capacities, and its questionable constructions of the public. The emphasis by social scientists[40] on ignorance as a state-of-affairs that matters to the public but which is neglected and denied by the institutional culture and its discourses is mistaken (see May earlier) as the reason for public rejection of the technology, when it is instead the *institutional denial* that is the main cause. Again, an institutional responsibility for public mistrust is seamlessly redefined as an *external* responsibility (public misunderstanding). Note how May's assertion of this also seamlessly reduces "uncertainty" to "balance of probabilities."[41]

Following the arguments by Latour and Stengers, we can also see how this conflict between public sensibilities and scientifically defined authority is a function of a problematic material institutional culture, yet is systematically misrepresented by its own agents as a purely (and closed) epistemic problem—"Why can't we get them to *understand* properly?" I refer to this as a material–cultural issue in that the neglect and denial of lack of predictive control through the epistemic–representational practices of "risk" and "risk assessment" are embodied in the material practices of those institutions themselves—they are not just epistemic and representational, but they also embody imaginations (of control or incipient control; and of publics) that are *performed* in regular practice.

An example is the 1999 U.K. government Public Consultation on the Biosciences. When qualitative fieldwork was being prepared for this, an advisory group recommended that background public information should also describe existing scientific disagreements and uncertainties. The responsible government official dismissed this, as the official minutes record:[42]

> [The OST official] accepted that scientific uncertainty is an important issue but stated that the very limited amount of scientific information being offered and its basic nature meant that to describe it as being uncertain would in fact be more confusing for participants.

Hence scientific disagreement and uncertainty was deleted as if the public were unaware of such conditions, and unable to do their own independent readings of them from various other public sources and occasions. The institutions imagined citizens not only as unable to face "uncertainty," but also as incapable of constructing their own independent meanings, including those deriving from experience of these kinds of institutional behavior. Moreover the institutions *performed* that imagination, rather than testing it rationally. The quality of existing knowledge, and of the public, were together effectively constructed by this official as *mutual* constructs. At most the public could have their own preferences; but not their own *meanings*. The idea that one might need to listen to and understand how publics make sense of their experience of such issues, including crucially of their actors, and as part of this what "uncertainties" actually mean to them, and why, never disrupted the official imagination.

Thus instead of trying to understand "the [typical] public," a particular normative construction of the public was imposed, and perversely, this creation would then be played back to that public as if it were their voice! Thus are "publics and their concerns" presumed and performed, whilst ostensibly being respectfully heard. The deafening noise of deep institutional insecurity and self-reference is silently imposed on the so-called dialogue between science and its publics.

The philosopher John Dewey recognized the fundamental processes through which notions of "the public" are created and given reality status:[43]

Is the public much more than what a cynical diplomat once called Italy: a geographical expression? Just as philosophers once imputed a substance to qualities and traits in order that the latter might have something in which to inhere and thereby gain a conceptual solidity which they lacked on their face, so perhaps our political "common-sense" philosophy imputes a public only to support and substantiate the behaviour of officials. How can the latter be public officers, we despairingly ask, unless there is a public?

However, this constructedness should not be misinterpreted as asserting the unreality of "the public." A more valid insight is that we should accept responsibility for how we understand and project constructs of "the public," and this applies to us social scientists pretending to research and to represent their attitudes and understandings, as much as it does to those purporting to represent them in policy. The point is that as philosophers such as Dewey, Taylor, Knott, and others like Kierkegaard have long noted, to really hear "the other" in a relationship, one has to place one's own self in question, as up for possible renegotiation. This is true of both individual or institutional interactions. This frightened refusal is the persistent obstacle crippling current attempts at mature public engagement with science. Here it is worth noting the deeper correspondences between the systematic, not deliberated, but more culturally entrenched inability of the scientific institutional culture to recognize and face "the other" in the form of ignorance and the forthcoming surprise it inevitably harbors, and "the other" in the form of the *human* culture of everyday public experience of "science" and its embodied but denied human-cultural commitments. Hastrup[44] has emphasized how an "other" in the form of suffering and pain challenges a central tenet of natural scientific epistemology, by being simply not knowable apart from subjective experience. As she notes, we can only experience this other through our own imaginative projection, which requires an imaginative capacity and readiness to use it. Of course science has imaginative capacity, otherwise it would not be so creative? However, this scientific capacity to imagine might be said to be highly channeled and structured by its own forms of strong socialization-training and a reinforcing practical culture that is very specific, control oriented, instrumental, and rote based, as well as by its social context that is increasingly commercial.

Despite apparently very different epistemic as well as emotional resonances, Hastrup's insight on suffering can also be applied to the ignorance problem. We cannot know it intellectually, except indirectly through imaginative allusion and practical cultural recognition of the limits in principle (the actual limits remaining constantly out of reach—that is the point) of our own culture. Polanyi[45] explained something close to this insight about the profound, unstated ambivalence of practical scientific research toward "what lies beyond" current knowledge. It cannot be pursued here, but the anthropological work by Hastrup[46] on hunger and suffering is highly suggestive of the parallels between the unseen cultural limitations of institutional self-consciously rational culture toward "the other" seen either as intellectual unknowns and shocks, or as cultural-human "others."

This digression into an area of anthropology, which at first sight appears to have nothing to do with science, public reactions, and risk, does help fill out some hidden dimensions that much of social science, let alone policy practice, has not really addressed. I raise these because they are inherent in what science has implicitly set out for itself in its aspirations, claims, and expectations of social authority. They have to be explicated and recognized as salient public questions for science as public knowledge. To ignore them will only be self-defeating.

Thus understanding others, in exercises like public dialogue with science and "listening-to-the-public" processes, as a precondition of avoiding failure, always has to involve awareness of the problematics, and finite limitations in principle of our own culture. As Hastrup says for an empathy with others' suffering, and I suggest for an effective awareness of our own scientific ignorance in risk assessment and rational decisions, a collective moral imagination and not just intellectual competence is essential. Otherwise we will be condemned to projecting our own buried and unacknowledged insecurities onto "the other" whom we are trying but failing to hear. This essential reflexive point has not yet been learnt in all the avalanche of attitudes surveys, public debate, "listening," dialogue, public deliberation, and confrontation that has marked the GEOs controversy, as well as other domains. It is not that there are no deficits of public understanding of science—there are too many, including those of scientifically well-qualified citizens. However, this does not mean that these "deficits" are *the cause* of public refusal to accept what are called "scientific" issue-definitions and ensuing policies, as if scientific propositions about risks were the only aspect. More salient is the public experience of the culture of institutional presumption, exaggeration, deafness, and denial. The fact that this is done in the name of science, more recently even a "listening science," only underlines how easy it is for this negative reaction to be expressed as if it were against "science," rather than as I suggest it is, a reaction against a science-policy culture that has overreached scientific knowledge and understanding into what is arrogant scientism.

In the context of the institutional culture's denial of ignorance, I suggest that this apparently purely intellectual shortcoming is much more than this; it is itself a repeated, habitual act of irresponsibility that also preemptively destroys the legitimacy and integrity of the cultural other. This is also, especially with the material powers of modern techno-science to hand which this culture legitimates, a culture of immodest if conveniently externalized *violence*.

Crucially these questions, deriving directly from the uncertainty issue, go beyond downstream impacts-oriented science and risk assessment, to focus instead on upstream *innovation,* and what human forces, purposes, and conditions drive it (e.g., who owns and controls it, for whose benefit?). This suggests a shift from instrumental issues of means, as in the case of concern about security (risk-centered meanings), to include neglected questions of human ends as essential matters of democratic attention.

CONSTRUCTED RISK: PERFORMING "PUBLICS" AND EXTERNALIZING RESPONSIBILITY

Risk is not just a matter of knowledge, but a material culture of modern scientific knowledge and policy. This is summarized in Exhibits 17.2 and 17.3, where I outline how the risk discourse, which is allowed not only to represent and inform, but to give the meaning of issues like GMOs, is both omitting key issues (Exhibit 17.2) but also silently interjecting, and then imposing, institutionally encultured

EXHIBIT 17.2. Some problems of risk as a defining discourse.

- Does not acknowledge—effectively denies—the reality of unpredicted effects, and wider unpredictability
- Ordinary people are concerned about this evident state of *denial:* of ignorance; of lack of control; and of responsibility
- Yet public refusal for this reason of mistrust is treated instead as risk-centered (but therefore mistaken—deficit-based)
- Thus public concerns are about institutional behavior, as much as about "risk-objects." This relational dimension is ignored
- Risk discourse obscures public questions about upstream human purposes, ends, visions driving innovation-oriented science
- This implicates private as well as public science, yet private upstream science's drivers are even more unaccountable
- *Imposes* the prescriptive assumption that the issue is risk, and only risk, as defined by institutional science
- Projects questionable and provocative constructions of the public as representations into the public domain

EXHIBIT 17.3. Risk discourse tacit constructions of the public.

- Universal, standard public meaning ("risk"; a scientific object)
- So, public deviation is misunderstanding, or willful anti-science
- Agency confined to impacts, risks—not innovation and its human aims and purposes
- Unknowns not "our" responsibility—leave it to others (marginalized, future people)
- Individual, and instrumental, self-centered motivations: no relational meanings
- No *autonomous* meanings—citizens dependent on science for these
- Behaviorism—behavior reflects singular attitude/feeling
- Zero-risk, certainty-obsessed
- But naïve and gullible to media and NGO misinformation
- Epistemic vacuity (ethics, (mis)trust, are emotive and individualized concerns only)

imaginations of its publics along with its representations of scientific knowledge. As already argued, representations are always implicit "performances," by the representer, of the identity of subject(s) being tacitly represented. There will always be a need for suppositions, or imaginations about those subjects as part of such representations (which encompass claims of public meaning as well as of "risks"). In theory, these imaginations could be provisional, hypothetical, and reflexively construed so as to be tested and maybe corrected by exposure in the arena of their projection, the public sphere. However, this is unduly optimistic, since these dimensions of representation such as I have given are not even necessarily explicit and visible to those they supposedly represent who could in theory act as their correctors, and maybe not either to their own authors. Instead they may take on an implicit normative role, influencing how people actually understand themselves and behave without this influence being an evident object of deliberation. This could simply be by dint of a passive influence, as part of the repertoire of human models available to people as cultural resources for thinking of themselves as human subjects in public. These dimensions of tacit projection, and construction on the part of institutional science and policy actors, even in the name of public dialogue, listening and engagement, are part of the culture of modern institutional science and policy, in the combined sense of being about meanings ("risk"), and also being cultural in the sense of gradually accreted, routinized, habituated, taken-for-granted, and "natural" ways of thinking and practice. Responsibility for this lies with the scientific world and the policy-political world together, the latter having by default attenuated its own agency and responsibility for cultivating democratic public deliberation over the accepted meanings of such public issues, instead allowing scientific reflexes to colonize and stand for the full range of possible public-sphere meanings.[47]

Interestingly enough, this reduction of public meanings to only propositional quasi-scientific questions has been reinforced by some social scientists.[48] The undemocratic imposition involved has not been deliberate, thus it is difficult to attribute responsibility for it. Indeed the dominant modern propensity to imagine forms of agency only as deliberate and strategically calculated makes this *cultural* syndrome, also encompassing much of mainstream social science, all the more difficult for its actors to recognize.

CONCLUSIONS: A LEARNING SCIENCE?

Different philosophers, anthropologists, and critical social scientists have emphasized how the more basic issues facing modern scientific institutions, which feel in crisis over their public legitimacy and trust, are repeatedly misdefined as if they were purely closed epistemic issues—"How to get *them* (the ignorant public) to understand what *we* know is true?"—when these confrontations are fundamentally issues of different cultures coming into confrontation with one another. That is, the first step that I have attempted here, of showing that different knowledges are in play reflecting different worldviews and ways of life, leads to a second, which exposes how this more open epistemic question, reflecting potentially different ideas of the human purposes or ends of knowledge-production, reflects different social relations or practical forms of life—that is, ontological, maybe metaphysical, as well as epistemological differences.

These differences are not at all monolithic, nor are they automatically mutually exclusive. They are also strongly imbued with *imagined* others, and imagined ends. In the GM issue, but in others too, where the demands of scientific-technological innovation are placing increasing demands on public quiescence, the one has become so asymmetrically powerful and overweening over the other that mutual recognition and respect, not to mention substantive negotiation, have become all but impossible between them.

In the GMO case we can see that the culture that is unable to recognize and hear other legitimate cultural actors and perspectives is the same one that cannot see its own ignorance as a condition betraying science's pretensions of control. As I have shown, in the GEOs controversy with anxious official attempts to reassert control over a rapidly escalating crisis of public disaffection and refusal, the closed epistemic definition of the public problems of science have held sway through repeated falsifications of the public deficit explanations involved.

In these respects the only differences between the North American and European aspects of this issue are that the institutional science-policy culture has not been forced by any overt civil society opposition to rationalize itself there, unlike its European counterparts; thus little articulation of supposed reasons for public opposition, in the form of deficit models, has been needed and performed in the elaborate way that these have in the United Kingdom and the EU. However, this does not mean that the same basic culture, nor that similar public concerns, do not prevail in the North American setting.

The association between an ignored or domesticated "cultural other" on one hand, and denial of the state of ignorance as a cultural syndrome on the other, as identical features of the science-policy culture, can explain to us why scientists such as the U.K. ACRE chair found it impossible to handle the ignorance question, because it cannot be represented in the subsuming parochial cognitivist terms of his domestic institutional culture. Interestingly, the only way he found to handle it involved his inadvertent spontaneous equation and dismissal of its putative recognition with its twin unthinkable, namely "imaginings" from the public's "fevered brow." Treating either seriously would be equally absurd to this culture. It further explains why the same systematic effective denial has been repeated time and again in different forms, in related places and occasions, as indicated earlier.

The cultural denial of scientific ignorance, which still prevails in the exclusive use of risk discourse to define such issues, can be seen almost as a cultural "necessity"—but only if we accept that culture as coherent, and not itself riddled with all manners of internal cleavages, fault-lines, contortions, and self-contradictions. This risk-discourse based scientistic culture preempts any mainstream democratic policy debate about the proper human purposes and cultural meanings of knowledge, because it is unable to imagine such questions as questions.

In his reflections on the sensibilities of Buddhism and psychotherapy toward ignorance, Mark Epstein[50] recognizes that the harder we struggle to know more powerfully (seen as a singularly positive indeed essential trait by modern rationality-culture) the less we are *able* to know in a deeper sense. As he put it,

> The concretisation of experience which the thinking mind is so expert at carrying out, is what [we] call ignorance. (p. 87)

In other words, he notes, modern forms of scientific knowing inadvertently but systematically falsify authentic knowledge—which necessarily includes *knowledge of oneself,* or self-reflexivity, as part of wisdom. They perform this aspect of falsification, because they constitutionally differentiate subject and object, knower and known. This separation is what defines proper knowing in Western scientifically shaped cultures. Thus paradoxically, the more rigorously (that is, one-dimensionally) we try to know, the more we intensify our own alienation, as knower-subject, from the known-object; thus, the less we are aware of and able to act on the limitations and parochialisms of our "more powerful" (but self-centered and instrumental-only) knowledge. Our knowledge-object always embodies a crucial element of ourselves as knower, since we have in some way humanly defined our object(s) of desired knowledge—we have been the silent (of course, never final) authors of meaning, and salience. "Risk" and GEOs is a key contemporary examples are of this, where the momentary experience of not knowing is seamlessly deleted immediately when the unpredicted and uncontrolled event occurs.

Therefore, it seems that trying to imagine appropriate recognition of ignorance, and performing this recognition responsibly in public policy, can never be a purely cognitive task—it has to be embodied in an appropriate practical public culture. This was the ACRE scientific chair's problem; he had no cultural context of practice to which he could relate the AEBC member's question, until on the spot he invented his own, "fevered brow" meaning. In so externalizing and banishing any potential for acknowledging the state of ignorance as endemically present "other," he also thus routinely patronized and alienated science's publics yet again, not as an individual but as an agent of a particular culture. In so doing he obliterated the self-regulatory cultural potential that our scientific-technological culture of hubris urgently needs, and which people seem inchoately to be calling for.

NOTES

1. Jasanoff, S., Busch, L., Grove-White, R., Winickoff, D., and Wynne, B. 2004. Amicus Curiae Brief submitted to the World Trade Organisation, Disputes Settlement Panel, in the case of EC: Measures Affecting the Approval and Marketing of Biotech Products, April 30. http://www.lancs.ac.uk/cesagen (Last accessed November 7, 2006); Winickoff, D., Busch, L., Grove-White. R., Jasanoff, S., and Wynne, B. 2005 Adjudicating the GM food wars: Science, risk, and democracy in world trade law. *Yale Journal of International Law* 30: 81-123.

2. For example, the EU Council of Ministers decision on June 24, 2005 to continue the ban on GMOs, which reflected the unflinching political refusal by five member states—Austria, Germany, Greece, France, and Luxemburg, was taken against official commitment by the European Commission to remove the ban on the grounds that there were no legal grounds for it because there was no definite scientific evidence of harm. The U.K. government stood alone amongst member states in affirming the Commission definition of the issue as a scientific one, where (lack of) scientific evidence of harm should override any other reasons for refusal, including solid European public rejection, not only in the five "refusenik" member states. This U.K. and EC position simply presumed to know the basis of public opposition, namely irrationality and misunderstanding.

3. Hoban, T. 1998. Trends in consumer attitudes about agricultural biotechnology. *AgBioForum* 1(Summer): 3-7. Thomas Hoban (1998) was director of the US food industry-funded International Food Information Council (IFIC), which produced surveys of supposed U.S. public attitudes to GM foods, which were rejected by professional bodies as crudely biased in favor of GM. See *GMOs—Change of heart for professor who got cold feet.* September 26, 2004, nlpwessex@btinternet. com last accessed 07 Nov 2006, and http://www.gmwatch.org/archive2.Asp? Arcid=4387. Last accessed 07 Nov 2006. According to Karen Charman in a PR Watch article on Hoban's surveys: "James Beniger, a communications professor at the University of Southern California and past president of the American Association for Public Opinion Research, reviewed the IFIC survey and said it is so biased with leading questions favoring positive responses that any results are meaningless." UCLA communications professor Michael Suman agreed, adding that the questions "only talk about the food tasting better, being fresher, protecting food from insect damage, reducing saturated fat and providing benefits. It's like saying 'Here's biotechnology, it does these great things for you, do you like it?'" After such criticism, Hoban had a change of heart and accepted that his survey results, which had been based upon leading questions such as: "All things being equal, how likely would you be to buy a variety of produce, like tomatoes or potatoes, if it had been modified by biotechnology to taste better or fresher?" were invalid. Yet these were the main bases for the assertions that U.S. consumers are perfectly happy with GM foods unlike their EU counterparts, even when Hoban's own data showed a systematic fall (which went unremarked) in levels of public acceptance as levels of awareness that their food contained GM products increased over the period of three two-yearly surveys; Levy. A., and Derby, B. 2001. *Report on consumer focus groups on biotechnology.* Paper to annual meeting of the Society for Social Studies of Science, University of Vienna, September 30, 2000. Washington, DC: Center for Food Safety and Applied Nutrition, US Food and Drug Administration.

4. Edelman. M. 1984. *Political language: Words that succeed and policies that fail.* Chicago, IL: University of Chicago Press.

5. Jasanoff, S. 2005. *Designs on nature: Science and democracy in the USA, Britain and Germany.* Princeton, NJ: Princeton University Press.

6. UK House of Lords. 2000 report, *Science and Society.* Select Committee on Science and Technology (chair, Patrick Jenkin). http://www.publications.parliament. uk/pa/ld199900/ldselect/ldsctech/38/3801.htm. Last accessed November 7, 2006.

7. Wynne, B. 2001. Creating public alienation: Expert discourses of risk and ethics on GMOs. *Science as Culture* 10: 1-40; Wynne. B. 2005. Public dialogue/engagement as a means of restoring public trust in science—hitting the notes, but missing the music? *Community Genetics* 10, 2006, 211-220.

8. Lynch, M., and Woolgar, S., eds. 1990. *Representation in scientific practice.* Cambridge, MA: MIT Press; Law, J., and Mol, A.-M., eds. 2002. *Complexities: Social studies of knowledge.* Durham, NC: University of North Carolina Press; Barad. K. 2003. Posthumanist performativity: Towards an understanding of how matter comes to matter. *Signs: Journal of Women in Culture and Society* 28: 801-831.

9. In another context, that of nanotechnology assessment, Jean-Pierre Dupuy and Alexei Grinbaum (2004. Living with uncertainty: Toward the on-going normative assessment of nanotechology. *Techne* 8(2): 4) draw upon Karl Popper to state that Popper defined the notion of metaphysical research as a set of ideas and worldviews that underlie any particular scientific research agenda. The positivist philosophy that drives most of modern science (and much of contemporary philosophy) takes "metaphysics" to be a meaningless quest for answers to unanswerable questions. However, Popper showed that there is no scientific research program that does not rest upon a set of general presuppositions about the structure of the world. To be sure, those metaphysical views are not empirically testable and they are not amenable to "falsification," However, this does not imply that they are of less importance or that they do not play a fundamental role in the advancement of science. I would just add that it is this systematic denial of such a metaphysical basis that leaves science being given the false extra public role I describe here—the default provider of meaning for issues involving science but not properly called "scientific issues." This shadow metaphysics to which Dupuy and Grinbaum refer also corresponds to some degree with the imaginations and visions that I describe as pervading scientific programs of research.

10. Power, M. *The risk management of everything.* London: Demos; Wynne, B. 2002. Risk and environment as legitimatory discourses of technology: reflexivity inside-out, *Current Sociology* 20: 459-477.

11. On this reduction of public policy issues to propositional meanings only, see the exchanges: Collins, H., and Evans, R. 2002. The third wave of science studies: studies of expertise and experience. *Social Studies of Science* 32: 235-296; Wynne, B. 2003. Seasick on the third wave? Subverting the hegemony of propositionalism. Social *Studies of Science* 33: 401-417; Jasanoff, S. 2005. See note 5.

12. Burke, D. GM food and crops: What went wrong in the UK? *European Molecular Biology Organisation Reports* 5: 2004, 432-436. http://www.emboreports.org.

13. Marris, C., Wynne, B., Simmons, P., and Weldon, S. *Public attitudes to agricultural biotechnologies in Europe.* 2001, Final report to D-G Research, European Commission, Brussels. www.pabe.net.

14. Wilsdon, J., and Willis, R. 2004. *See-Through science: Why public engagement needs to move upstream.* London: Demos.

15. Wynne.B., Wilsdon. J., and Stilgoe, J. 2005. *The public value of science: Or how to ensure that science really matters.* London: Demos. http://www.demos.co.uk/catalogue/publicvalueofscience. (Last accessed November 7, 2006).

16. Levy. A., and Derby, B. 2001. See note 3.

17. UK Agriculture and Environment Biotechnology Commission. 2001. *Crops on Trial.* London: UK Department of Environment, Food and Rural Affairs.

18. Ibid.

19. Firbank, L. Personal communication.

20. Jasanoff, S. 1990. *The fifth branch.* Cambridge, MA: Harvard University Press.

21. See the footnotes in Jasanoff, S., et al. 2004. (cited in note 1) for some more broad-based and realistic definitions in-principle.

22. Latour, B. 1998. *Pandora's hope: Essays on the reality of science studies.* Cambridge MA: Harvard University Press.

23. Stengers, I. 1997. *Power and invention: Situating science.* Minneapolis: University of Minnesota Press.

24. Wynne, B. 2001. See note 7.

25. Blair, Rt. Hon Tony, MP. 2002. Science matters. UK Prime Minister in a speech at The Royal Society, London, May.

26. Marris, C., et al., 2001. See note 13.

27. Sir Robert May, then chief-scientific adviser to the UK Government. 2000. Lecture at Hanover Expo, Germany, July 11.

28. I refer here to May's evidence in 1999 to the UK House of Lords (see note 6). Select Committee on Science and Technology, Science and Society Inquiry Vol. 2, Minutes of Evidence, March 2000, UK Parliament papers.

29. Wynne. B. 1992. Uncertainty and environmental learning: Reconceiving science and policy in the preventive paradigm. *Global Environmental Change* 2: 111-137; Stirling, A. 1998. Risk at a turning point? *Journal of Risk Research* 2: 87-103.

30. Schon, D. 1982. Uncertainty and risk. In *Science in context.* Edited by B. Barnes and D. Edge. Milton Keynes, UK: Open University Press, pp. 57-71.

31. Rumsfeld, D., U.S. Secretary for Defense. 2004. US Department of Defense news briefing, Washington, DC, February 12.

32. At the time of the negotiation of the EU 1990/220 Directive on the deliberate release of GMOs to the environment, a fourth hurdle in regulation was proposed, namely a question about social benefit of whichever innovation is in question; but this was rejected. The only place where it has prevailed as far as I am aware is as an Article in the 1995 Norwegian Gene Technology Act.

33. Wynne. B, 1980. Technology risk and participation: The social treatment of uncertainty. In *Society Risk and Technology.* Edited by J. Conrad. London: Academic Press, pp. 83-107.

34. As I have noted elsewhere, (see note 7) this places an interesting light on the conventional dismissal, or if recognized at all, the individualization, of public concerns about "Playing God" over commercializing GEOs.

35. UK Treasury/DTI/DfEE. 2004. A ten-year framework for science and innovation. London: Treasury, Department of Trade and Industry and Department for Education and Employment.

36. Sir Robert May. 2002. Anniversary Presidential Address. The Royal Society, London, November.

37. UK Government. 2001. *Interim response to the report of the BSE inquiry.* Presented to Parliament by the Minister of Agriculture Fisheries and Food, February. London: UK Government, CM 5049. Quote, p. 23.

38. European Commission. 2000. *Communication on the precautionary principle.* European Commission, Brussels, March.

39. Wilsdon, J., and Willis, R. 2004. See note 14.

40. Marris, C., et al. 2001. See note 13. Grove-White, R., Mayer, S., Macnaghten, P. and Wynne, B. 1997. *Uncertain world: Genetically modified organisms, food and public attitudes in Britain.* Lancaster University, CSEC; Levy, A., and Derby, B. 2001. See note 3.

41. Latour, B. 1998. See note 22; Stengers, I. 1997. See note 23.

42. UK Government Office for Science and Technology. 1999. Public Consultation on the Biosciences, Minutes of the Advisory Group, January 14. London: Department of Trade and Industry.

43. Dewey. J, 1927. *The public and its problems.* Chicago, IL: University of Chicago Press, reprint 1997. quote p. 73.

44. Hastrup, K. 1993. Hunger and the hardness of facts. *Man* 28: 727-739.

45. Polanyi, M. 1958. *Personal knowledge.* London: Routledge and Kegan Paul.

46. Hastrup, K. 1993. See note 44.

47. Not that I would wish to exonerate the political–economic context of that science from responsibility too. It clearly coincides with the interests of global corporate commercial powers to preemptively instrumentalize such democratic arenas (sadly it seems, in harness with the state) so that questions about the proper human purposes and priorities, thus appropriate criteria of knowledge-production, are not addressed but instead subordinated to the definition of innovation trajectories that are thought to maximize extraction (and externalization of uncontrolled dimensions) but then defined as due only to "sound science" and without alternatives.

48. Collins, H., and Evans, R. 2002. See note 11.

49. See footnotes in Jasanoff, S., 2004, cited in note 1; Latour, B. 1998. See note 22; Haraway, D. 1997. *Modest_Witness@Second_Millenium. FemaleCman meets OncoMouseTM: Feminism and Technoscience.* New York: Routledge.

50. Epstein, M. 1996. *Thoughts without a thinker: Psychotherapy from a Buddhist perspective.* London: Duckworth and New York, Basic Books.

Index

Genetically Engineered Crops
© 2007 by The Haworth Press, Inc. All rights reserved.
doi:10.1300/5880_18

Printed in the United States
by Baker & Taylor Publisher Services